T0192129

Communications in Computer and Information Science 1803

Rationale

The CCIS series is devoted to the publication of proceedings of computer science conferences. Its aim is to efficiently disseminate original research results in informatics in printed and electronic form. While the focus is on publication of peer-reviewed full papers presenting mature work, inclusion of reviewed short papers reporting on work in progress is welcome, too. Besides globally relevant meetings with internationally representative program committees guaranteeing a strict peer-reviewing and paper selection process, conferences run by societies or of high regional or national relevance are also considered for publication.

Topics

The topical scope of CCIS spans the entire spectrum of informatics ranging from foundational topics in the theory of computing to information and communications science and technology and a broad variety of interdisciplinary application fields.

Information for Volume Editors and Authors

Publication in CCIS is free of charge. No royalties are paid, however, we offer registered conference participants temporary free access to the online version of the conference proceedings on SpringerLink (http://link.springer.com) by means of an http referrer from the conference website and/or a number of complimentary printed copies, as specified in the official acceptance email of the event.

CCIS proceedings can be published in time for distribution at conferences or as post-proceedings, and delivered in the form of printed books and/or electronically as USBs and/or e-content licenses for accessing proceedings at SpringerLink. Furthermore, CCIS proceedings are included in the CCIS electronic book series hosted in the SpringerLink digital library at http://link.springer.com/bookseries/7899. Conferences publishing in CCIS are allowed to use Online Conference Service (OCS) for managing the whole proceedings lifecycle (from submission and reviewing to preparing for publication) free of charge.

Publication process

The language of publication is exclusively English. Authors publishing in CCIS have to sign the Springer CCIS copyright transfer form, however, they are free to use their material published in CCIS for substantially changed, more elaborate subsequent publications elsewhere. For the preparation of the camera-ready papers/files, authors have to strictly adhere to the Springer CCIS Authors' Instructions and are strongly encouraged to use the CCIS LaTeX style files or templates.

Abstracting/Indexing

CCIS is abstracted/indexed in DBLP, Google Scholar, EI-Compendex, Mathematical Reviews, SCImago, Scopus. CCIS volumes are also submitted for the inclusion in ISI Proceedings.

How to start

To start the evaluation of your proposal for inclusion in the CCIS series, please send an e-mail to ccis@springer.com.

Alexander Dudin · Anatoly Nazarov ·
Alexander Moiseev
Editors

Information Technologies and Mathematical Modelling

Queueing Theory and Applications

21st International Conference, ITMM 2022
Karshi, Uzbekistan, October 25–29, 2022
Revised Selected Papers

Editors
Alexander Dudin
Belarusian State University
Minsk, Belarus

Anatoly Nazarov
Tomsk State University
Tomsk, Russia

Alexander Moiseev
Tomsk State University
Tomsk, Russia

ISSN 1865-0929 ISSN 1865-0937 (electronic)
Communications in Computer and Information Science
ISBN 978-3-031-32989-0 ISBN 978-3-031-32990-6 (eBook)
https://doi.org/10.1007/978-3-031-32990-6

This Springer imprint is published by the registered company Springer Nature Switzerland AG
The registered company address is: Gewerbestrasse 11, 6330 Cham, Switzerland

Preface

The series of scientific conferences Information Technologies and Mathematical Modelling (ITMM) was started in 2002. In 2012, the series acquired an international status, and selected revised papers have been published in *Communications in Computer and Information Science* since 2014. The conference series was named after Alexander Terpugov, one of the first organizers of the conference, an outstanding scientist of the Tomsk State University and a leader of the famous Siberian school on applied probability, queueing theory, and applications.

Traditionally, the conference has about ten sections in various fields of mathematical modelling and information technologies. Throughout the years, the sections on probabilistic methods and models, queueing theory, and communication networks have been the most popular ones at the conference. These sections gather many scientists from different countries. Many foreign participants come to this Siberian conference every year because of our warm welcome and serious scientific discussions. The ITMM conference has now been held internationally: in 2022, for the first time, the conference was held not in Russia, but in one of the most ancient cities of the Republic of Uzbekistan - the city of Karshi. The conference was organized by the National Research Tomsk State University of Russia, Karshi State University of Uzbekistan, Peoples' Friendship University of Russia (RUDN University), Trapeznikov Institute of Control Sciences of Russian Academy of Sciences, and Romanovsky Institute of Mathematics of the Academy of Sciences of the Republic of Uzbekistan.

This volume presents selected papers from the 21th ITMM conference. The conference received 89 submissions, from which 19 were selected to be published in the current collection. Papers have passed single-blind peer review and each of them had at least three reviewers.

The papers are devoted to new results in queueing theory and its applications. Its target audience includes specialists in probabilistic theory, random processes, and mathematical modelling as well as engineers engaged in logical and technical design and operational management of data processing systems, communication, and computer networks.

October 2022

Alexander Dudin
Anatoly Nazarov
Alexander Moiseev

Organization

International Program Committee Chairs

Alexander Dudin	Belarusian State University, Belarus
Shavkat Ayupov	Uzbekistan Academy of Sciences, Uzbekistan
Azam Imomov	Karshi State University, Uzbekistan
Svetlana Moiseeva	Tomsk State University, Russia
Anatoly Nazarov	Tomsk State University, Russia
Alexander Moiseev	Tomsk State University, Russia
Utkir Rozikov	Romanovsky Institute of Mathematics of the Academy of Sciences of the Republic of Uzbekistan, Uzbekistan

International Program Committee

Abdurahim Abdushukurov	Romanovsky Institute of Mathematics of the Academy of Sciences of the Republic of Uzbekistan, Uzbekistan
Khalid Al-Begain	Kuwait College of Science and Technology, Kuwait
Mohammad Ahmad Alia	Al-Zaytoonah University, Jordan
Ravshan Ashurov	Romanovsky Institute of Mathematics of the Academy of Sciences of the Republic of Uzbekistan, Uzbekistan
Ivan Atencia	University of Malaga, Spain
Abdulla Azamov	Romanovsky Institute of Mathematics of the Academy of Sciences of the Republic of Uzbekistan, Uzbekistan
Dilmurod Bojtillaev	Romanovsky Institute of Mathematics of the Academy of Sciences of the Republic of Uzbekistan, Uzbekistan
Pedro Cabral	Universidade Nova de Lisboa, Portugal
Pau Fonseca i Casas	Universitat Politècnica de Catalunya, Spain
Srinivas Chakravarthy	Kettering University, USA
Bong Dae Choi	National Institute for Mathematical Sciences, South Korea

Dmitri Moltchanov	Tampere University, Finland
Paulo Montezuma-Carvalho	Universidade Nova de Lisboa, Portugal
Evsey Morozov	Institute of Applied Mathematical Research, Karelian Research Centre of Russian Academy of Sciences, Russia
Valery Naumov	Service Innovation Research Institute, Finland
Rein Nobel	Vrije Universiteit Amsterdam, The Netherlands
Michele Pagano	Pisa University, Italy
Tuan Phung-Duc	University of Tsukuba, Japan
Muzaffar Rahmatullaev	Romanovsky Institute of Mathematics of the Academy of Sciences of the Republic of Uzbekistan, Uzbekistan
Gul'nara Raimova	Romanovsky Institute of Mathematics of the Academy of Sciences of the Republic of Uzbekistan, Uzbekistan
Jacques Resing	Eindhoven University of Technology, The Netherlands
Vladimir Rykov	Gubkin Russian State University of Oil and Gas, Russia
Konstantin Samouylov	Peoples' Friendship University of Russia (RUDN University), Russia
Olimzhon Sharipov	Romanovsky Institute of Mathematics of the Academy of Sciences of the Republic of Uzbekistan, Uzbekistan
Stanislav Shidlovskiy	Tomsk State University, Russia
Ahmadzhan Soleev	Samarkand State University, Uzbekistan
Sergey Suschenko	Tomsk State University, Russia
János Sztrik	University of Debrecen, Hungary
Zhozil Tahirov	Romanovsky Institute of Mathematics of the Academy of Sciences of the Republic of Uzbekistan, Uzbekistan
Henk Tijms	Vrije Universiteit Amsterdam, The Netherlands
Oleg Tikhonenko	Cardinal Stefan Wyszyński University in Warsaw, Poland
Gurami Tsitsiashvili	Institute of Applied Mathematics, Far Eastern Branch of Russian Academy of Sciences, Russia
Vladimir Vishnevsky	Institute of Control Sciences, Russian Academy of Sciences, Russia
Anton Voitishek	Institute of Computational Mathematics and Mathematical Geophysics, Siberian Branch, Russian Academy of Sciences, Russia
Vladimir Zadorozhny	Omsk State Technical University, Russia

Alexander Zamyatin	Tomsk State University, Russia
Andrey Zorin	Lobachevsky State University of Nizhni Novgorod, Russia
Amdjed Zraiqat	Al-Zaytoonah University, Jordan

Local Organizing Committee

Azam Imomov, Chair
Svetlana Moiseeva, Co-chair
Svetlana Paul, Co-chair
Golibzhan Botirov, Co-chair
Abdimumin Abdurahmanov
Mumin Abulov
Zhahongir Azimov
Valentina Broner
Elena Danilyuk
Ekaterina Fedorova
Nodir Holmirzaev
Abdulhamid Holmurodov
Yakubzhan Husanbaev
Irina Kochetkova
Ivan Lapatin
Andrey Larionov
Ekaterina Lisovskaya
Olga Lizyura
Abror Mejliev
Anna Morozova
Ekaterina Pankratova
Svetlana Rozhkova
Utkir Rozikov
Daria Semenova
Dmitry Shashev
Maria Shklennik
Alexey Shkurkin
Bajramali Tursunov
Konstantin Voytikov
Hurshid Zhumakulov

Contents

Limit Theorems for the Positive Recurrent Q-process

Azam A. Imomov[1,2(✉)] and Zuhriddin A. Nazarov[2]

[1] Karshi State University, Karshi, Uzbekistan
imomov_azam@mail.ru

[2] Romanovskiy Institute of Mathematics, Tashkent, Uzbekistan

Abstract. We examine the population growth system called Q-processes. This is defined by the Galton-Watson Branching system conditioned on non-extinction of its trajectory in the remote future. We find an explicit form of the generating function of the distribution of limit states of the system. We prove that this generating function produces an invariant distribution. As a consequence of this result, a local limit theorem with a tail part estimate is established. In addition to all this, at the end we observe the structural and asymptotic behavior of the total progeny participating in the evolution of the Q-process. We establish an analogue of the Central Limit Theorem for the total progeny.

Keywords: Branching system · Extinction Time · Q-process · Markov chain · Monotone Ratio Convergence theorem · Invariant Distribution · Local Limit Theorem · Total Progeny · Central Limit Theorem

1 Introduction

1.1 Preliminaries

Models of stochastic branching systems form an essential part of the general theory of random processes. The dynamic interest in these models is due to many factors. The first of these, which became the main impetus for the creation of the theory of branching models, is the possibility of estimating, with their help, the probabilities of survival of a population of monotypic individuals. In this regard, the most primitive branching model, initiated by famous English statisticians H.Watson and F.Galton in the second half of the 19th century, is now called the Galton-Watson Branching (GWB) system; see [1–3,7] and [12]. Among the random trajectories of branching systems, there are those that continue a long time. In the case of the GWB model, the class of such trajectories forms another stochastic model called Q-process; see [2] and [5]. In the case of continuous-time Markov branching systems, a similar model, called the *Markov Q-process*, was first introduced in [6].

Consider GWB system $\{Z(n), n \in \mathbb{N}_0\}$ with branching rates $\{p_k, k \in \mathbb{N}_0\}$, where $\mathbb{N}_0 = \{0\} \cup \mathbb{N}$ and $\mathbb{N} = \{1, 2, \ldots\}$, the variable $Z(n)$ denote the population

A. Dudin et al. (Eds.): ITMM 2022, CCIS 1803, pp. 1–15, 2023.
https://doi.org/10.1007/978-3-031-32990-6_1

size at the moment n in the system. The evolution of the system occurs according to the following mechanism. Each individual lives a unit length life time and then gives $k \in \mathbb{N}_0$ descendants with probability p_k. This process is a reducible, homogeneous-discrete-time Markov chain with a state space consisting of two classes: $\mathcal{S}_0 = \{0\} \cup \mathcal{S}$, where $\{0\}$ is absorbing state, and $\mathcal{S} \subset \mathbb{N}$ is the class of possible essential communicating states. Throughout the paper assume that $p_0 > 0$ and $p_0 + p_1 > 0$ which called the Schröder case. We suppose that $p_0 + p_1 < 1$ and $m := \sum_{k \in \mathcal{S}} k p_k < \infty$.

The system n-step transition probabilities

$$P_{ij}(n) := \mathsf{P}\left\{Z(n+k) = j \mid Z(k) = i\right\} \qquad \textit{for any} \quad k \in \mathbb{N}_0$$

are defined in terms of branching rates $\{p_k, k \in \mathbb{N}_0\}$. In fact, we observe that a probability generating function (GF)

$$\sum_{k \in \mathcal{S}_0} P_{ij}(n) s^k = \left[f_n(s)\right]^i, \tag{1}$$

where

$$f_n(s) := \sum_{k \in \mathcal{S}_0} \mathsf{p}_k(n) s^k,$$

therein $\mathsf{p}_k(n) := P_{1k}(n)$ and, in the same time $f_n(s)$ is n-fold iteration of the offspring GF $f(s) := \sum_{k \in \mathcal{S}_0} p_k s^k$. Needless to say that $f_n(0) = \mathsf{p}_0(n)$ is a vanishing probability of the system starting from one individual. Note that this probability tends as $n \to \infty$ monotonously to q, which called an extinction probability of the system, i.e. $\lim_{n\to\infty} \mathsf{p}_0(n) = q$; see [2]. The extinction probability $q = 1$ if $m \leq 1$ and $q < 1$ when $m > 1$. Based on this, the system is called sub-critical, critical and supercritical if $m < 1$, $m = 1$ and $m > 1$ respectively.

Further we are dealing with the GWB system conditioned on the event $\{n < \mathcal{H} < \infty\}$, where $\mathcal{H} := \min\{n \in \mathbb{N} : Z(n) = 0\}$ is the extinction time. Let $\mathsf{P}_i\{*\} := \mathsf{P}\left\{* \mid Z(0) = i\right\}$ and define conditioned probability measure

$$\mathsf{P}_i^{\mathcal{H}(n+k)}\{*\} := \mathsf{P}_i\left\{* \mid n + k < \mathcal{H} < \infty\right\} \qquad \textit{for any} \quad k \in \mathbb{N}.$$

In [2, p. 58] proved, that

$$\mathcal{Q}_{ij}(n) := \lim_{k\to\infty} \mathsf{P}_i^{\mathcal{H}(n+k)}\left\{Z(n) = j\right\} = \frac{jq^{j-i}}{i\beta^n} P_{ij}(n), \tag{2}$$

where $\beta := f'(q)$. Observe that $\sum_{j \in \mathbb{N}} \mathcal{Q}_{ij}(n) = 1$ for each $i \in \mathbb{N}$. Thus, the probability measure $\mathcal{Q}_{ij}(n)$ can determine a new population growth system – a discrete-homogeneous time irreducible Markov chain with the state space $\mathcal{E} \subset \mathbb{N}$ which we denote by $\{W(n), n \in \mathbb{N}_0\}$. This is called in the monograph [2, p. 58] *the Q-process*. Undoubtedly that $W(0) \stackrel{d}{=} Z(0)$ and transition probabilities

$$\mathcal{Q}_{ij}(n) = \mathsf{P}\left\{W(n) = j \mid W(0) = i\right\} = \mathsf{P}_i\left\{Z(n) = j \mid \mathcal{H} = \infty\right\},$$

so that the Q-process can be interpreted as a "long-living" GWB system.

Put into consideration a GF

$$w_n^{(i)}(s) := \sum_{j\in\mathcal{E}} \mathcal{Q}_{ij}(n)s^j.$$

Then from (1) and (2) we obtain

$$w_n^{(i)}(s) = \left[\frac{f_n(qs)}{q}\right]^{i-1} w_n(s), \tag{3}$$

where the GF $w_n(s) := w_n^{(1)}(s) = \mathsf{E}\left[s^{W(n)} \mid W(0) = 1\right]$ has a form of

$$w_n(s) = s\frac{f_n'(qs)}{\beta^n} \quad \textit{for all} \quad n \in \mathbb{N}. \tag{4}$$

Application of iteration for $f(s)$ in the relation (3) leads us to the following functional equation:

$$w_{n+1}^{(i)}(s) = \frac{w(s)}{f_q(s)} w_n^{(i)}(f_q(s)), \tag{5}$$

where $w(s) := w_1(s)$ and $f_q(s) = f(qs)/q$. Thus, Q-process is completely defined by setting the GF

$$w(s) = s\frac{f'(qs)}{\beta}. \tag{6}$$

An evolution of the Q-process is in essentially regulated by the structural parameter $\beta > 0$. In fact, as it has been shown in [2, p. 59, Theorem 2], that if $\beta < 1$ then $\mathcal{E}$ is positive recurrent and, $\mathcal{E}$ is transient if $\beta = 1$. On the other hand, it is easy to be convinced that positive recurrent case $\beta < 1$ of Q-process corresponds to the non-critical case $m \neq 1$ of GWB system. Note that $\beta \leq 1$ and nothing but.

The paper considers the positive recurrent case. Let $\alpha := w'(1-) < \infty$. Then differentiating (6) on the point $s = 1$ we obtain $\alpha = 1 + (1 - \beta)\gamma_q$, where

$$\gamma_q := \frac{qf''(q)}{\beta(1-\beta)}.$$

It follows from (3) and (4) that

$$\mathsf{E}_i W(n) = (i-1)\beta^n + \mathsf{E}W(n),$$

where $\mathsf{E}W(n) = 1 + \gamma_q \cdot (1 - \beta^n)$.

It is obvious, that when initial GWB system is sub-critical, then the condition $\alpha < \infty$ is nothing but the condition of finiteness of the second factorial moment of offspring distribution, i.e. $f''(1-) < \infty$. Further we everywhere will be accompanied by this condition by default.

Now, denote $R_n(s) := q - f_n(s)$ for all $s \in \mathrm{U}_q(0,1]$, where

$$\mathrm{U}_q(0,1] := \{[0,q) \cup (q,1)\}.$$

First we recall the fact that in our conditions

$$\sup\left\{\left|\frac{\beta^n}{R_n(s)} - \frac{1}{\mathcal{A}_q(s)}\right| : s \in \mathrm{U}_q[0,1)\right\} = o(1) \qquad as \quad n \to \infty \tag{7}$$

which proved in [4], where

$$\mathcal{A}_q(s) = \frac{q-s}{1+\gamma \cdot (q-s)}, \tag{8}$$

for all $s \in \mathrm{U}_q(0,1]$, and

$$\gamma := \frac{\gamma_q}{2q} = \frac{f''(q)}{2\beta\,(1-\beta)}. \tag{9}$$

In [2, p. 40] proved that $\lim_{n\to\infty} R_n'(s)/\beta^n = \mathcal{A}_q'(s)$. Thus, we obtain an assertion similar to statement (6) for the function $R_n'(s)$. Therefore, combination of (6) and (8) immediately yields the following result.

Theorem 1. *Let $\beta < 1$. Then*

$$\rho_n(s) := \left|w_n(s) - \mathcal{M}_q(s)\right| \longrightarrow 0 \tag{10}$$

as $n \to \infty$ uniformly in $s \in [0,1)$, where $\mathcal{M}_q(s) = -\mathcal{A}_q'(qs)$ and

$$\mathcal{M}_q(s) = \frac{s}{\left(1+q\gamma \cdot (1-s)\right)^2}, \tag{11}$$

at that γ is defined in (9).

Our principal progress on this issue is that the limit function $\mathcal{M}_q(s)$ was explicitly found in the Theorem 1. Subsequent results appear in the next step. Here we improve the formulation of Theorem 1, by specifying the explicit approximation rate of the function $w_n(s)$ to $\mathcal{M}_q(s)$ in Theorem 2. Next we state an analogue of the Monotone Ratio Convergence theorem for the transition probabilities $\mathcal{Q}_{i1}(n)$. This certainly gives the local limit theorem for the positive recurrent Q-processes.

1.2 Main Results

We begin by stating our first result which improves Theorem 1, specifying the decreasing rate of the asymptotic estimator $\rho_n(s)$ in (10).

Theorem 2. *Let $\beta < 1$. Then*

$$\left|w_n(s) - \mathcal{M}_q(s)\right| = \mathcal{O}(\beta^n) \tag{12}$$

as $n \to \infty$ uniformly in $s \in [0,1)$, where $\mathcal{M}_q(s)$ is defined in (11).

Since $\mathcal{M}_q(s)$ is a limit of power series, it may be written as follows:

$$\mathcal{M}_q(s) = \sum_{j\in\mathcal{E}} \mu_j s^j, \tag{13}$$

where $\mu_j > 0$. The following result asserting the direct convergence of transition probabilities $\mathcal{Q}_{ij}(n)$ to μ_j immediately follows from Theorem 2.

Let's observe now an analogue of the Monotone Ratio Convergence theorem for the probabilities $\mathcal{Q}_{1j}(n)$.

Theorem 3. *Let $\beta < 1$. Then for all $j \in \mathcal{E}$*

$$\frac{\mathcal{Q}_{1j}(n)}{\mathcal{Q}_{11}(n)} \uparrow \pi_j < \infty \qquad as \quad n \to \infty,$$

where the numbers $\{\pi_j\}$ are the invariant distribution for the Q-process, i.e.

$$\pi_j = \sum_{i\in\mathcal{E}} \pi_i \mathcal{Q}_{ij}(1) \qquad and \qquad \sum_{j\in\mathcal{E}} \pi_j < \infty;$$

at that $\sum_{j\in\mathcal{E}} \pi_j = \left(1 + q\gamma\right)^2$.

The following theorem generalizes the assertion of the last Theorem 3 for all states $i, j \in \mathcal{E}$, while establishing the rate of convergence of the ratios.

Theorem 4. *Let $\beta < 1$. Then*

$$\frac{\mathcal{Q}_{ij}(n)}{\mathcal{Q}_{11}(n)} = \pi_j \cdot \left(1 + r_n\right) \qquad for\ all \quad i, j \in \mathcal{E}, \tag{14}$$

where $r_n = \mathcal{O}\left(\beta^n\right)$ as $n \to \infty$.

Remark 1. The arguments of the proof of the last two theorems implies that

$$\pi(s) = \frac{\mathcal{M}_q(s)}{\mu_1} \qquad and \qquad \mu_1 = \frac{1}{\left(1 + q\gamma\right)^2}$$

at that the local probabilities have a limiting expansion

$$\mathcal{Q}_{ij}(n) = \mu_1 \pi_j \cdot \left(1 + \mathcal{O}\left(\beta^n\right)\right) \qquad as \quad n \to \infty$$

for all $i, j \in \mathcal{E}$.

2 Proof of the Results

Proof of Theorem 2. Our proof discussions are fully based on the following auxiliary statement asserting the geometric decay of an approximation rate of the function $R'_n(s)/\beta^n$ to $\mathcal{A}'_q(s)$ as $n \to \infty$.

Lemma 1. *The following assertion holds:*

$$\sup\left\{\left|\frac{R_n'(s)}{\beta^n} - \mathcal{A}_q'(s)\right| : s \in U_q[0,1)\right\} = \mathcal{O}(\beta^n) \qquad as \quad n \to \infty.$$

Proof. First, using Taylor expansion formula, we write

$$\frac{R_{n+1}'(s)}{R_n'(s)} = \beta - f''\big(\xi_n(s)\big)R_n(s), \tag{15}$$

where $\xi_n(s)$ is an intermediate value between q and the function $R_n(s)$ and this can easily be estimated by

$$q_r(n) = q + (r-q)\beta^n \quad \text{for} \quad r = 0, 1.$$

It follows

$$\begin{cases} q_0(n) < \xi_n(s) < q & \text{if} \quad s \in [0, q), \\ q < \xi_n(s) < q_1(n) & \text{if} \quad s \in (q, 1); \end{cases}$$

for details on estimations of $\xi_n(s)$ in the last step, we refer to [4]. Repeatedly using the relation (15) produces that

$$\frac{R_n'(s)}{\beta^n} = -\prod_{k=0}^{n-1}\left[1 - \frac{f''\big(\xi_k(s)\big)}{\beta}R_k(s)\right],$$

since $R_0'(s) = -1$. Now we write

$$\frac{R_n'(s)}{\beta^n} = \frac{\mathcal{A}_q'(s)}{\prod_{k=n}^{\infty}\left[1 - A_k(s)\right]}, \tag{16}$$

where

$$A_n(s) := \frac{f''\big(\xi_n(s)\big)}{\beta}R_n(s).$$

In view of asymptotic behaviour of the function $R_n(s)$, the function $A_n(s)$ becomes much less than 1 for sufficiently large n. Then using the well-known inequality $\ln(1-x) \geq -x - x^2/(1-x)$ available for $s \in [0, q)$, allows us to write

$$\ln\prod_{k=n}^{\infty}\left[1 - A_k(s)\right] = -\Sigma_n(s) + \sigma_n(s), \tag{17}$$

where

$$\Sigma_n(s) = \sum_{k=n}^{\infty} A_k(s),$$

and

$$0 \leq -\sigma_n(s) \leq \sum_{k=n}^{\infty}\frac{A_k^2(s)}{1 - A_k(s)}.$$

It is convincing, that

$$\sigma_n(s) = o\big(\Sigma_n(s)\big) \quad as \quad n \to \infty \tag{18}$$

provided that the infinite series $\sum_{k\in\mathbb{N}_0} A_k(s)$ converges.

For $R_n(s)$ we have the following uniform estimates:

$$\begin{cases} 0 < R_n(s) < q\beta^n & \text{if } s \in [0, q), \\ 0 < -R_n(s) < (1-q)\beta^n & \text{if } s \in (q, 1) \end{cases}$$

for all $n \in \mathbb{N}_0$; see [4]. Then, since $q_r(n) < 1$ for all $n \in \mathbb{N}_0$ and both $r = 0, 1$,

$$\big|A_n(s)\big| < \frac{f''\big(\xi_n(s)\big)}{\beta}\big|R_n(s)\big| < \frac{f''(1-)}{\beta}\mu_q\beta^n,$$

where $\mu_q = \max\{q, 1-q\}$, hence

$$\big|\Sigma_n(s)\big| < \frac{f''(1-)}{\beta\,(1-\beta)}\mu_q\beta^n \tag{19}$$

uniformly in $s \in \mathbb{U}_q[0,1)$. Moreover, it follows that $\big|1 - A_n(s)\big| < \infty$ and consequently we have the following uniform estimate:

$$0 \le \big|\sigma_n(s)\big| \le \mathcal{O}\big(\beta^n\big) \cdot \Sigma_n(s)$$

for all $s \in \mathbb{U}_q[0,1)$ and $n \to \infty$. Thus (18) holds.

Considering the relation $e^{-y} \sim 1 - y$ as $y \to 0$, equations (17)–(19) yield

$$\prod_{k=n}^{\infty}\big[1 - A_k(s)\big] = 1 - \Sigma_n(s) + o\big(\Sigma_n(s)\big) \qquad as \quad n \to \infty. \tag{20}$$

Now statement of the lemma readily follows from (16), (19) and (20). □

Lemma 1 implies

$$\frac{R'_n(s)}{\beta^n} = \mathcal{A}'_q(s)\left(1 + \mathcal{O}(\beta^n)\right) \quad as \quad n \to \infty. \tag{21}$$

To get to the assertion (12) we rewrite (4) as

$$w_n(s) = -s\frac{R'_n(qs)}{\beta^n} \qquad for\ all \quad n \in \mathbb{N} \tag{22}$$

and recall $\mathcal{M}_q(s) = -\mathcal{A}'_q(qs)$. Then we complete the proof of the theorem by combining formulas (21) and (22). □

***Proof of Theorem* 3.** First we will prove the convergence of ratios $\mathcal{Q}_{1j}(n)/\mathcal{Q}_{11}(n)$ to the limit π_j. It follows from relation (4) that

$$w_{n+1}(s) = w_n(s)\frac{f'\big(f_n(qs)\big)}{\beta}; \tag{23}$$

iterations for $f_n(s)$ used in last step. We see that $\mathcal{Q}_{11}(n) = \lim_{s\downarrow 0}\left[w_n(s)/s\right]$ and

$$\mathcal{Q}_{1j}(n) = \frac{1}{j!}\lim_{s\downarrow 0}\left[\mathrm{D}^j w_n(s)\right],$$

where $\mathrm{D}^j := \partial^j/\partial s^j$ is jth derivatives operator. By repeated differentiation of (23) one can verify the validity of the following relation for the j-th derivative of the function $w_n(s)$:

$$\mathrm{D}^j w_{n+1}(s) = \mathrm{D}^j w_n(s)\frac{f'\left(f_n(s)\right)}{\beta} + a_{j,n}(s),$$

where $a_{j,n}(s)$ is a power series with nonnegative coefficients. Thus we have

$$\begin{aligned}\frac{\mathcal{Q}_{1j}(n+1)}{\mathcal{Q}_{11}(n+1)} &= \frac{1}{j!}\lim_{s\downarrow 0}\frac{\mathrm{D}^j w_n(s)f'\left(f_n(s)\right) + \beta a_{j,n}(s)}{w_n(s)f'\left(f_n(s)\right)/s} \\ &\geq \frac{1}{j!}\lim_{s\downarrow 0}\frac{\mathrm{D}^j w_n(s)}{w_n(s)/s} = \frac{\mathcal{Q}_{1j}(n)}{\mathcal{Q}_{11}(n)}.\end{aligned}$$

So the ratios $\mathcal{Q}_{1j}(n)/\mathcal{Q}_{11}(n)$ are monotonically increasing. They are also bounded uniformly for all $n \in \mathbb{N}$. Therefore, they increasing converge to a finite limit, which we will denote by π_j. Since $f_n(s) \to q$ uniformly in $s \in [0,1)$ as $n \to \infty$, it follows from relation (3) that $w_n^{(i)}(s)/w_n(s) \to 1$, i.e. $\mathcal{Q}_{ij}(n)/\mathcal{Q}_{1j}(n) \to 1$ as $n \to \infty$ and for all $i, j \in \mathcal{E}$. Then

$$\lim_{n\to\infty}\frac{\mathcal{Q}_{ij}(n)}{\mathcal{Q}_{11}(n)} = \pi_j.$$

Since the Q-process is the discrete-homogeneous time Markov chain, the probabilities $\mathcal{Q}_{ij}(n)$ satisfy the Kolmogorov-Chapman equation. Then we can transform this equation to the following one:

$$\frac{\mathcal{Q}_{ij}(n+1)}{\mathcal{Q}_{11}(n+1)}\frac{\mathcal{Q}_{11}(n+1)}{\mathcal{Q}_{11}(n)} = \sum_{k\in\mathcal{E}}\frac{\mathcal{Q}_{ik}(n+1)}{\mathcal{Q}_{11}(n)}\mathcal{Q}_{kj}(1). \tag{24}$$

Now, taking limits as $n \to \infty$ of both sides of equation (24) we have

$$\pi_j = \sum_{k\in\mathcal{E}}\pi_k\mathcal{Q}_{kj}(1) \qquad \textit{for all} \quad j \in \mathcal{E}$$

since $\mathcal{Q}_{11}(n+1)/\mathcal{Q}_{11}(n) \to 1$ as $n \to \infty$. Repeatedly using last relation with itself for any $n \in \mathbb{N}$ we write the functional equation $\pi_j = \sum_{k\in\mathcal{E}}\pi_k\mathcal{Q}_{kj}(n)$. This equation point out to the invariant property of the numbers π_j with respect to probabilities $\mathcal{Q}_{ij}(n)$. To show that $\sum_{j\in\mathcal{E}}\pi_j = \left(1+q\gamma\right)^2$ we write

$$\pi(s) := \sum_{j \in \mathcal{E}} \pi_j s^j = \lim_{n \to \infty} \sum_{j \in \mathcal{E}} \frac{\mathcal{Q}_{1j}(n)}{\mathcal{Q}_{11}(n)} s^j = \lim_{n \to \infty} \frac{w_n(s)}{\mathcal{Q}_{11}(n)}$$

$$= \mathcal{M}_q(s) \lim_{n \to \infty} \frac{1}{\mathcal{Q}_{11}(n)} = \mathcal{M}_q(s) \lim_{n \to \infty} \lim_{s \downarrow 0} \frac{s}{w_n(s)}$$

$$= \mathcal{M}_q(s) \lim_{n \to \infty} \lim_{s \downarrow 0} \left(- \frac{\beta^n}{R_n'(qs)} \right) = \mathcal{M}_q(s) \big(1 + q\gamma\big)^2.$$

We used (8), (11), (21) and (22) in the last step, while taking into account the fact that $\mathcal{M}_q(s) = -\mathcal{A}_q'(qs)$. Since $\mathcal{M}_q(1) = 1$, we get to the desired. □

Proof of Theorem **4.** We will show now that the ratios $\mathcal{Q}_{ij}(n)/\mathcal{Q}_{11}(n)$ approach the limit π_j at the speed rate $\mathcal{O}(\beta^n)$. We have

$$\pi_n^{(i)}(s) := \sum_{j \in \mathcal{E}} \frac{\mathcal{Q}_{ij}(n)}{\mathcal{Q}_{11}(n)} s^j = \frac{1}{\mathcal{Q}_{11}(n)} \left[\frac{f_n(qs)}{q} \right]^{i-1} w_n(s) \tag{25}$$

Theorem 2 implies that

$$w_n(s) = \mathcal{M}_q(s) \left(1 + \mathcal{O}(\beta^n)\right) \qquad as \quad n \to \infty. \tag{26}$$

Then by the generating function continuity theorem

$$\mathcal{Q}_{11}(n) = \mu_1 \cdot \left(1 + \mathcal{O}(\beta^n)\right) \qquad as \quad n \to \infty. \tag{27}$$

Therewith integrating (26) over the interval $(1, s)$, taking into account (4), gives

$$f_n(qs) - q = \mathcal{O}(\beta^n) \int_1^s \mathcal{M}_q(x) dx \left(1 + \mathcal{O}(\beta^n)\right) \qquad as \quad n \to \infty.$$

Last integral converges since $\mathcal{M}_q(s)$ is bounded in $s \in [0, 1)$. Thus combination of relations (25)–(27) entail

$$\pi_n^{(i)}(s) = \frac{\mathcal{M}_q(s)}{\mu_1} \left(1 + \mathcal{O}(\beta^n)\right) \qquad as \quad n \to \infty.$$

This is more than the statement (14).

The proof of Theorem 4 is completed. □

Appendix: Total progeny

In the final part, we observe a random variable

$$S_n = W_0 + W_1 + \cdots + W_{n-1},$$

denoting the total state up to time n in the Q-process. By analogy with branching systems, this variable is of great interest in studying the deep properties of the

Q-process. For details on total progeny in GWB systems and related models results see e.g. [8–11].

At the first stage we need the joint GF of the variables W_n and S_n

$$J_n(s;x) = \sum_{j\in\varepsilon}\sum_{l\in\mathbb{N}} \mathsf{P}\{W_n = j, S_n = l\}\, s^j x^l$$

on a two-dimensional space

$$\mathbb{K} = \Big\{(s;x) \in \mathbb{R}^2 : \; s \in [0,1], x \in [0,1], \; \sqrt{(s-1)^2 + (x-1)^2} \neq 0\Big\}.$$

Lemma 2. *For $(s;x) \in \mathbb{K}$ and any $n \in \mathbb{N}$, the following recursive relation holds:*

$$J_{n+1}(s;x) = \frac{w(s)}{f_q(s)} J_n\big(xf_q(s);x\big), \tag{28}$$

where $w(s) = sf'(qs)/\beta$ and $f_q(s) = f(qs)/q$.

Proof. Consider a two-dimensional process $\{W_n, S_n\}$, which is a Markov chain with transition probabilities

$$\mathsf{P}\big\{W_{n+1} = j, S_{n+1} = l \;\big|\; W_n = i, S_n = k\big\} = \mathsf{P}_i\{W_1 = j, S_1 = l\}\, \delta_{l,i+k},$$

where δ_{ij} is the Kronecker's delta function. Therefore, we have

$$\begin{aligned}\mathsf{E}_i\left[s^{W_{n+1}}x^{S_{n+1}} \mid S_n = k\right] &= \sum_{j\in\mathcal{E}}\sum_{l\in\mathbb{N}} \mathsf{P}_i\{W_1 = j, S_1 = l\}\,\delta_{l,i+k} s^j x^l \\ &= \sum_{j\in\mathcal{E}} \mathsf{P}_i\{W_1 = j\}\, s^j x^{i+k} = w^{(i)}(s)\cdot x^{i+k}.\end{aligned}$$

From here, using the formula of total probabilities, we find

$$\begin{aligned}J_{n+1}(s;x) &= \mathsf{E}\left[\mathsf{E}\left[s^{W_{n+1}}x^{S_{n+1}} \mid W_n, S_n\right]\right] = \mathsf{E}\left[w^{(W_n)}(s)\cdot x^{W_n+S_n}\right] \\ &= \mathsf{E}\left[\big(f_q(s)\big)^{W_n-1}\cdot w(s)\cdot x^{W_n+S_n}\right] = \frac{w(s)}{f_q(s)}\cdot \mathsf{E}\left[\big(xf_q(s)\big)^{W_n}\cdot x^{S_n}\right].\end{aligned}$$

In the last line we have used formula (3). From here we obtain formula (28).

The proof of Lemma 2 is completed. □

Using relation (28), we can now obtain an explicit expression for the GF $J_n(s;x)$. Indeed, applying (28) consistently, taking into account (6) and, after simple transformations, we have

$$J_n(s;x) = \frac{s}{\beta^n}\cdot\frac{\partial H_n(s;x)}{\partial s}, \tag{29}$$

where the function $H_n(s;x)$ is defined for any $(s;x) \in \mathbb{K}$ by the following recursive relations:

$$\begin{cases} H_0(s;x) = s; \\ H_{n+1}(s;x) = xf_q\big(H_n(s;x)\big). \end{cases} \tag{30}$$

Since $\partial J_n(s;x)/\partial x\big|_{(s;x)=(1;1)} = \mathsf{E}S_n$, from (29) and (30), we find that

$$\mathsf{E}S_n = (1+\gamma)n - \gamma_q \frac{1-\beta^n}{1-\beta}, \tag{31}$$

where $\gamma_q = qf''(q)/\beta(1-\beta)$.

Let $\{Z_q(n), n \in \mathbb{N}_0\}$ be the GWB system generated by GF $f_q(s) = f(qs)/q$.

Remark 2. Needles to say that $\{Z_q(n), n \in \mathbb{N}_0\}$ is sub-critical GWB system. Then $V_n = \sum_{k=0}^{n-1} Z_q(k)$ is the total number of individuals that have existed up to the n-th generation in the system $\{Z_q(n)\}$. It is known that the GF of the joint distribution $\big(Z_q(n), V_n\big)$ satisfies the recursive equation (30); see [10, p. 126]. Thus, the function $H_n(s;x)$ is a two-dimensional GF for all $n \in \mathbb{N}$ and $(s;x) \in \mathbb{K}$ and has all the properties of the GF $\mathsf{E}\left[s^{Z_q(n)}x^{V_n}\right]$.

By virtue of what said in Remark 2, in studying $H_k(s;x)$ we use the properties of the GF $\mathsf{E}\left[s^{Z_q(n)}x^{V_n}\right]$. Since $\{Z_q(n), n \in \mathbb{N}_0\}$ is subcritical, it goes extinct with probability 1. Therefore, there exists a proper random variable $V = \lim_{n\to\infty} V_n$, which means the total number of descendants participating in the whole evolution of the system $\{Z_q(n)\}$. So

$$h(x) := \mathsf{E}x^V = \lim_{n\to\infty} \mathsf{E}x^{V_n} = \lim_{n\to\infty} H_n(1;x)$$

and, according to (30) it satisfies the functional equation

$$h(x) = xf_q\big(h(x)\big). \tag{32}$$

Further, we note that

$$\mathsf{P}\big\{Z_q(n) = 0, V_n = k\big\} = \mathsf{P}\big\{Z_q(n) = 0, V = k\big\}.$$

Then, due to the monotonicity of the probabilistic GF, we find

$$\mathsf{P}\big\{V = k\big\} - \sum_{i\in\mathbb{N}} \mathsf{P}\big\{Z_q(n) = i, V_n = k\big\}s^i \le \mathsf{P}\big\{V = k, Z_q(n) > 0\big\}.$$

Therefore, denoting

$$R_n(s;x) := h(x) - H_n(s;x)$$

for $(s;x) \in \mathbb{K}$, we have

$$R_n(s;x) \le \sum_{k\in\mathbb{N}} \mathsf{P}\left\{V = k, Z_q(n) > 0\right\} x^k = R_n(0;x).$$

It is easy to see $R_n(0;x) \le R_n(0;1) = \mathsf{P}\{Z_q(n) > 0\}$. Then

$$\big|R_n(s;x)\big| \le \mathsf{P}\{Z_q(n) > 0\} \longrightarrow 0 \qquad as \quad n \to \infty. \tag{33}$$

On the other hand, due to the fact that $|h(x)| \le 1$ and $|H_n(s;x)| \le 1$ we have

$$R_n(s;x) = x\left[f_q\big(h(x)\big) - f_q\big(H_{n-1}(s;x)\big)\right]$$

$$= x\mathsf{E}\big[h(x) - H_{n-1}(s;x)\big]^{Z_q(n)} \le \beta R_{n-1}(s;x)$$

for all $(s;x) \in \mathbb{K}$. This implies that

$$\big|R_n(s;x)\big| \le \beta^{n-k}\big|R_k(s;x)\big| \tag{34}$$

for any $n \in \mathbb{N}$ and $k = 0, 1, \ldots, n$.

In what follows, where the function $R_n(s;x)$ will be used, we deal with the domain $\mathbb{K}$, where this function does not vanish. By virtue of (33), taking into account (30), (32), we obtain the asymptotic formula

$$R_{n+1}(s;x) = xf'_q\big(h(x)\big)R_n(s;x) - x\frac{f''_q\big(h(x)\big) + \eta_n(s;x)}{2}R_n^2(s;x), \tag{35}$$

where $|\eta_n(s;x)| \to 0$ as $n \to \infty$ uniformly in $(s;x) \in \mathbb{K}$. Since $R_n(s;x) \to 0$, it follows from (35) that

$$R_n(s;x) = \frac{R_{n+1}(s;x)}{xf'_q\big(h(x)\big)}\big(1 + o(1)\big) \qquad as \quad n \to \infty.$$

By virtue of the last equality, we transform (35) to the form

$$R_{n+1}(s;x) = xf'_q\big(h(x)\big)R_n(s;x) - \left[\frac{f''_q\big(h(x)\big)}{2f'_q\big(h(x)\big)} + \epsilon_n(s;x)\right]R_n(s;x)R_{n+1}(s;x)$$

and, therefore

$$\frac{u(x)}{R_{n+1}(s;x)} = \frac{1}{R_n(s;x)} + v(x) + \epsilon_n(s;x), \tag{36}$$

where

$$u(x) = xf'_q\big(h(x)\big) \qquad and \qquad v(x) = x\frac{f''_q\big(h(x)\big)}{2u(x)}$$

and $\sup_{(s;x)\in\mathbb{K}}\big|\epsilon_n(s;x)\big| \le \epsilon_n \to 0$ as $n \to \infty$. By successively applying (36), we find the following representation for $R_n(s;x)$:

$$\frac{u^n(x)}{R_n(s;x)} = \frac{1}{R_0(s;x)} + \frac{v(x)\big[1 - u^n(x)\big]}{1 - u(x)} + \sum_{k=1}^{n}\epsilon_k(s;x)u^k(x). \tag{37}$$

Note that in the monograph [10, p. 136] formula (37) was stated for the critical GWB system.

Further, for convenience, we use the following form of the GF $J_n(s;x)$, which follows from relations (29) and (30):

$$J_n(s;x) = s\prod_{k=0}^{n-1}\frac{xf_q'(H_k(s;x))}{\beta}. \tag{38}$$

If we assume that $w''(1-) < \infty$, then using the the formula (38) by standard calculation we can find the following asymptotic relation as $n \to \infty$ for the covariance of W_n and S_n:

$$\mathsf{cov}(W_n, S_n) \sim \begin{cases} \dfrac{(\alpha-1)^2}{6}n^3, & \text{if } \beta = 1, \\ \mathcal{O}(1), & \text{if } \beta < 1. \end{cases}$$

Next we obtain the following asymptotes as $n \to \infty$ for the variances:

$$\mathsf{D}W_n \sim \begin{cases} \dfrac{(\alpha-1)^2}{2}n^2, & \text{if } \beta = 1, \\ \mathcal{O}(1), & \text{if } \beta < 1, \end{cases}$$

and

$$\mathsf{D}S_n \sim \begin{cases} \dfrac{(\alpha-1)^2}{12}n^4, & \text{if } \beta = 1, \\ \mathcal{O}(n), & \text{if } \beta < 1. \end{cases}$$

Therefore, we obtain the following asymptotic formula for the correlation coefficient of W_n and S_n:

$$\lim_{n\to\infty}\rho(W_n, S_n) = \begin{cases} \dfrac{\sqrt{6}}{3}, & \text{if } \beta = 1, \\ 0, & \text{if } \beta < 1. \end{cases}$$

The last statement indicates that in the case $\beta < 1$ there is an "asymptotic independence" between the variables W_n and S_n. In this case, we study the asymptotic properties of S_n.

Consider GF $T_n(x) := \mathsf{E}x^{S_n} = J_n(1;x)$. According to (38) we have

$$T_n(x) = \prod_{k=0}^{n-1} u_k(x), \tag{39}$$

where

$$u_n(x) = \frac{xf_q'(h_n(x))}{\beta},$$

and GF $h_n(x) = \mathsf{E}x^{V_n}$ for which $h_{n+1}(x) = xf_q(h_n(x))$.

Consider the function

$$\Delta_n(x) := h(x) - h_n(x).$$

This is a special case for $s = 1$ of the function $R_n(s;x)$ considered above. The following inequality is a consequence of inequality (34):

$$\left|\Delta_n(x)\right| \le \beta^{n-k}\left|\Delta_k(x)\right|$$

for $x \in \mathbb{K}$ and any $n \in \mathbb{N}$ and $k = 0, 1, \ldots n$. Successive application of the last inequality gives

$$\left|\Delta_n(x)\right| = \mathcal{O}\left(\beta^n\right) \to 0 \qquad as \quad n \to \infty \tag{40}$$

uniformly in $x \in \mathbb{K}$. Similarly to the case $R_n(s;x)$, taking into account (40) we find the following representation:

$$\frac{u^n(x)}{\Delta_n(x)} = \frac{1}{h(x)-1} + \frac{v(x)\left[1-u^n(x)\right]}{1-u(x)} + \sum_{k=1}^{n} \epsilon_k(x)u^k(x), \tag{41}$$

where

$$u(x) = xf_q'\left(h(x)\right) \qquad and \qquad v(x) = x\frac{f_q''\left(h(x)\right)}{2u(x)}$$

and $\sum_{x\in\mathbb{K}}\left|\epsilon_n(x)\right| \le \epsilon_n \to 0$ as $n \to \infty$.

Above reasonings imply the following analogue of the central limit theorem.

Theorem 5. *Let $\beta < 1$. Then*

$$\left|P\left\{\frac{S_n - ES_n}{\sqrt{2Dn}} < x\right\} - \Phi(x)\right| \le \frac{L(n)}{n^{1/4}},$$

where

$$D = \gamma\frac{2+\beta\gamma}{2(1-\beta)},$$

$\Phi(x)$–standard normal distribution function and $L(x)$–slowly varies at infinity.

The proof of this theorem and related details we leave for our subsequent papers.

References

1. Asmussen, S., Hering, H.: Branching Processes. Birkhäuser, Boston (1983)
2. Athreya, K.B., Ney, P.E.: Branching processes. Springer, New York (1972). https://doi.org/10.1007/978-3-642-65371-1
3. Harris, T.E.: The Theory of Branching Processes. Springer, Berlin (1963)
4. Imomov, A.A., Murtazaev, M.: On explicit form of the Kolmogorov constant in the theory of Galton-Watson Branching Processes. ArXiv: https://arxiv.org/abs/2205.03024, 2022, 9 pages

5. Imomov, A.A.: Limit Theorem for the Joint Distribution in the Q-processes. J. Siberian Federal Univ. Math. Phys. **7**(3), 289–296 (2014)
6. Imomov, A.A.: On Markov continuous time analogue of Q-processes. Theory Prob. and Math. Stat. **84**, 57–64 (2012)
7. Jagers, P.: Branching Progresses with Biological applications. Pitman Press, GB, JW & Sons (1975)
8. Karpenko, A.V., Nagaev, S.: V Limit theorems for the total number of descendents for the Galton-Watson branching process. Theory Probab. Appl. **38**, 433–455 (1994)
9. Kennedy, D.P.: The Galton-Watson process conditioned on the total progeny. Jour. Appl. Prob. **12**, 800–806 (1975)
10. Kolchin, V.F.: Random mappings. Nauka, Moscow (1984). (Russian)
11. Pakes, A.G.: Some limit theorems for the total progeny of a branching process. Adv. App. Prob. **3**(1), 176–192 (1971)
12. Sevastyanov, B.A.: Branching Processes. Nauka, Moscow (1971). (Russian)

Numerical Analysis of Shortest Queue Problem for Time-Scale Queueing System with a Small Parameter

Sergey A. Vasilyev(✉), Mohamed A. Bouatta, Shahmurad K. Kanzitdinov, and Galina O. Tsareva

Peoples' Friendship University of Russia (RUDN University), 6 Miklukho-Maklaya St, Moscow 117198, Russian Federation
vasilyev-sa@rudn.ru
http://www.rudn.ru

Abstract. 5G networks are a new technological step in the field of telecommunications. 5G networks provide the implementation of the required quality of communication with the growth of subscriber devices and lack of frequency bands. The application of queuing theory methods to analyze network performance is very important at the design, implementation and operation stages, as it is necessary to ensure a high return on investment that will be directed to the introduction of this new technology. Consequently, the attention of 5G researchers is particularly focused on the analysis of the shortest queue problem which is widely used as balancing mechanisms in time-scale queueing system (TSQS).

In this paper we employ simulation analysis of the TSQS evolution dynamics under the supposition that there are the large number of identical single-service devices and it is suppose this number increases indefinitely. It is assumed that all single-service devices have identical exponentially distributed service time with a finite mean value and a finite service intensity. It is supposed that there is a Poisson incoming stream of arriving requests with a finite intensity and TSQS fulfills a service discipline so that for each incoming request is provided a random selection a server device from random selected m-set server devices that has the s-th shortest queue size. The evolution of TSQS states can be represent by solutions of a system of differential equations of infinite degree. We formulate the singularly perturbed Cauchy problem for this system of differential equations with a small parameter. We apply the truncation procedure for this singularly perturbed Cauchy problem and formulate the finite order system of differential equations. We use a high-order non-uniform grid scheme for numerical solving of the truncated Cauchy problem. We implement the numerical scheme with different sets which allows to evaluate the impact of a small parameter in time-scaling processes for TSQS. The grid scheme demonstrates good convergence of solutions of the singularly perturbed Cauchy problem when a small parameter tend

S. A. Vasilyev—This paper has been supported by the RUDN University Strategic Academic Leadership Program.

A. Dudin et al. (Eds.): ITMM 2022, CCIS 1803, pp. 16–28, 2023.
https://doi.org/10.1007/978-3-031-32990-6_2

to zero. The results of the numerical analysis demonstrate that this TSQS keep services with a high incoming flow of requests.

Keywords: Countable Markov chains · Time-scale network analysis · Singular perturbed systems of differential equations · Numerical analysis of the Cauchy problem · Layer-adapted piecewise uniform Shishkin-type meshes

1 Introduction

The simulation of time-scale queueing systems (TSQS) is very important because of the implementation of 5G/6G networks and Internet of Things (IoT) requires to solve queuing theory problems using not only analytical methods [6,10,17,20] but also simulation ones [2,9].

The researching of TSQS with a huge number of identical server devices and shortest queue disciplines [5,11,19] associated with the Caushy problems for infinite degree systems of ordinary differential equations.

The problem of the time-scaling TSQS especially attracts attention among the methods of TSQS analysis [7,10,17] since it is an effective technique for analyzing large-scale complex systems. If there is a purpose to investigate the scale in time invariance of TSQS, then there is an opportunity to study the transformation properties of solutions of differential equations which define the dynamics of the system changes over time. The time-scaling change is similarity transformations of the solutions of a system of differential equations with a time-scaling parameter and it form a group of time-scale transformations of TSQS. The time-scaling methods often identify with the use of a small time-scaling parameter in TSQS models [3,4,12,13].

In recent times, it is proposed layer-adapted methods for the numerical simulation analysis of singularly perturbed systems differential equations which are modifications of non-uniform mesh methods [1,8,14,18].

In the paper [16] we considered the shortest queue system model and demonstrated the system evolution dynamics which was described by the solutions of the Cauchy problem of the Tikhonov type for singularly perturbed systems of differential equations of the infinite order for which the truncation method was applied.

In this paper we implement the numerical method for analysis of the evolution dynamics of TSQS with $n \to \infty$ identical single-service devices. We suppose that each single-service device has exponentially distributed service times of mean $\bar{t} = 1/\mu$, where $\mu > 0$ is a service intensity. We allow that there are a Poisson incoming requests with the intensity $n\lambda > 0$. We presume that TSQS implements the service discipline so that it is possible provide a randomly selection for each input request from any m-set server devices the one server device that has the s-th shortest (or equivalently, the $(m - s)$-th longest) queue size $(1 \leq s \leq m)$. The evolution dynamics of TSQS can be obtained as a solution $u_k^{s,m}(t)$ of a system of differential equations of infinite degree. We consider the singularly perturbed Cauchy problem for this system of differential equations with a small parameter. We use the truncation procedure for this singularly perturbed Cauchy problem and formulate the finite order system of differential equations. We apply a high-order non-uniform grid scheme of the Shishkin-type for numerical solving of the truncated Cauchy problem. We investigate of the solutions of singularly perturbed Cauchy problem for truncated system of differential equations with a small parameter and consider time-scaling processes for TSQS with different sets of small parameters. The grid scheme show good convergence of solutions of the singularly perturbed Cauchy problem when a small parameter $\varepsilon \to 0$. The results of the numerical simulation demonstrate that this TSQS can keep services with a high incoming flow of requests.

2 Time-Scale Queueing Systems Model

We consider a time-scale queueing system (TSQS), which consists of a large number $n \to \infty$ of infinite-buffer FCFS identical single service devices. We suppose that each device has exponentially distributed service time of mean $\bar{t} = 1/\mu$, where $\mu > 0$ is a service intensity. It is assumed the Poisson arrival process with rate of requests with $n\lambda > 0$.

We study TSQS which randomly selects m service devices for each arrived request and we assume that there is at least one device of this set m that has the s-th shortest queue length in the choice moment. If this condition holds then the request is sent to one of the selected service device immediately. If there happen to be more than one service devices with the s-th shortest queue size among m devices then one of the set is chosen randomly.

Let $\mathbf{y}^{s,m} = \{y_k^{s,m}(t)\}_{k=0}^{\infty}$ be shares of the service devices that have the queues lengths not less than k at time t ($y_0^{s,m}(t) = 1, y_k^{s,m}(t) \geq y_{k+1}^{s,m}(t)$, $k \in \mathcal{Z}_+$, $\mathcal{Z}_+ = \{0,1,2,\ldots\}$, $s,m \in \mathcal{N}$, $\mathcal{N} = \{1,2,\ldots\}$) are non-negative integers and $\sum_{k=1}^{\infty} y_k^{s,m}(t) < \infty$ for any $t \geq 0$.

If we assume that $n \to \infty$, the functions of the sequence $\mathbf{y}^{s,m}(t)$ become deterministic and the TSQS dynamics is described by solutions of an infinite system of differential equations

$$\dot{y}_k^{s,m}(t) = \mu\left(y_{k+1}^{s,m}(t) - y_k^{s,m}(t)\right) + \lambda \Delta h_{s,m}(y_k^{s,m}(t)),\ t \geq 0, \tag{1}$$

where $k \geq 1$, $1 \leq s \leq m$, $k,s,m \in \mathcal{N}$ and the function $\Delta h_{s,m}(y_k^{s,m}(t))$ has the form for $1 \leq s \leq m$ ($s,m \in \mathcal{N}$),

$$\Delta h_{s,m}(y_k^{s,m}(t)) = \left(h_{s,m}(y_{k-1}^{s,m}(t) - h_{s,m}(y_k^{s,m}(t))\right),$$

$$h_{s,m}(y_k^{s,m}(t)) = \sum_{l=0}^{s-1} C_m^l (1 - y_k^{s,m}(t))^l (y_k^{s,m}(t))^{m-l} =$$

$$= \sum_{l=0}^{s-1}\sum_{p=0}^{l} \frac{(-1)^{l-p} m!}{p!(m-l)!(l-p)!} (y_k^{s,m}(t))^{m-p},$$

$$\Delta h_{s,m}(y_k^{s,m}(t)) = \sum_{l=0}^{s-1}\sum_{p=0}^{l}(-1)^{l-p}C_m^l C_l^p[(y_{k-1}^{s,m}(t))^{m-p} - (y_k^{s,m}(t))^{m-p}] =$$

$$= \sum_{l=0}^{s-1}\sum_{p=0}^{l}\frac{(-1)^{l-p}m!}{p!(m-l)!(l-p)!}[(y_{k-1}^{s,m}(t))^{m-p} - (y_k^{s,m}(t))^{m-p}].$$

3 Time-Scale Queueing Systems Model with a Small Parameter

We investigate time-scaling properties of a solution of the Cauchy problem for the singularly perturbed infinite system of the differential equations (1) with small parameter $\varepsilon > 0$ in the such form:

$$\begin{cases} \varepsilon^{\rho_k}\dot{y}_k^{s,m}(t) + \mu\left(y_k^{s,m}(t) - y_{k+1}^{s,m}(t)\right) = \lambda\Delta h_{s,m}, k \geq 1,\ t \geq 0, \\ y_0^{s,m}(t) = 1,\ y_k^{s,m}(0) = y_k^0 \geq 0,\ k \geq 1,\ y_k^0 \geq y_{k+1}^0,\ 1 \leq s \leq m,\ s, m \in \mathcal{N}, \end{cases} \quad (2)$$

where $\rho = \{\rho_k\}_{k=1}^{\infty}$, $(\rho_k \geq 0)$ is a numerical sequence of real numbers. Hence, we use the time transformation in this form $\bar{t}_k = \varepsilon^{-\rho_k}t$. It is possible to consider the rapid changes processes with the help of such multi-scaling transformations and investigate TSQS evolution at different time intervals. A characteristic feature of singularly perturbed equations is that their solutions and/or derived solutions have special narrow zones (inner and boundary layers) in which there is a sharp transition from one stable state to another or to specified boundary values. Such situations arise, for example, in hydrodynamics in the flow of a viscous liquid in the region of boundary layers, where the viscous liquid passes from the boundary values set by the adhesion condition to an inviscid liquid, or in the vicinity of shock waves, where the gas passes from subsonic to supersonic condition. Chemical reactions are also characterized by a rapid transition from one state to another. In biology, such drastic changes occur in population genetics.

As a rule, the derivative solutions of singularly perturbed equations in the center of the inner and/or boundary layers reach the order of ε^{-k}, $k > 0$, i.e. tend to infinity when the small parameter tends to zero. Outside the layers, the derivatives are evaluated by some positive constant independent of ε. Due to these features, singularly perturbed equations are quite difficult to solve by standard methods in the case when the parameter ε is very small. If we consider such problems, it is used a method that is a set of analytical and numerical approaches. The analytical approach consists in qualitative analysis of solutions, namely, in obtaining suitable estimates of derivative solutions in the vicinity of zones with features. At the same time, complex multidimensional problems are reduced to simpler ordinary differential equations (amenable to analytical study) with the same structure of the qualitative behavior of the solution in layers. The obtained estimates are used to construct coordinate transformations that eliminate layers. Using the latter, we find an approximate solution of the problem we are interested in using standard numerical schemes, approximating either the transformed problem in a new computational domain in the resulting coordinate system on a uniform grid, or an initial problem in the physical domain on an uneven grid, on which the grid thickening in layers is described by a coordinate transformation. Approach, described in the monograph, it is mainly aimed at solving nonlinear and bisingular equations. Bisingular equations are equations whose solutions have additional features when reducing (ε =) the problem. Their solutions have various types of layers—both well-known exponential and more frequently occurring power and combined ones. Such layers appear, for example, in solutions of singularly perturbed second -order equations whose coefficients before the first derivatives are not separated from zero (equations with turning points).

The Cauchy problem (2) can be represented as:

$$\begin{cases} \dot{\mathbf{y}} + \mathbf{M}(\mu, \varepsilon, \rho)\mathbf{y}(t) = \mathbf{F}(\mathbf{y}(t), \lambda, \varepsilon, \mathbf{R}), \\ \mathbf{y}(0) = \mathbf{y}^0, \end{cases} \tag{3}$$

$$\mathbf{y}(t) = (y_1^{s,m}(t), y_2^{s,m}(t), \ldots, y_n^{s,m}(t), \ldots),\ y_0^{s,m}(t) = 1,$$
$$\mathbf{M} = (M_1, M_2, \ldots, M_n, \ldots),\ \mathbf{F} = (F_1, F_2, \ldots, F_n, \ldots),$$
$$M_k = \varepsilon^{-\rho_k} \mu \left(y_k^{s,m}(t) - y_{k+1}^{s,m}(t)\right),\ F_k = \varepsilon^{-\rho_k} \lambda \Delta h_{s,m},\ k \ge 1,$$
$$\mathbf{y}(0) = (y_1^{s,m}(0), \ldots, y_n^{s,m}(0), \ldots),\ \mathbf{y}^0 = (y_1^0, y_2^0, \ldots, y_n^0, \ldots),\ y_k^0 \ge y_{k+1}^0,\ k \ge 1$$
$$\mathbf{R} = (\rho_1, \rho_2, \ldots, \rho_n, \ldots).$$

4 Truncated System of Differential Equations and Numerical Analysis

The Cauchy problem (3) for a truncated system of differential equations with a finite order n and n initial conditions has the form:

$$\begin{cases} \dot{\tilde{y}} + \tilde{\mathbf{M}}(\mu, \varepsilon, \tilde{\mathbf{R}})\tilde{\mathbf{y}}(t) = \tilde{\mathbf{F}}(\tilde{\mathbf{y}}(t), \lambda, \varepsilon, \tilde{\mathbf{R}}), \\ \tilde{\mathbf{y}}(0) = \tilde{\mathbf{y}}^0, \end{cases} \tag{4}$$

where

$$\tilde{\mathbf{y}}(t) = (y_1^{s,m}(t), y_2^{s,m}(t), \ldots, y_n^{s,m}(t)),\ y_0^{s,m}(t) = 1,$$
$$\tilde{\mathbf{M}} = (M_1, M_2, \ldots, M_n),\ \tilde{\mathbf{F}} = (F_1, F_2, \ldots, F_n),$$
$$M_k = \varepsilon^{-\rho_k} \mu \left(y_k^{s,m}(t) - y_{k+1}^{s,m}(t)\right),\ k \in \overline{1, n-1},\ M_n = \varepsilon^{-\rho_k} \mu y_k^{s,m}(t),$$
$$F_k = \varepsilon^{-\rho_k} \lambda \Delta h_{s,m},\ k \in \overline{1, n},$$
$$\tilde{\mathbf{y}}(0) = (y_1^{s,m}(0), \ldots, y_n^{s,m}(0)),\ \tilde{\mathbf{y}}^0 = (y_1^0, y_2^0, \ldots, y_n^0),\ y_k^0 \ge y_{k+1}^0,\ k \in \overline{1, n-1},$$
$$\tilde{\mathbf{R}} = (\rho_1, \rho_2, \ldots, \rho_n).$$

This Cauchy problem (4) is the Tikhonov problem, if we assume that $b_k = 0$, $k = \overline{1, l}$, $b_k > 0$, $k = \overline{l+1, n}$ $(2 \le l \le n-1)$ [15].

We use a piecewise-uniform grid $\bar{\Xi}_\tau[0,T_0]$ $(t_0 = 0, t_i < t_{i+1}, i = \overline{0, N-1}, t_N = T_0)$ for numerical analysis of the problem (4)

$$\bar{\Xi}_\tau[0,T_0] = (t_i | t_i = i\tau_1;\ i = \overline{0,K};\ t_i = t_K + i\tau;\ i = \overline{1,M};$$
$$t_i = t_{K+M} + i\tau_2;\ i = \overline{1,2L};\ t_i = t_{K+M+2L} + i\tau;\ i = \overline{1, N-K-M-2L}),$$
$$\tau_1 = \delta_1/K,\ \tau_2 = \delta_2/2L,\ \delta_j = \bar{D}_j \varepsilon\, ln(\varepsilon^{-1})),\ j = 1,2;$$
$$\tau = (T_0 - \delta_1 - \delta_2)/(N - K - M - 2L),$$

where parameters $\bar{D}_j$ $(j = 1,2)$ are determined analytically by asymptotic estimates of solutions of the Tikhonov problem (4) and $T_{IBL} = t_{K+M} + L\tau_2$ is a point where there is an inner boundary layer. Thus, this piecewise-uniform grid $\bar{\Xi}_\tau$ has K small steps τ_1, $2L$ small steps τ_2, and $(N-K-M-2L)$ big steps τ on the segment $[0, T_0]$.

We use a finite-difference approximation of the problem (4) in the form:

$$\begin{cases} \tilde{\mathbf{y}}_{i+1} = \mathbf{B}_i, i = \overline{1, N-1}, \\ \tilde{\mathbf{y}}_0 = \tilde{\mathbf{y}}^0, \end{cases} \tag{5}$$

where $h_i = t_i - t_{i-1}$, $(i = \overline{1, N-1})$, $y_{k,i}^{s,m} = y_k^{s,m}(t_i)$, $y_{0,i}^{s,m} = 1$ $(i = \overline{0,N})$,

$$\mathbf{B}_i = (B_{1i}, B_{2i}, \ldots, B_{ni}),\ \mathbf{B}_i = \tilde{\mathbf{y}}_i + h_i[\tilde{\mathbf{F}}_i - \tilde{\mathbf{M}}\tilde{\mathbf{y}}_i],$$

$$F_{ki} = \varepsilon^{-\rho_k} \lambda \sum_{l=0}^{s-1} \sum_{p=0}^{l} \frac{(-1)^{l-p} m!}{p!(m-l)!(l-p)!} [(y_{k-1,i}^{s,m})^{m-p} - (y_{k,i}^{s,m})^{m-p}],$$
$$k \in \overline{1,n},\ i = \overline{0,N}$$
$$B_{ki} = y_{k,i}^{s,m} + h_i \varepsilon^{-\rho_k} [\mu \left(y_{k+1,i}^{s,m} - y_{k,i}^{s,m} \right) +$$
$$+ \lambda(h_{s,m}(y_{k-1,i}^{s,m}) - h_{s,m}(y_{k,i}^{s,m}))],\ k = \overline{1,n},\ i = \overline{0,N},$$
$$\tilde{\mathbf{y}}_0 = (y_{1,0}^{s,m}, y_{2,0}^{s,m}, \ldots, y_{n,0}^{s,m}), \tilde{\mathbf{y}}^0 = (y_1^0, y_2^0, \ldots, y_n^0),\ y_k^0 \geq y_{k+1}^0,\ k \in \overline{1, n-1}.$$

We apply the fourth-order Runge-Kutta method for problem (5):

$$\mathbf{g}_i^1 = \mathbf{B}(\mathbf{y}_i, t_i),\ \mathbf{g}_i^2 = \mathbf{B}(\mathbf{y}_i + \frac{h_i}{2}\mathbf{g}_i^1, t_i + h_i/2),$$
$$\mathbf{g}_i^3 = \mathbf{B}(\mathbf{y}_i + \frac{h_i}{2}\mathbf{g}_i^2, t_i + h_i/2),\ \mathbf{g}_i^4 = \mathbf{B}(\mathbf{y}_i + h_i \mathbf{g}_i^3, t_i + h_i),$$
$$\mathbf{y}_{i+1} = \mathbf{y}_i + \frac{h_i}{6}(\mathbf{g}_i^1 + 2\mathbf{g}_i^2 + 2\mathbf{g}_i^3 + \mathbf{g}_i^4),$$

where $\mathbf{g}_i^j \in R^n$ $(i = \overline{0,N},\ j = \overline{1,4})$ are vectors.

The numerical analysis of the solutions $w_k^{s,m}(t)$ $(k = \overline{1,25};\ m = 2,3,4;\ 1 \leq s \leq m)$ is presented in the figures (see Fig. 1, 2, 3, 4, 5, 6, 7, 8 and 9).

Table 1. Simulation parameters

Parameters	Values of the parameters
λ_l	4
λ_h	6
μ	5
n	25
$l\ (1 \le k \le 9)$	9
$b_k(1 \le k \le 9)$	0
$b_k(10 \le k \le 25)$	$1/k$
N	10^4
μ	10^{-5}

The parameters of the simulation such as the low incoming mode of arrival request rates λ_l, the high incoming mode of arrival request rates λ_h, the service intensity μ, the number of steps of the grid N, the permissible error μ and so on are shown in the Table 1.

The values of the initial conditions are the following sequence of numbers $y_k^0 = (28 - k)/30$, $k = \overline{1, 25}$.

The values of the parameters s, m, ε are presented in the captions under the figures (see Fig. 1, 2, 3, 4, 5, 6, 7, 8 and 9). In the Fig. 1 we can see the results of the numerical stimulation of the solution $y_k^{1,2}(t)$. It is shown that TSQS has an unstable service mode under overload conditions. There are the left and inner transition layers. The transitions become more sharper when $\varepsilon \to 0$. In the Fig. 2 we can see the results of the numerical stimulation of the solution $y_k^{2,2}(t)$. It is shown that TSQS has a stable service mode under overload conditions. Thus, an increase in the parameter s leads to the stably service. There are only the left

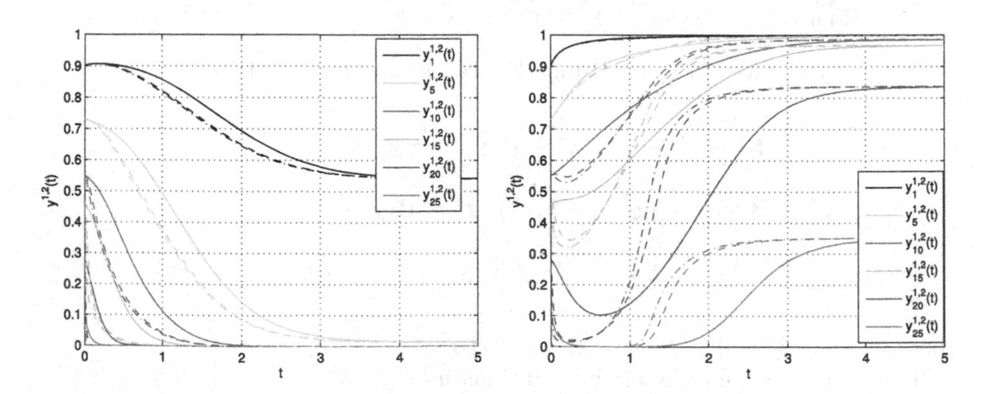

Fig. 1. The function $y_k^{1,2}(t)$, $\lambda = 4$ for the left graph, $\lambda = 6$ for the right graph, $\mu = 5$, $\varepsilon = 0.1$ solid line, $\varepsilon = 0.01$ long dash line, $\varepsilon = 0.001$ short dash line.

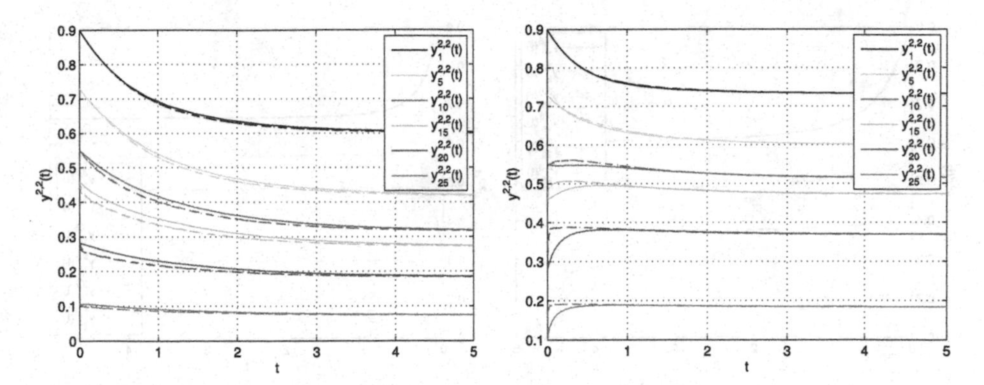

Fig. 2. The function $y_k^{2,2}(t)$, $\lambda = 4$ for the left graph, $\lambda = 6$ for the right graph, $\mu = 5$, $\varepsilon = 0.1$ solid line, $\varepsilon = 0.01$ long dash line, $\varepsilon = 0.001$ short dash line.

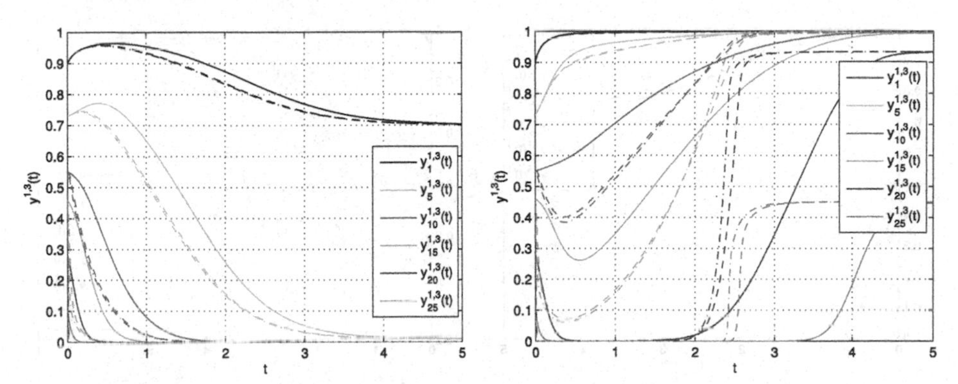

Fig. 3. The function $y_k^{1,3}(t)$, $\lambda = 4$ for the left graph, $\lambda = 6$ for the right graph, $\mu = 5$, $\varepsilon = 0.1$ solid line, $\varepsilon = 0.01$ long dash line, $\varepsilon = 0.001$ short dash line.

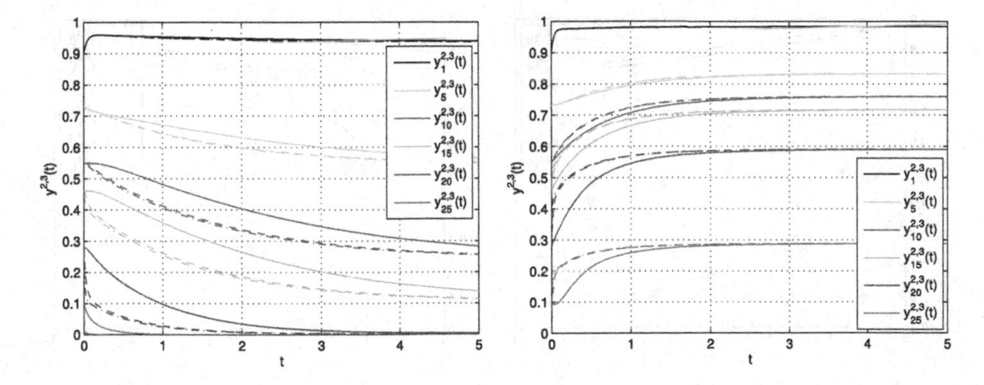

Fig. 4. The function $y_k^{2,3}(t)$, $\lambda = 4$ for the left graph, $\lambda = 6$ for the right graph, $\mu = 5$, $\varepsilon = 0.1$ solid line, $\varepsilon = 0.01$ long dash line, $\varepsilon = 0.001$ short dash line.

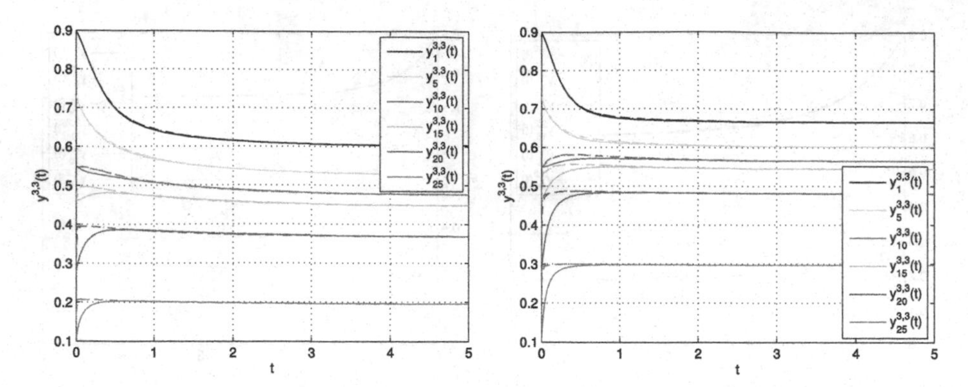

Fig. 5. The function $y_k^{3,3}(t)$, $\lambda = 4$ for the left graph, $\lambda = 6$ for the right graph, $\mu = 5$, $\varepsilon = 0.1$ solid line, $\varepsilon = 0.01$ long dash line, $\varepsilon = 0.001$ short dash line.

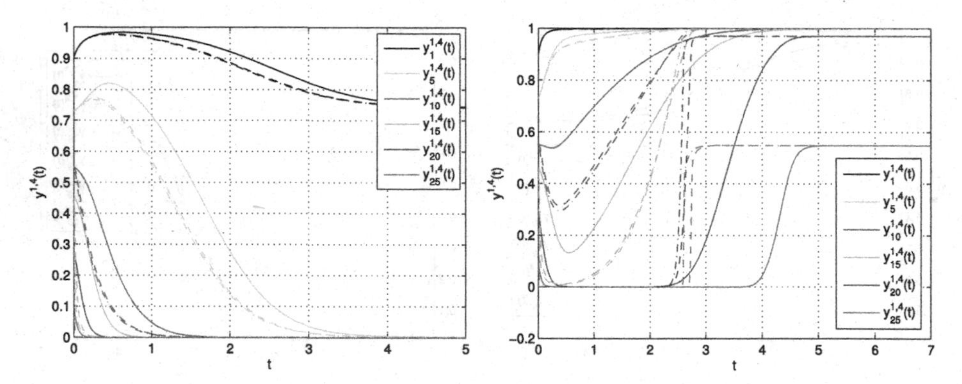

Fig. 6. The function $y_k^{1,4}(t)$, $\lambda = 4$ for the left graph, $\lambda = 6$ for the right graph, $\mu = 5$, $\varepsilon = 0.1$ solid line, $\varepsilon = 0.01$ long dash line, $\varepsilon = 0.001$ short dash line.

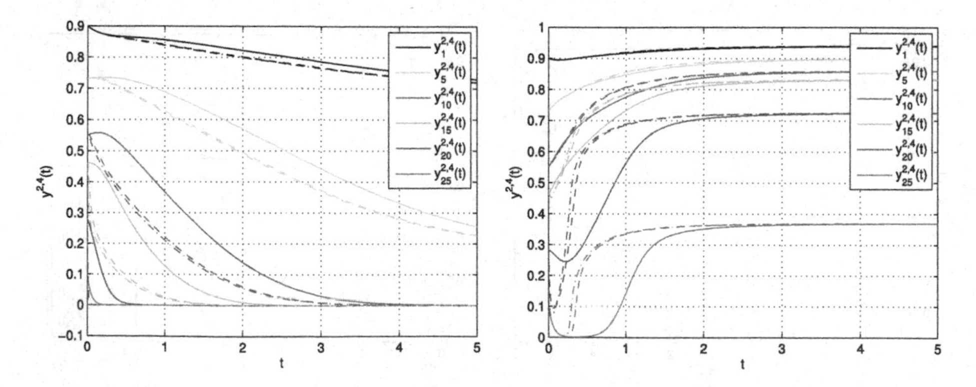

Fig. 7. The function $y_k^{2,4}(t)$, $\lambda = 4$ for the left graph, $\lambda = 6$ for the right graph, $\mu = 5$, $\varepsilon = 0.1$ solid line, $\varepsilon = 0.01$ long dash line, $\varepsilon = 0.001$ short dash line.

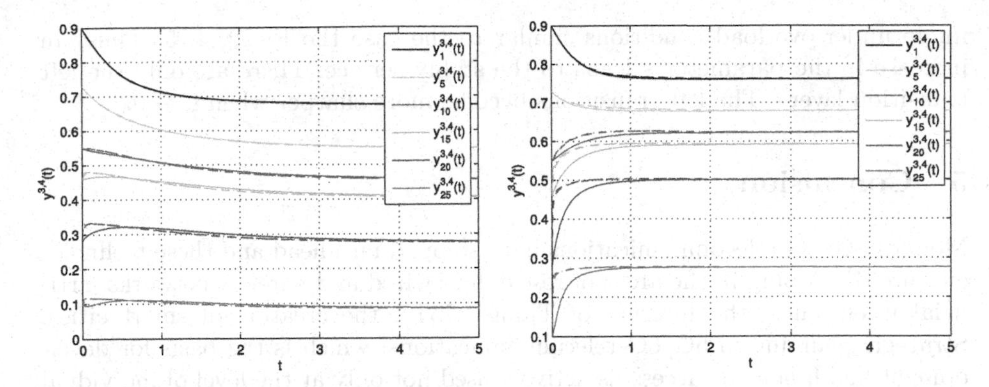

Fig. 8. The function $y_k^{3,4}(t)$, $\lambda = 4$ for the left graph, $\lambda = 6$ for the right graph, $\mu = 5$, $\varepsilon = 0.1$ solid line, $\varepsilon = 0.01$ long dash line, $\varepsilon = 0.001$ short dash line.

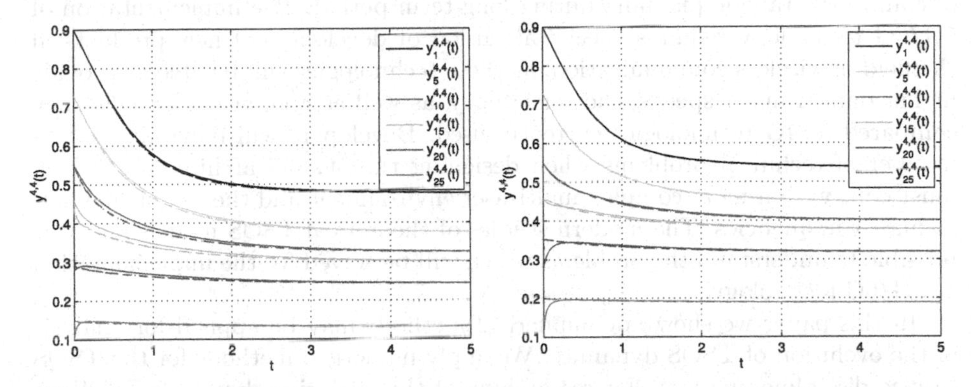

Fig. 9. The function $y_k^{4,4}(t)$, $\lambda = 4$ for the left graph, $\lambda = 6$ for the right graph, $\mu = 5$, $\varepsilon = 0.1$ solid line, $\varepsilon = 0.01$ long dash line, $\varepsilon = 0.001$ short dash line.

transition layers. The left transitions become more sharper when $\varepsilon \to 0$. In the Fig. 3 we can see the results of the numerical stimulation of the solution $y_k^{1,3}(t)$. It is shown that TSQS has an unstable service mode under overload conditions similar to the case in the Fig. 1. There are the left and inner transition layers. The transitions become more sharper when $\varepsilon \to 0$. In the Fig. 4, Fig. 5 we can see the results of the numerical stimulation of the solutions $y_k^{2,3}(t)$, $y_k^{3,3}(t)$. It is shown that TSQS has an stable service mode under overload conditions similar to the case the Fig. 2. Thus, an increase in the parameter s leads to the stably service. There are only the left transition layers. The left transitions become more sharper when $\varepsilon \to 0$. In the Fig. 6 we can see the results of the numerical stimulation of the solution $y_k^{1,4}(t)$. It is shown that TSQS has an unstable service mode under overload conditions similar to the case in the Fig. 1 and Fig. 3. There are the left and inner transition layers. The transitions become more sharper when $\varepsilon \to 0$. In the Fig. 7, 8 and 9 we can see the results of the numerical stimulation of the solutions $y_k^{2,4}(t)$, $y_k^{3,4}(t)$, $y_k^{4,4}(t)$. It is shown that TSQS has an stable service mode under overload conditions similar to the case the Fig. 2, 4, 5. Thus, an increase in the parameter s leads to the stably service. There are only the left transition layers. The left transitions become more sharper when $\varepsilon \to 0$.

5 Conclusion

Modern 5G/6G telecommunications have stepped far ahead and these technologies are developing in the areas of research of wired and wireless networks, artificial intelligence, the Internet of Things (IoT), the creation of smart cities. Strategic planning in 5G/6G telecommunications, which is the basis for development and financial success, is actively used not only at the level of individual companies, but also entire countries. It is necessary to predict the innovative changes for successful development of the 5G/6G telecommunications technology and for strategic planning during long-term period. The implementation of 5G/6G technology requires special attention of developers of new products in the field of wireless communication, as these technologies will be able to provide higher data transfer speeds, shorter latency, as well as greater energy efficiency compared to 4G technologies currently used. Developers will inevitably face a number of technical problems when designing the 5G/6G architecture, which must cope with a more complex multi-user environment and the use of channels at higher frequencies. The modern studies of theoretical TSQS models makes it possible to understand the problems which will be solved for the implementation of 5G/6G technology.

In this paper we show how numerical methods may be applied for analysis of the evolution of TSQS dynamics. We apply numerical methods for the TSQS service discipline analysis. For researching of the evolution dynamics TSQS we investigate the function $u_k^{s,m}(t)$ which can be found by numerically solving the system of differential equations finite degree. We use a high-order non-uniform grid scheme of the Shishkin-type for numerical solving of the Cauchy problem for the system of differential equations finite degree. We use different sets of small parameters for time-scaling processes analysis for TSQS. The high-order non-uniform grid scheme demonstrates good convergence of solutions of the singularly perturbed Cauchy problem when a small parameter $\varepsilon \to 0$. In addition, our proposed methods can serve as an approximation of queuing systems with a large but finite number of servers.

This paper has been supported by the RUDN University Strategic Academic Leadership Program (recipient S.A. Vasilyev, mathematical model development, simulation model development, numerical analysis).

References

1. Andreev, V., Savin, I.: The uniform convergence with respect to a small parameter of A. A. Samarskii's monotone scheme and its modification. Comput. Math. Math. Phys. **35**, 581–591 (1995). http://mi.mathnet.ru/eng/zvmmf/v35/i5/p739

2. Baddour, A., Malykh, M., Sevastianov, L.: On periodic approximate solutions of dynamical systems with quadratic right-hand side. J. Math. Sci. **261**, 698–708 (2022). https://doi.org/10.1007/s10958-022-05781-4
3. Bushkova, T., Moiseeva, S., Moiseev, A., Sztrik, J., Lisovskaya, E., Pankratova, E.: Using infinite-server resource queue with splitting of requests for modeling two-channel data transmission. Methodol. Comput. Appl. Probabil. 1–20 (2021). https://doi.org/10.1007/s11009-021-09890-6
4. Danilyuk, E., Moiseeva, S., Nazarov, A.: Asymptotic diffusion analysis of an retrial queueing system M/M/1 with impatient calls. In: Vishnevskiy, V.M., Samouylov, K.E., Kozyrev, D.V. (eds.) Distributed Computer and Communication Networks. DCCN 2021. Communications in Computer and Information Science, vol. 1552. Springer, Cham (2022) https://doi.org/10.1007/978-3-030-97110-6_18
5. Dimitriou, I.: Analysis of the symmetric join the shortest orbit queue. Oper. Res. Lett. **49**(1), 23–29 (2021). https://doi.org/10.1016/j.orl.2020.10.011
6. Dudin, A., Dudina, O., Dudin, S., Gaidamaka, Y.: Self-service system with rating dependent arrivals. Mathematics **10**(3), 297 (2022). https://doi.org/10.3390/math10030297
7. Hu, K., Wang, B., Cao, S., Li, W., Wang, L.: A novel model predictive control strategy for multi-time scale optimal scheduling of integrated energy system. Energy Rep. **8**, 7420–7433 (2022). https://doi.org/10.1016/j.egyr.2022.05.184
8. Kaushik, A., Choudhary, M.: A higher-order uniformly convergent defect correction method for singularly perturbed convection-diffusion problems on an adaptive mesh. Alexandria Eng. J. **61**(12), 9911–9920 (2022). https://doi.org/10.1016/j.aej.2022.03.005
9. Kondratyeva, A., et al.: Characterization of dynamic blockage probability in industrial millimeter wave 5G deployments. Future Internet **14**(7), 193 (2022). https://doi.org/10.3390/fi14070193
10. van Kreveld, L.R., Boxma, O.J., Dorsman, J.L., Mandjes, M.R.H.: Scaling limits for closed product-form queueing networks. Perform. Eval. **151**, 102220 (2021). https://doi.org/10.1016/j.peva.2021.102220
11. Liu, X., Gong, K., Ying, L.: Steady-state analysis of load balancing with Coxian-2 distributed service times. Naval Res. Logist. **69**(1), 57–75 (2022). https://doi.org/10.1002/nav.21986
12. Nazarov, A., Dudin, A., Moiseev, A.: Pseudo steady-state period in non-stationary infinite-server queue with state dependent arrival intensity. Mathematics **10**(15), 2661 (2022). https://doi.org/10.3390/math10152661
13. Polkhovskaya, A., Moiseeva, S., Danilyuk, E.: Asymptotic analysis of retrial queueing system $M/M/1$ with non-persistent customers and collisions. In: Dudin, A., Nazarov, A., Moiseev, A. (eds.) Information Technologies and Mathematical Modelling. Queueing Theory and Applications. ITMM 2021. Communications in Computer and Information Science, vol. 1605. Springer, Cham (2022) https://doi.org/10.1007/978-3-031-09331-9_27
14. Roul, P.: A fourth-order non-uniform mesh optimal B-spline collocation method for solving a strongly nonlinear singular boundary value problem describing electrohydrodynamic flow of a fluid. Appl. Numer. Math. **153**, 558–574 (2020). https://doi.org/10.1016/j.apnum.2020.03.018
15. Tihonov, A.N.: Systems of differential equations containing small parameters in the derivatives. Mat. Sbornik N. S. **31**(73), pp. 575–586 (1952). http://mi.mathnet.ru/eng/msb/v73/i3/p575

16. Vasilyev, S.A., Bouatta, M.A., Tsareva, G.O.: High-order non-uniform grid scheme for numerical analysis of queueing system with a small parameter. In: Silhavy, R., Silhavy, P., Prokopova, Z. (eds.) Data Science and Algorithms in Systems. CoMeSySo 2022. Lecture Notes in Networks and Systems, vol. 597, pp. 785–797. Springer, Cham (2023) https://doi.org/10.1007/978-3-031-21438-7_66
17. Yang, P., et al.: Hierarchical multiple time scales cyber-physical modeling of demand-side resources in future electricity market. Int. J. Electrical Power Energy Syst. **133**, 107184 (2021) https://doi.org/10.1016/j.ijepes.2021.107184
18. Zhang, J., Liu, X.: Uniform convergence of a weak Galerkin finite element method on Shishkin mesh for singularly perturbed convection-diffusion problems in 2D. Appl. Math. Comput. **432**, 127346 (2022). https://doi.org/10.1016/j.amc.2022.127346
19. Zhou, X., Shroff, N., Wierman, A.: Asymptotically optimal load balancing in large-scale heterogeneous systems with multiple dispatchers. Perform. Eval. **145**, 102146 (2021). https://doi.org/10.1016/j.peva.2020.102146
20. Zisgen, H.: An approximation of general multi-server queues with bulk arrivals and batch service. Oper. Res. Lett. **50**(1), 57–63 (2022). https://doi.org/10.1016/j.orl.2021.12.006

Mathematical Modeling of Virtual Machine Life Cycle Using Branching Renewal Process

Ekaterina Fedorova, Ivan Lapatin, Olga Lizyura(✉), Alexander Moiseev, Anatoly Nazarov, and Svetlana Paul

Institute of Applied Mathematics and Computer Science, National Research Tomsk State University, 36 Lenina ave., Tomsk 634050, Russia

oliztsu@mail.ru

Abstract. In this paper, we propose the branching renewal process as a model of virtual machine life cycle in cloud node. We assume that virtual machines have several working modes with different resource consumption. Each machine begins with zero phase and continues performing by probabilistic change of other phases. Virtual machines can leave the cloud node after any phase with some probability. For this model, we obtain probability distribution of the number of simultaneously performing virtual machines in each phase and the distribution function of resource consumption in cloud node.

Keywords: Branching process · Virtual machine life cycle · Cloud system · Resource consumption

1 Introduction

Cloud services are rapidly growing area of services that provides to the customers storage, computing resources, and communications. The cloud server consists of physical machines that must be distributed among customers according to their requests and tariffs. To do this, virtual machines are used, which can be located on one physical machine of the cloud node and isolate the services provided to the user [3,9,13].

Mathematical models of cloud services usually aim to reflect the whole cloud network. The most studies of cloud models consider multiserver systems with one or more queues and some kind of traffic manager, which distribute the requests between cloud servers to satisfy the QoS requirements [1,2,15].

The most popular model for the cloud node is classical queue with finite or infinite queue length. In paper [7], authors describe Markovian single-server model of cloud node with correlated reneging. They consider stationary and transient probability characteristics of queue length in the system.

This work has been supported by Huawei Cloud.

A. Dudin et al. (Eds.): ITMM 2022, CCIS 1803, pp. 29–39, 2023.
https://doi.org/10.1007/978-3-031-32990-6_3

Paper [5] consider the multiserver model of cloud service with finite queue length, correlated reneging and resubmission. The model is markovian and the study is devoted to the steady state distribution of queue length.

Many papers are devoted to the predicting of virtual machine resource consumption and load balancing in cloud networks using neural networks [4,6,10]. Sometimes, instead of predicting workload, the scheduling algorithms for virtual machine migration are used to improve performance [8,11,12,14].

Thus, the problem of evaluation of workload generated by virtual machines is relevant for predicting resource consumption (for instance, CPU utilization) in a cloud node.

The main idea of our research is to take into account the heterogeneity of service by considering multiple phases of virtual machine (VM) life cycle. We assume that VM occupies some random amount of resource in any phase of performance. Thus, the model can help us to obtain the distribution of the total amount of the resource, which is occupied in the server. By resource, we mean any significant supplies that virtual machines compete for, such as memory, CPU utilization, energy, etc.

The rest of the paper is organized as follows. In Sect. 2, we describe mathematical model of branching renewal process based on Poisson arrivals. Each event of Poisson process generates a branch of additional events. Section 3 is devoted to the investigation of the number of simultaneously performed virtual machines, which is also the number of branches of considered branching process. In Sect. 4, we consider the probability distribution of the total amount of occupied resource in the cloud node based on the results of Sect. 3. Section 5 is dedicated for the numerical experiments and obtaining of performance metrics of the cloud system. In Sect. 6, we present some concluding remarks.

2 Mathematical Model

In the considered branching process, the number of phases is $N+1$. Each VM begins with zero phase, which means the interval of time from definition of VM to the first moment of phase switching. After zero phase, VM can leave the server or switch to another phase. Each phase can be completed by the VM leaving the server or by a probabilistic phase change.

We propose mathematical model of VM life cycle as branching renewal process based on $N+1$ types of VM activity phases (Fig. 1).

The moments of arrivals of virtual machines in the flow are marked with asterisks on the time axis t. We will call such a flow as **generating (main)**. Each event of the main flow generates a sequence of a virtual machine arriving and probabilistic repetitions of its phases until it departures from the node.

Let the generating (main) flow be a stationary Poisson process with parameter λ. At the moment of arrival, zero phase of the branching flow begins. Its duration has an exponential distribution with the parameter μ_0. After the end of zero phase, the flow cease to exist with probability v_0 or, which is the same, the virtual machine departures from the node, and with probability r_{0n}, its n-th

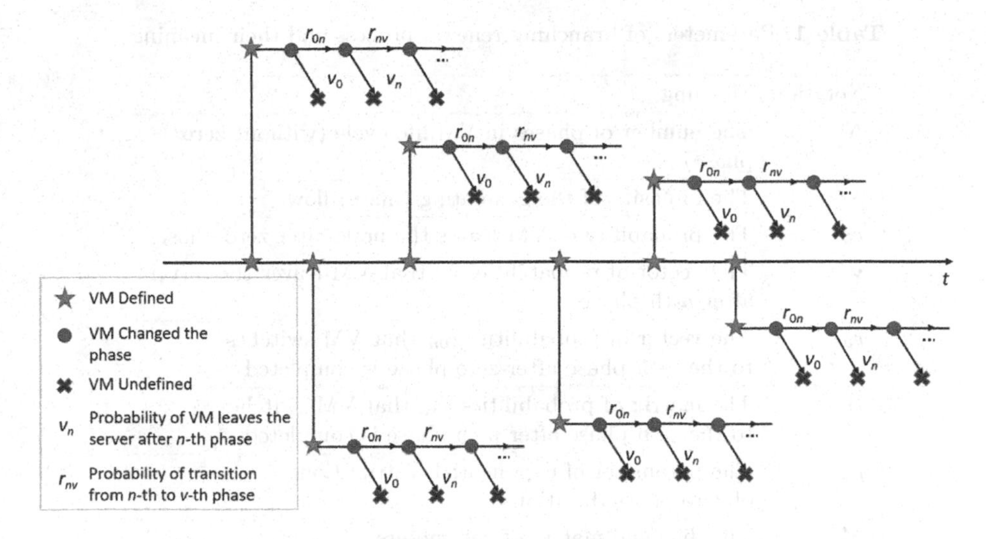

Fig. 1. Branching renewal process with $N+1$ types of VM activity

phase begins, $n = 1, 2, \ldots, N$. The n-th phase's duration has an exponential distribution with parameter μ_n, $n = 1, 2, \ldots, N$. At the end of n-th phase, the branching flow can cease to function with probability v_n or go into another v-th phase with probability r_{nv}, $n, v = 1, 2, \ldots, N$.

In this model, we will assume that the cloud system allocates resources of volume V_n, $n = 0, 1, \ldots, N$ to the virtual machine in the corresponding phase. Volumes V_n are defined by characteristic functions $h_n(u), n = 0, 1, \ldots, N$.

We seek for the distribution of the total volume V of resources occupied by all virtual machines. To do this, it is necessary to know the probability distribution of the numbers $i_0, i_1, \ldots, i_N$ of simultaneously implemented phases of the branching flow and to define characteristic functions $h_n(u)$ of resource volumes V_n, $n = 0, 1, \ldots, N$ allocated to virtual machines in the corresponding phase.

In Table 1, we introduce the required parameters and notations for the model.

3 Probability Distribution of the Number of Simultaneously Performed Phases

Let $i_n(t)$ be the number of virtual machines operating in n-th phase, $n = 0, 1, \ldots, N$. Denote probability distribution

$$\mathbb{P}\{i_0(t) = i_0,\ i_1(t) = i_1,\ ...,\ i_N(t) = i_N\} = P(i_0,\ i_1,\ ...,\ i_N)$$

and write the distribution in the vector form

$$P(i_0, i_1, ..., i_N) = P(\mathbf{i}).$$

Table 1. Parameters of branching renewal process and their meaning

Notation	Meaning
N	The number of phases in VM life cycle (without zero phase)
λ	The intensity of the generating (main) flow
v_0	The probability of VM leaves the node after zero phase
$\mathbf{v}$	The vector of probabilities v_n that VM leaves the server after n-th phase
$\mathbf{r}_0$	The vector of probabilities r_{0n} that VM switches to the n-th phase after zero phase is completed
$\mathbf{R}$	The matrix of probabilities r_{nv} that VM switches to the v-th phase after n-th phase is completed
μ_0	The parameter of exponential distributions of zero phase duration
$\mathbf{M}$	The diagonal matrix of parameters of exponential distributions of phases durations μ_n

Here $\mathbf{i}$ is a vector of i_n. We also denote identity matrix $\mathbf{I}$ and unit vectors $\mathbf{e}_n$, whose elements are zeroes except for n-th element, which is equal to one. For this distribution, we derive the Kolmogorov equation

$$-\left(\lambda + \sum_{n=0}^{N} i_n\mu_n\right) P(\mathbf{i}) + \lambda P(\mathbf{i}-\mathbf{e}_0) +$$

$$+\sum_{n=0}^{N} P(\mathbf{i}+\mathbf{e}_n)(i_n+1)\mu_n v_n + \sum_{n=0}^{N}\sum_{k=1}^{N} P(\mathbf{i}+\mathbf{e}_n-\mathbf{e}_k)(i_n+1)\mu_n r_{nk} = 0. \quad (1)$$

After that, we introduce characteristic function

$$H(\mathbf{u}) = \sum_{i_0,i_1,\ldots,i_N=0}^{\infty} e^{ju_0i_0+ju_1i_1+\ldots+ju_Ni_N} P(\mathbf{i}),$$

and rewrite the equation

$$H(\mathbf{u})(e^{ju_0}-1)\lambda - j\sum_{n=0}^{N} \frac{\partial H(\mathbf{u})}{\partial u_n}\mu_n e^{-ju_n}\left\{v_n + \sum_{k=1}^{N} e^{ju_k} r_{nk} - e^{ju_n}\right\} = 0.$$

Since the following is true

$$v_n + \sum_{k=1}^{N} r_{nk} = 1, n = 0, 1, ..., N,$$

we can transform the equation

$$H(\mathbf{u})(e^{ju_0}-1)\lambda +$$

$$+j\sum_{n=0}^{N}\frac{\partial H(\mathbf{u})}{\partial u_n}\mu_n e^{-ju_n}\left\{(e^{ju_n}-1)-\sum_{k=1}^{N}(e^{ju_k}-1)r_{nk}\right\}=0.$$

We propose to seek the solution in the following form:

$$H(\mathbf{u})=\exp\left\{\sum_{k=0}^{N}(e^{ju_k}-1)\rho_k\right\},$$

which we substitute to the equation and obtain

$$(e^{ju_0}-1)(\lambda-\rho_0\mu_0)+\sum_{n=1}^{N}(e^{ju_n}-1)\left\{\sum_{k=0}^{N}\rho_k\mu_k r_{kn}-\rho_n\mu_n\right\}=0.$$

The equality above is true when the coefficients of $(e^{ju_n}-1)$ are equal to zero. Thus, we obtain the following system:

$$\lambda-\rho_0\mu_0=0,$$

$$\sum_{k=0}^{N}\rho_k\mu_k r_{kn}-\rho_n\mu_n=0, n=1,2,...,N.$$

Having the system of equations, we solve the first equation

$$\rho=\frac{\lambda}{\mu_0}.$$

After this, we introduce vector

$$\boldsymbol{\rho}=\{\rho_1,...,\rho_N\},$$

and write the second equation of the system in matrix form

$$\boldsymbol{\rho}\mathbf{M}(\mathbf{I}-\mathbf{R})=\lambda\mathbf{r}_0.$$

Then the solution of the matrix equation is as follows:

$$\boldsymbol{\rho}=\lambda\mathbf{r}_0(\mathbf{I}-\mathbf{R})^{-1}\mathbf{M}^{-1}.$$

The characteristic function $H(\mathbf{u})$ of $(N+1)$-dimensional stationary distribution $P(\mathbf{i})$ of the number of simultaneously performed phases has the form

$$H(\mathbf{u})=\exp\left\{\sum_{n=0}^{N}(e^{ju_n}-1)\rho_n\right\}, \tag{2}$$

where

$$\rho_0=\frac{\lambda}{\mu_0},\ \boldsymbol{\rho}=\lambda\mathbf{r}_0(\mathbf{I}-\mathbf{R})^{-1}\mathbf{M}^{-1}. \tag{3}$$

The next section of the paper is devoted to the derivation of probability distribution of the volume of resource occupied by all VMs in the cloud system.

4 Total Volume V of Resources Occupied by All Virtual Machines

We assume that characteristic functions $h_n(u)$ of the amount of resource V_n, which is occupied by a virtual machine in the corresponding n-th phase are given, $n = 1, 2, \ldots, N$. Let us write the characteristic function of the total amount of the resource occupied by all virtual machines in the following form:

$$H(u) = \exp\left\{\sum_{n=0}^{N}(h_n(u) - 1)\rho_n\right\}, \quad (4)$$

where ρ_n are given by Eq. (3).

The distribution function of the total amount of resource occupied by all virtual machines can be derived from the previous formula using the inverse Fourier transform as follows:

$$F(x) = \frac{1}{2\pi}\int_{-\infty}^{\infty}\frac{1 - e^{-jux}}{ju}H(u)du, \quad (5)$$

5 Performance Metrics of Cloud System

5.1 Special Case: Branching Process with Three Phases

We consider the branching process for a particular case with only three phases: zero, first and second phase.

The flow is given by the following parameters:

- $N = 2$ is the number of phases (without zero phase);
- $\lambda = 7 \cdot 10^{-5}$ (events/sec) is the number of VMs arrived per second;
- $\mu_0 = 7 \cdot 10^{-3}$ (1/sec) is the reciprocal to the mean duration of zero phase;
- $\mathbf{M} = \begin{bmatrix} \mu_1 & 0 \\ 0 & \mu_2 \end{bmatrix}$, where
  $\mu_1 = 4 \cdot 10^{-3}$ (1/sec) is the reciprocal to the mean duration of the first phase,
  $\mu_2 = 2 \cdot 10^{-5}$ (1/sec) is the reciprocal to the mean duration of the second phase;
- $v_0 = 0.001$ is the probability of VM leaving after a zero phase;
- $\mathbf{v} = \begin{bmatrix} v_1 \\ v_2 \end{bmatrix}$, where
  $v_1 = 0.01$ is the probability of VM leaving after the first phase,
  $v_2 = 0$ is the probability of VM leaving after the second phase;
- $\mathbf{r}_0 = \begin{bmatrix} r_{01} & r_{02} \end{bmatrix}$, where
  $r_{01} = 0.999$ is the probability of VM switching to the first phase after zero phase,
  $r_{02} = 0$ is the probability of VM switching to the second phase after zero phase;

– $\mathbf{R} = \begin{bmatrix} 0 & 0.99 \\ 1 & 0 \end{bmatrix}$ is the matrix of probabilities of VM switching between phases.

The parameters of the distribution of the number of VMs performing in different phases are determined as follows:

$$\rho_0 = 0.01, \ \rho_1 = 1.748, \ \rho_2 = 346.154.$$

In Table 2, we show the distributions of volumes V_n and their parameters.

Table 2. Distributions of the volume of resources occupied by single VM in different phases

Notation	Distribution type	Parameters	Mean	Variance
V_0	Gamma	$shape_0 = 2$, $rate_0 = 3$	0.67	0.22
V_1	Uniform	$a = 0$, $b = 3$	1.5	0.75
V_2	Gamma	$shape_2 = 3$, $rate_2 = 4$	0.75	0.1875

In Fig. 2, we show the graph of the distribution function of the total volume V of the resource occupied by all virtual machines. Dash lines show the position of quantiles.

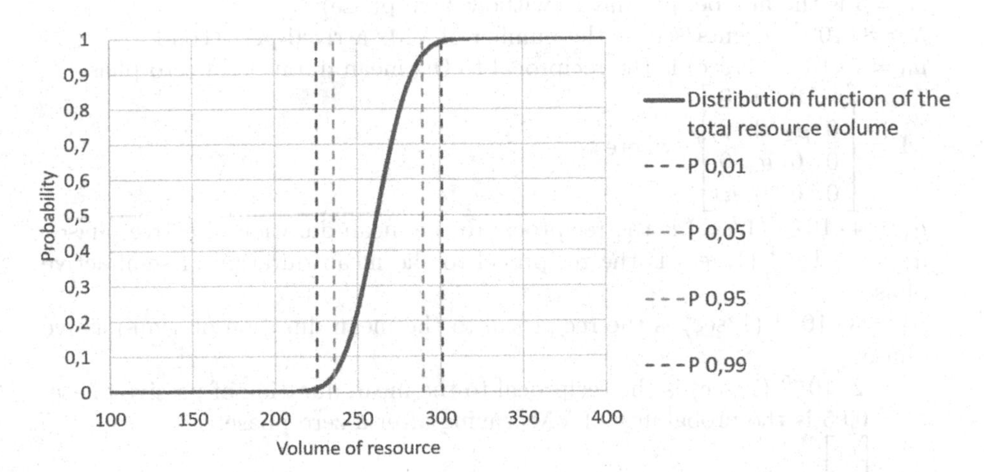

Fig. 2. Graph of the distribution function $F(x)$ of the total volume V of occupied resource

Mean and standard deviation of the total volume V of the resource occupied by all virtual machines are given by

$$m_V = 262.244, \ \sigma_V = 16.594.$$

The quantiles of significance levels 0.01, 0.05, 0.95, 0.99, are equal to

$$p_{0.01} = 225.327, \; p_{0.05} = 235.84, \; p_{0.95} = 289.37, \; p_{0.99} = 301.029.$$

Using this distribution, we also can obtain the probability that the total volume V of occupied resource exceed some critical level $\mathbb{P}\{V > x\} = 1 - F(x)$. The probabilities for different critical levels of resource are shown in Table 3.

Table 3. Probabilities that the occupied resource V exceeds the critical level

Critical level	200	225	250	275	300	325
Probability	$1 - 3 \cdot 10^{-5}$	0.991	0.772	0.215	0.012	$1.1 \cdot 10^{-4}$

The results in Table 3 show the probability of system overload in case when the resource is bounded by the critical level.

5.2 Special Case: Branching Process with Five Phases

Let us consider the case when VM life cycle consists of five phases. Besides zero phase, VM have four phases of activity: active, semi-active, semi-passive and passive phase. The parameters of the flow are the following:

- $N = 4$ is the number of phases (without zero phase);
- $\lambda = 8 \cdot 10^{-5}$ (events/sec) is the number of VMs arrived per second;
- $\mu_0 = 7 \cdot 10^{-3}$ (1/sec) is the reciprocal to the mean duration of zero phase;
- $\mathbf{M} = \begin{bmatrix} \mu_1 & 0 & 0 & 0 \\ 0 & \mu_2 & 0 & 0 \\ 0 & 0 & \mu_3 & 0 \\ 0 & 0 & 0 & \mu_4 \end{bmatrix}$, where
  $\mu_1 = 4 \cdot 10^{-3}$ (1/sec) is the reciprocal to the mean duration of active phase,
  $\mu_2 = 5 \cdot 10^{-4}$ (1/sec) is the reciprocal to the mean duration of semi-active phase,
  $\mu_3 = 3 \cdot 10^{-4}$ (1/sec) is the reciprocal to the mean duration of semi-passive phase,
  $\mu_4 = 2 \cdot 10^{-5}$ (1/sec) is the reciprocal to the mean duration of passive phase;
- $v_0 = 0.05$ is the probability of VM leaving after a zero phase;
- $\mathbf{v} = \begin{bmatrix} v_1 \\ v_2 \\ v_3 \\ v_4 \end{bmatrix}$, where
  $v_1 = 0.01$ is the probability of VM leaving after the first phase,
  $v_2 = 0.02$ is the probability of VM leaving after the second phase,
  $v_3 = 0.03$ is the probability of VM leaving after the first phase,
  $v_4 = 0.1$ is the probability of VM leaving after the second phase;

- $\mathbf{r}_0 = \begin{bmatrix} r_{01} \; r_{02} \; r_{03} \; r_{04} \end{bmatrix}$, where
  $r_{01} = 0.9$ is the probability of VM switching to the first phase after zero phase,
  $r_{02} = 0.03$ is the probability of VM switching to the second phase after zero phase,
  $r_{03} = 0.02$ is the probability of VM switching to the first phase after zero phase,
  $r_{04} = 0.05$ is the probability of VM switching to the second phase after zero phase,
- $\mathbf{R} = \begin{bmatrix} 0 & 0.75 & 0.23 & 0.01 \\ 0.3 & 0 & 0.63 & 0.05 \\ 0.07 & 0.2 & 0 & 0.7 \\ 0 & 0.4 & 0.5 & 0 \end{bmatrix}$ is the matrix of probabilities of VM switching between phases.

The parameters of the distribution of the number of VMs performing in different phases are determined as follows:

$$\rho_0 = 0.011, \;\; \rho_1 = 0.065, \;\; \rho_2 = 0.998, \;\; \rho_3 = 2.009, \;\; \rho_4 = 22.669.$$

In Table 4, we show the distributions of volumes V_n and their parameters.

Table 4. Distributions of the volume of resources occupied by single VM in different phases

Notation	Distribution type	Parameters	Mean	Variance
V_0	Gamma	$shape_0 = 1.5$, $rate_0 = 0.25$	6	24
V_1	Uniform	$a_1 = 0$, $b_1 = 18$	9	27
V_2	Gamma	$shape_2 = 4$, $rate_2 = 0.5$	8	16
V_3	Gamma	$shape_3 = 3$, $rate_3 = 0.5$	6	12
V_4	Uniform	$a_4 = 0$, $b_4 = 9$	4.5	6.75

In Fig. 3, we show the graph of the distribution function of the total volume V of the resource occupied by all virtual machines. Dash lines show the position of quantiles.

Mean and standard deviation of the total volume V of the resource occupied by all virtual machines are given by

$$m_V = 122.703, \; \sigma_V = 76.236.$$

The quantiles of significance levels 0.01, 0.05, 0.95, 0.99, are equal to

$$p_{0.01} = 62.997, \; p_{0.05} = 78.586, \; p_{0.95} = 171.223, \; p_{0.99} = 193.837.$$

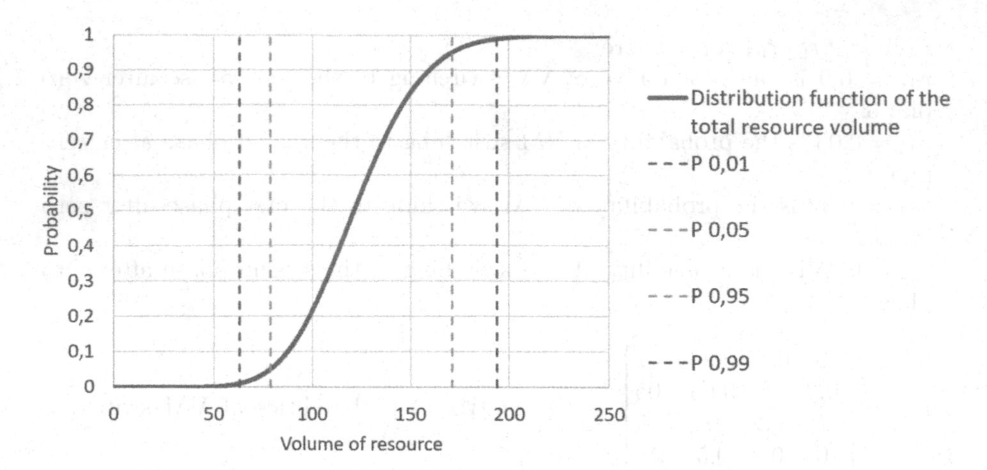

Fig. 3. Graph of the distribution function $F(x)$ of the total volume V of occupied resource

Using this distribution, we also can obtain the probability that the total volume V of occupied resource exceed some critical level $\mathbb{P}\{V > x\} = 1 - F(x)$. The probabilities for different critical levels of resource are shown in Table 5.

Table 5. Probabilities that the occupied resource V exceeds the critical level

Critical level	25	50	100	125	150	175	200	250
Probability	$1 - 7 \cdot 10^{-6}$	0.998	0.785	0.450	0.166	0.039	0.006	$5 \cdot 10^{-5}$

The results in Table 5 show the probability of system overload in case when the resource is bounded by the critical level.

6 Conclusion

We have considered multi-phase branching renewal process with Poisson generating process and exponentially distributed phase durations. The obtained probability distribution of the number of simultaneously performed phases is multi-dimensional Poisson. We also have derived the distribution of occupied resource in the node and shown the calculation of its main characteristics.

Considered model is a base for future investigation of the mutual influence of virtual machines on the resource consumption of a cloud system node. Numerical examples show that the utilization of any resource can be predicted as random variable having the rate of VMs arrivals, service rates and probabilities of VM transitions between phases. In future research, we plan to take into account the effect of service degradation due to the high number of VMs in the node.

References

1. Bai, W.H., Xi, J.Q., Zhu, J.X., Huang, S.W.: Performance analysis of heterogeneous data centers in cloud computing using a complex queuing model. In: Mathematical Problems in Engineering 2015 (2015)
2. Goswami, V., Patra, S.S., Mund, G.B.: Performance analysis of cloud with queue-dependent virtual machines. In: 2012 1st International Conference on Recent Advances in Information Technology (RAIT), pp. 357–362. IEEE (2012)
3. Hedhli, A., Mezni, H.: A survey of service placement in cloud environments. J. Grid Comput. **19**(3), 1–32 (2021)
4. Karim, M.E., Maswood, M.M.S., Das, S., Alharbi, A.G.: Bhyprec: a novel bi-LSTM based hybrid recurrent neural network model to predict the CPU workload of cloud virtual machine. IEEE Access **9**, 131476–131495 (2021)
5. Kuaban, G.S., Soodan, B.S., Kumar, R., Czekalski, P.: Analysis of the performance of a cloud computing processing queue with correlated reneging of tasks and resubmission. In: 2021 International Conference on Electrical, Computer and Energy Technologies (ICECET), pp. 1–8. IEEE (2021)
6. Kumar, J., Saxena, D., Singh, A.K., Mohan, A.: Biphase adaptive learning-based neural network model for cloud datacenter workload forecasting. Soft. Comput. **24**(19), 14593–14610 (2020)
7. Kumar, R., Soodan, B.S., Kuaban, G.S., Czekalski, P., Sharma, S.: Performance analysis of a cloud computing system using queuing model with correlated task reneging. J. Phys. Conf. Ser. **2091**, 012003. IOP Publishing (2021)
8. Liaqat, M., Naveed, A., Ali, R.L., Shuja, J., Ko, K.M.: Characterizing dynamic load balancing in cloud environments using virtual machine deployment models. IEEE Access **7**, 145767–145776 (2019)
9. Marinescu, D.C.: Cloud Computing: Theory and Practice. Morgan Kaufmann (2022)
10. Ouhame, S., Hadi, Y., Ullah, A.: An efficient forecasting approach for resource utilization in cloud data center using CNN-LSTM model. Neural Comput. Appl. **33**(16), 10043–10055 (2021)
11. Qi, L., Chen, Y., Yuan, Y., Fu, S., Zhang, X., Xu, X.: A QoS-aware virtual machine scheduling method for energy conservation in cloud-based cyber-physical systems. World Wide Web **23**(2), 1275–1297 (2020)
12. Ragmani, A., Elomri, A., Abghour, N., Moussaid, K., Rida, M.: Faco: a hybrid fuzzy ant colony optimization algorithm for virtual machine scheduling in high-performance cloud computing. J. Amb. Intell. Hum. Comput. **11**(10), 3975–3987 (2020)
13. Rashid, A., Chaturvedi, A.: Cloud computing characteristics and services: a brief review. Int. J. Comput. Sci. Eng. **7**(2), 421–426 (2019)
14. Supreeth, S., Patil, K.K.: Virtual machine scheduling strategies in cloud computing-a review. Int. J. Emerging Technol. **10**(3), 181–188 (2019)
15. Vetha, S., Devi, K.V.: Dynamic resource allocation in cloud using queueing model. J. Ind. Pollut. Control **33**(2), 1547–1554 (2017)

On Modeling a Section of a Single-Track Railway Network Based on Queuing Networks

Alexander Kazakov, Anna Lempert, and Maxim Zharkov(✉)

Matrosov Institute for System Dynamics and Control Theory of Siberian Branch of Russian Academy of Sciences (IDSTU SB RAS), Irkutsk, Russia
{kazakov,lempert}@icc.ru, zharkm@mail.ru
http://idstu.irk.ru

Abstract. The paper considers the problem of estimating the capacity of a railway network section with single-track lines, the distinctive feature of which is a package (group) train schedule. We propose a methodology for the mathematical modeling of trains running along the railway network using the queuing theory. Based on the methodology, models consist of two parts: a description of incoming train flows and a simulation of the train passing through the system. The arrival of train flows from different directions is described by several BMAP flows, which makes it possible to take into account the category of trains, the intensity of their arrival, and the size of the incoming train package. The trains running is simulated using a queuing network. It allows us to consider the non-linear structure of the railway network; describe the various parameters of the stations and the capacity of the railway lines between them; take into account the presence of train packages, as well as the influence of random factors. We consider a section of the Ulaanbaatar railway located in Mongolia to test the proposed methodology. Based on the results of a numerical study of the constructed model, we draw conclusions about the permissible capacity of this section and the bottlenecks in its structure.

Keywords: Mathematical model · queuing network · railway network section · simulation · computational experiment

1 Introduction

Queuing theory is widely applied to simulate various technical systems, the operation of which has a stochastic character [1]. It is mainly used to describe computer and telecommunication networks [2–4], less often in transport and logistics [5]. As a rule, researchers use single-phase Markov queuing systems (QSs) [4,5].

The study was funded by the Ministry of Science and Education of the Russian Federation in the framework of the basic part, project No 121041300065-9.

A. Dudin et al. (Eds.): ITMM 2022, CCIS 1803, pp. 40–54, 2023.
https://doi.org/10.1007/978-3-031-32990-6_4

However, QSs are inefficient for describing complex systems that involve multilevel non-linear servicing of requests, which, in particular, include transport objects [6]. In this case, multiphase QSs and queuing networks (QNs) appear to be more convenient tools [1]. In [7], the authors use this mathematical apparatus for modeling urban train networks. Article [8] proposes a model for estimating delays in the movement of trains on the urban railway network, papers [9,10] develop a method for analyzing the capacity of a railway line with unknown schedules, and [11] presents a model of commuter train movement. In the mentioned works, the authors use QNs, whose nodes are single-channel QSs with an infinite one.

The scientific literature also contains examples of using more complex QNs, whose nodes are multichannel QSs. In article [12], the authors apply a similar method to describe the run of freight trains along the railway network. In [13], it helps to simulate the operation of a marshalling yard, and in [14] - a large railway junction.

Previously, we developed a methodology for modeling marshalling and freight railway stations, in which QN nodes are multichannel QSs with finite queues and describe the operation of station yards. In contrast to the works presented above, we use the BMAP for the mathematical description of the incoming train flow with a complex structure. It allows us to combine several transport sub-flows into a single pattern, which can be correlated and grouped [15–17].

In this paper, we propose a methodology for modeling larger objects: single-track sections of the railway network, on which the train running between neighboring stations is carried out in two directions along one track. As a result, such sections have a more complex operation technology compared to conventional double-track lines. Such railways are widely met in Asian countries. In particular, these include the Ulaanbaatar Railway (UBRW), the shortest railway corridor from the central regions of Russia to China. We apply the proposed methodology to simulate its operation.

2 Problem Statement

We mean a section of the railway network is several stations of different types [6] interacting with each other in a relatively small area. Trains arrive at the section from two or more directions, each of which includes several categories of trains.

Stations consist of one or more yards with different track amounts and capacities. There are specific tracks for loading/unloading trains called freight yards at the stations. The train running between neighboring stations is carried out by one or more railway lines. With a single-track line, trains arrive (depart) at the station in packages (groups), the size of which, as a rule, does not exceed three trains. There may be sidings - short sections of the railroad with two or three tracks, which allow passing oncoming trains by a single line.

Stations and railway lines connecting them form a non-linear hierarchical structure with several routes. The capacity of the railway network is limited. Therefore, trains are forced to stop and await vacating the line or a place at the station.

The purpose of the study of the railway section is to estimate its capacity, to determine "bottlenecks" in the structure, and to give recommendations for improving its operation. To achieve it, we apply the methods of mathematical modeling and simulation.

3 Methodology

At the stations and lines of the railway section, the same type of operations with trains are regularly performed: arrivals, servicing, passing, and departure. Although their duration is regulated, it is influenced by many random factors, such as weather conditions, equipment breakdowns, and human factors. At the same time, trains visit stations and pass through lines sequentially, following a given route; so, there is a multistage nonlinear service in the system. In this case, a queueing network is an effective tool for describing the operation of such systems.

Queuing network (QN) is the union of a finite number $S \geq 2$ of QSs (hereinafter referred to as nodes), in which requests move from one node to another in accordance with the routing matrix P [1]. We deal with open QN. It means that requests come from an external source. Matrix $P_{(S+1)\times(S+1)}$ consists of probabilities of request transition from one node to another node.

When constructing the model, we distinguish four types of elements: incoming train flows, station yards, railway lines, and train routes on the railway network. We use a separate BMAP flow to describe the arrival of train packages from one direction [4]. One train is considered a request, then a group of requests is a package.

We describe the operation of the yards by multi-channel QSs without queues, in which the number of channels is equal to the number of tracks in the corresponding yard. The service time T for the channels is the sum $T = t_1 + t_2 + t_3$, where t_1 is the duration of receiving a train to the yard, t_2 is the time of its servicing (or parking), t_3 is the duration of operations for departure. In modeling freight yards, the number of channels depends on the capacity – the number of trains allowed simultaneously. The service time $T = t_1 + t_3 + t_4$, where t_4 is the loading/unloading time.

To simulate the operation of railway lines, we use two single-channel QSs without a queue according to the following rules.

a) In the case of a double-track line, we consider it as two separate QSs, each QS serves single requests one at a time.
b) If there is single-track line and the siding has three or more tracks, then both QSs serve requests in groups whose size is the maximum size of the train package.
c) If the siding has only two tracks, then the corresponding QS serves requests in groups, and another QS - one at a time.

Rules ***b***) and ***c***) help us to describe the situation when one train (or a package) stops at a siding and lets an oncoming train package pass. The service time for requests in these QSs means the time of train movement along the corresponding track.

We take into account various train routes using several types of requests. For each type, its own route matrix is constructed. We describe the stop of a train or a package due to busy lines and stations by temporary blocking the channels of the previous QS until enough space appears in the further QS for request receiving.

The constructed BMAP and QSs, as well as routing matrices, form the QN. Its study is carried out using a previously developed simulation model [16]. The purpose of the numerical research is to determine the performance indicators of the QN, such as the loss probability, the average sojourn time of a request in each node and the system, the average number of busy channels, and the time of channel blocking. Based on these indicators, we draw conclusions about the efficiency of operation.

4 The Northern Section of the Ulaanbaatar Railway

To test the proposed methodology, we consider the section of the Ulaanbaatar railway from Darkhan-1 station, a large freight station located 40 km from the border with Russia, to Ulaanbaatar 1 station, the main marshalling yard in Mongolia (hereinafter referred to as the trunk). The trunk has a length of about 283 km and includes 12 stations: Darkhan-1, Darkhan-2, Salkhit, Erkhet, Baruunkharaa, Zuunkharaa, Tunkh, Mandal, Arshaant, Tolgoit, Ulaanbaatar 2 and Ulaanbaatar 1. Figure 1 shows the station locations, where black dots mark the sidings.

There are double-track lines between stations Darkhan-1 and Darkhan-2, as well as Tolgoit and Ulaanbaatar 1; therefore, there are no sidings here. Table 1 shows the length of railway lines, as well as the average train running time. The running time is obtained from the data of standard train speed for these lines for 2018.

Let us describe the stations in more detail. We obtained their technical parameters based on the analysis of satellite images, operating regulations and train schedules.

First, we consider the way stations **Salkhit, Erkhet, Baruunkharaa, Zuunkharaa, Tunkh, and Mandal**, whose main task is passing through and briefly stopping all types of trains. Stations consist of one receiving and departure yard (RDY): **Baruunkharaa** has 4 tracks; **Erkhet**, **Tunkh**, and **Mandal** have 5 tracks each; **Salkhit** has 6 tracks; **Zuunkharaa** has 7 tracks. The transit time through the station is 3–8 mins, the stop of passenger trains takes 9–16 mins, and freight trains - 20–60 mins.

Darkhan-1, Darkhan-2, Arshaant and **Tolgoit** are freight stations where passenger trains also can stop. These stations have one or two RDY and cargo yard: **Darkhan-1** has one RDY with 10 tracks; **Darkhan-2** has two RDYs with 6 tracks each; **Arshaant** has one RDY with 6 tracks; **Tolgoit** has two RDYs with 6 and 7 tracks. The capacity of the cargo yards for all stations is one train,

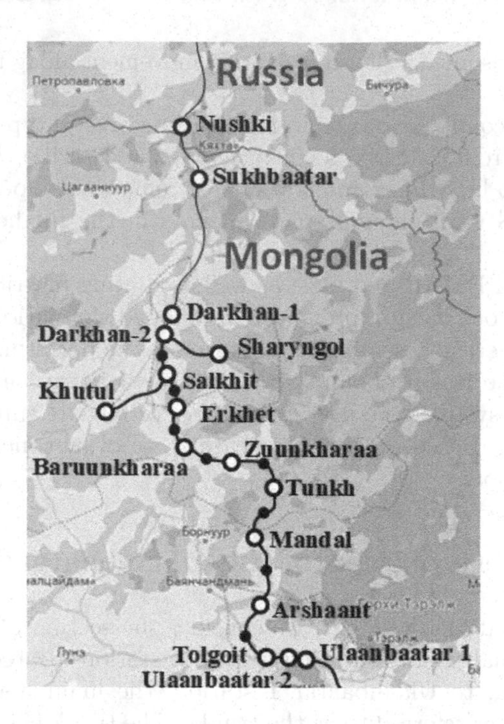

Fig. 1. Scheme of the trunk Darkhan-1 – Ulaanbaatar 1

the average loading/unloading time of the train is one day. The parameters of RDY operation at these stations are similar to those at the way stations.

Ulaanbaatar 2 is an auxiliary station to Ulaanbaatar 1, it is also a way station. It has one RDY with 7 tracks and perform two classes of operations with different durations. The first is passing and stopping trains (operation parameters are similar to way stations); the second is train length change or long stop before transfer to Ulaanbaatar 1 (service time is 50–80 mins).

Ulaanbaatar 1 is the largest marshalling yard in Mongolia. It is located within the city, therefore, in addition to handling freight trains, it also receives passenger ones. The station includes one RDY with nine tracks, a marshaling yard with 11 tracks, and a cargo yard that can receive one train as a whole. The average train service time at the marshaling yard is 4 h, and at the cargo yard - 12 h.

Train traffic on the considered trunk is unevenly distributed and follows a partially batch schedule. Freight trains (local and transit) run in packages. Their size does not exceed two trains due to the small capacity of way stations (4-5 tracks). Passenger trains are not included in the packages, as they have priority over freight trains. Note that incoming packages from other railway sections can reach three trains.

According to the plan, 15-16 trains pass through Darkhan-1 station in each direction. Then the train flow gradually increases and amounts to 74 trains at

Table 1. Train running time on railway lines

Railway lines	Length of Railway line (km)	Average running time (min)	
		Passenger train	Freight train
Darkhan-1 – Darkhan-2	5	7	9
Darkhan-2 – Salkhit	30	28	34
Salkhit – Erkhet	11	12	14
Erkhet – Baruunkharaa	32	29	34
Baruunkharaa – Zuunkharaa	31	29	34
Zuunkharaa – Tunkh	39	40	44
Tunkh – Mandal	39	40	44
Mandal – Arshaant	45	41	45
Arshaant – Tolgoit	38	40	47
Tolgoit – Ulaanbaatar 2	7	11	14
Ulaanbaatar 2 – Ulaanbaatar 1	3	3	5

the Ulaanbaatar 1 station (forecast for 2025). Figure 2 shows the scheme of train flows, where pairs of numbers indicate the number of trains per day in 2022 (field data) and the forecast for 2025. At the station, the train can turn around, then the route of its movement changes to the opposite. Oblique arrows mark the reversals.

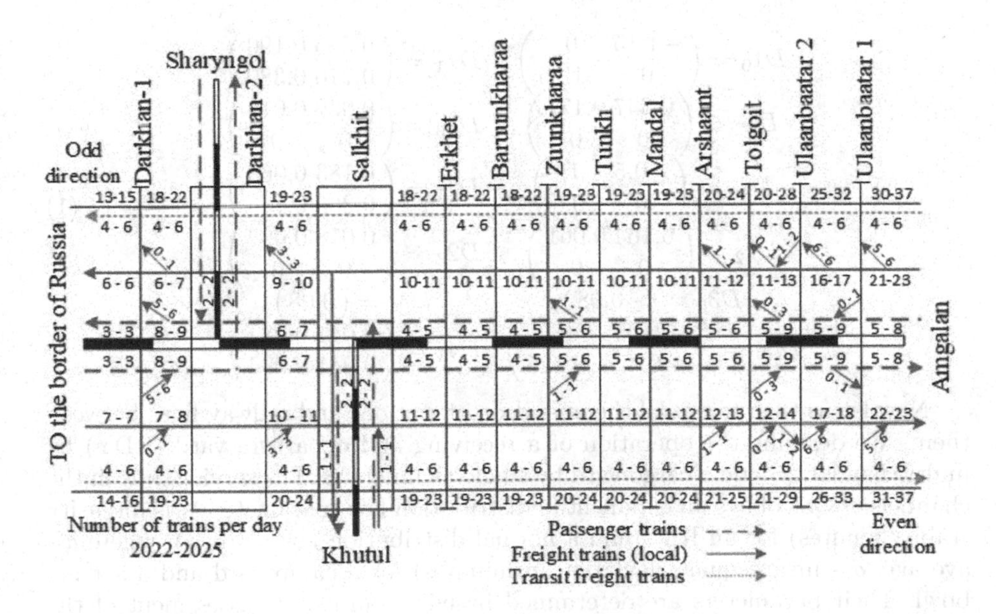

Fig. 2. Train flow diagram

5 Mathematical Model

Trains enter the trunk from four directions, each of which we describe as a separate BMAP. Main directions are following: $BMAP_1$ – 32 trains/day arrive from Amgalan (1.15 packages/h); $BMAP_2$ – from the border with Russia (Sukhbaatar station) – 15 trains/day (0.5 packages/h). Additional directions are $BMAP_3$ – from Sharyngol – 2 trains/day (0.08 packages/h) and $BMAP_4$ – from Khutul stations – 3 trains/day (0.125 packages/h).

$BMAP_1, BMAP_2$, and $BMAP_4$ simulate the arrival of freight (local and transit) and passenger trains, so they have two sub-flows each. $BMAP_3$ describes the arrival of passenger trains only. We determine the probabilities of the arrival of a train of a certain category as relative frequencies. Freight trains arrive in packages; their size obeys the discrete distribution (Table 2):

Table 2. The distribution of packages size

Trains in a package	1	2	3
Probability	0.5	0.45	0.05

These parameters were obtained based on expert assessments. The BMAP matrices have the following form, where the upper lines correspond to the arrival of freight trains, and the lower ones describe passenger trains.

$$\begin{gathered} D1_0 = \begin{pmatrix} -1.15 & 0 \\ 0 & -1.15 \end{pmatrix}, \quad D1_1 = \begin{pmatrix} 0.385 & 0.190 \\ 0.770 & 0.380 \end{pmatrix}, \\ D1_2 = \begin{pmatrix} 0.347 & 0.171 \\ 0 & 0 \end{pmatrix}, \quad D1_3 = \begin{pmatrix} 0.039 & 0.019 \\ 0 & 0 \end{pmatrix}, \\ D2_0 = \begin{pmatrix} -0.5 & 0 \\ 0 & -0.5 \end{pmatrix}, \quad D2_1 = \begin{pmatrix} 0.183 & 0.068 \\ 0.365 & 0.135 \end{pmatrix}, \\ D2_2 = \begin{pmatrix} 0.164 & 0.061 \\ 0 & 0 \end{pmatrix}, \quad D2_3 = \begin{pmatrix} 0.018 & 0.007 \\ 0 & 0 \end{pmatrix}, \\ D3_0 = \left(-0.08 \right), \quad D3_1 = \left(0.08 \right), \\ D4_0 = \begin{pmatrix} -0.125 & 0 \\ 0 & -0.125 \end{pmatrix}, \quad D4_1 = \begin{pmatrix} 0.041 & 0.084 \\ 0.041 & 0.084 \end{pmatrix}. \end{gathered} \tag{1}$$

Now let us turn to model the operation of stations and railway lines between them. We describe the operation of a receiving and departure yard (RDY) by multi-channel QS and a cargo yard by single-channel QS. The service time in the channels of QSs obeys an exponential distribution law (parameter λ_i is intensity trains/minutes) for an RDY and a normal distribution law ($N(a, \sigma)$, where a - average, σ - mean square deviation in minutes) for a cargo yard and a sorting bowl. Their parameters are determined based on an expert assessment of the train schedule and the station's operating regulations. In terms of queuing theory, the models have the following form:

Darkhan-1: RDY (node 1) - $BMAP_2/M/9/0$, $\lambda_1 = 0.08$; the cargo yard (node 2) - $*/G/1/0$, $N(1440, 60)$;
Darkhan-2 has two yards of the same type, so we describe them with one node (node 5) - $BMAP_3/M/12/0$, $\lambda_5 = 0.11$; and it also has the cargo yard (node 6) - $*/G/1/0$, $N(420, 50)$;
Salkhit has only one RDY (node 9) - $BMAP_4/M/6/0, \lambda_9 = 0.11$;
Erkhet (node 12), **Tunkh** (node 21), and **Mandal** (node 24) - $*/M/5/0$, $\lambda_{12} = \lambda_{21} = \lambda_{24} = 0.11$;
Baruunkharaa (node 15) - $*/M/4/0$, $\lambda_{15} = 0.11$;
Zuunkharaa (node 18) - $*/M/7/0$, $\lambda_{18} = 0.11$;
Arshaant: RDY (node 27) - $*/M/6/0$, $\lambda_{27} = 0.08$; the cargo yard (node 28) - $*/G/1/0$, $N(1440, 60)$;
Tolgoit: RDY (node 31) - $*/M/13/0$, $\lambda_{31} = 0.11$; the cargo yard (node 32) - $*/G/1/0$, $N(1440, 60)$.

We describe **Ulaanbaatar 2** by two QSs that perform different classes of operations: node 35 simulates the passing of trains - $*/M/5/0$, $\lambda_{35} = 0.11$; and node 36 describes the operation to change the length of the train - $*/G/2/0$, $N(65, 5)$.

Ulaanbaatar 1 has three subsystems: RDY (node 39) - $BMAP_1/M/9/0$, $\lambda_{39} = 0.11$; the sorting bowl (node 40) - $*/G/11/0$, $N(240, 35)$; the cargo yard (node 41) - $*/G/1/0$, $N(720, 60)$.

The trunk includes 11 sections:
a) Lines Darkhan-1 – Darkhan-2, Tolgoit – Ulaanbaatar 2, Ulaanbaatar 2 – Ulaanbaatar 1 have double-track line, therefore, we simulate their operating by two single-channel QSs (nodes 3, 4, 33, 34, 37, and 38) – $*/G/1/0$.
b) The Mandal – Arshaant single-track line has a siding with three tracks, then both QSs (nodes 25 and 26) service requests in packages – $*/G^X/1/0$, where $X = 2$ is the maximum size of serviced package.
c) The sidings on the remaining lines include two tracks. Then one QS services requests in packages (nodes 7, 10, 13, 16, 19, 22, and 29) – $*/G^X/1/0$, and the second system processes requests one at a time (nodes 8, 11, 14, 17, 20, 23, and 30) – $*/G/1/0$.
We assume the service time in these nodes has a normal distribution [15,16], the parameters of which are determined based on the data from Table 1: a is the average running time; we choose σ according to the "three sigma" rule. Then the service time at nodes 3 and 4 obeys $N(8; 0.33)$, at nodes 7 and 8 – $N(31; 1)$, at nodes 10 and 11 – $N(13; 0.5)$, at nodes 13, 14, 16 and 17 – $N(31.5; 1)$, at nodes 19, 20, 22 and 23 – $N(42; 2)$, at nodes 25 and 26 – $N(43; 2)$, at nodes 29 and 30 – $N(43.5; 2)$, at nodes 33 and 34 – $N(12.5; 0, 5)$, at nodes 37 and 38 – $N(4; 0.33)$.

The parameters of all normal distributions presented above satisfy the property $a - 8\sigma > 0$. Then the probability of a negative value appearance is $P(N(a, \sigma) > 8\sigma) < 10^{-15}$. Therefore, to solve practical problems, this event can be neglected, and all normal distributions can be used in their original form rather than their truncated versions [15].

Thus, the model of train traffic along the trunk has the form of the QN with four BMAP and 41 nodes, of which 19 nodes simulate the operation of 12 stations, and 22 nodes describe 11 railway lines. We add to these nodes four fictitious ones, according to the number of incoming request flows [1]: node 0 corresponds to $BMAP_1$, node 42 – $BMAP_2$, node 43 – $BMAP_3$, and node 44 – $BMAP_4$. Figure 3 shows the QN scheme as an oriented graph, where ellipses marks yards at stations and circles correspond to railway lines.

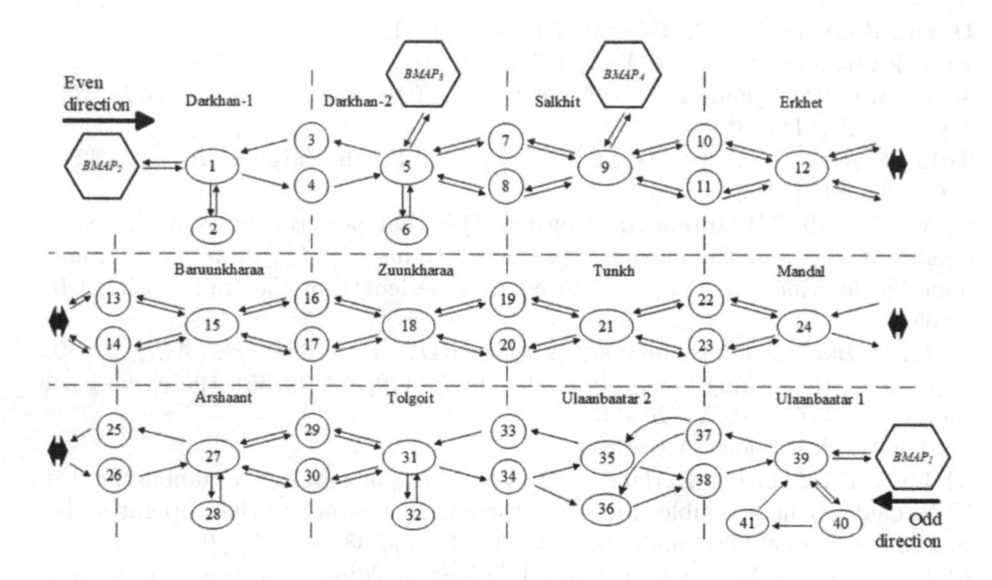

Fig. 3. Queuing network diagram

Recall that trains run in two main directions along the considered trunk. At the same time, the routes of passenger and freight trains are different; in particular, passenger trains cannot be serviced at the cargo yard. Therefore, four train routes are possible. In the model, we take them into account using various types of requests: 1 and 3 describe the movement of odd freight and passenger trains, 2 and 4 – even freight and passenger trains, respectively. For each type of request, its own route matrix P_z is constructed. Their dimension is 45 by 45, but they are very sparse. Therefore, below we describe only the nonzero elements $p_{i,j}$ of matrix P_z, which are probabilities of the transition from node i to node j.

Elements of the route matrix P_1:

$$p_{2,1} = p_{3,1} = p_{6,5} = p_{7,5} = p_{8,5} = p_{10,9} = p_{11,9} = p_{13,12} = p_{14,12} = p_{16,15} = \\ = p_{17,15} = p_{19,18} = p_{20,18} = p_{22,21} = p_{23,21} = p_{25,24} = p_{28,27} = p_{29,27} = p_{30,27} = \\ = p_{32,31} = p_{33,31} = p_{35,33} = p_{41,39} = p_{42,39} = p_{43,5} = p_{44,9} = 1;$$

$$p_{12,10} = p_{15,13} = p_{18,16} = p_{21,19} = p_{24,22} = p_{37,35} = 0.8;$$

$$p_{9,8} = p_{12,11} = p_{15,14} = p_{18,17} = p_{21,20} = p_{24,23} = p_{31,30} = p_{37,36} = 0.2;$$

$$p_{1,0} = 0.85; p_{1,2} = 0.15; p_{5,3} = 0.6; p_{5,6} = 0.4; p_{9,7} = 0.74; p_{9,44} = 0.06;$$

$$p_{27,25} = 0.94; p_{27,28} = 0.06; p_{31,29} = 0.74; p_{31,32} = 0.06;$$

$$p_{39,37} = 0.87; p_{39,40} = 0.13; p_{40,39} = 0.7; p_{40,41} = 0.3.$$

Elements of the route matrix P_2:

$$p_{0,1} = p_{1,4} = p_{2,1} = p_{4,5} = p_{6,5} = p_{7,9} = p_{8,9} = p_{10,12} = p_{11,12} = p_{13,15} = \\ = p_{14,15} = p_{16,18} = p_{17,18} = p_{19,21} = p_{20,21} = p_{22,24} = p_{23,24} = p_{24,26} = \\ = p_{26,27} = p_{28,27} = p_{29,31} = p_{30,31} = p_{31,34} = p_{32,31} = p_{35,38} = p_{36,38} = \\ = p_{38,39} = p_{39,42} = p_{41,39} = 1;$$

$$p_{5,7} = p_{9,10} = p_{12,13} = p_{15,16} = p_{18,19} = p_{21,22} = p_{27,29} = 0.8;$$

$$p_{5,8} = p_{9,11} = p_{12,14} = p_{15,17} = p_{18,20} = p_{21,23} = p_{27,30} = 0.2;$$

$$p_{34,35} = 0.9; p_{34,36} = 0.1; p_{40,39} = 0.7; p_{40,41} = 0.3.$$

Elements of the route matrix P_3:

$$p_{3,1} = p_{5,3} = p_{8,5} = p_{9,8} = p_{11,9} = p_{12,11} = p_{14,12} = p_{14,12} = p_{15,14} = \\ = p_{17,15} = p_{18,17} = p_{20,18} = p_{21,20} = p_{23,21} = p_{24,23} = p_{25,24} = p_{27,25} = \\ = p_{30,27} = p_{31,30} = p_{33,31} = p_{35,33} = p_{37,35} = p_{39,37} = p_{42,39} = 1.$$

Elements of the route matrix P_4:

$$p_{0,1} = p_{1,4} = p_{4,5} = p_{7,9} = p_{8,9} = p_{11,12} = p_{12,14} = p_{14,15} = p_{15,17} = \\ = p_{17,18} = p_{18,20} = p_{20,21} = p_{21,23} = p_{23,24} = p_{24,26} = p_{26,27} = p_{27,30} = \\ = p_{30,31} = p_{31,34} = p_{34,35} = p_{35,38} = p_{36,38} = p_{38,39} = p_{39,42} = 1;$$

$$p_{5,8} = 0.85; p_{5,43} = 0.15; p_{9,44} = 0.33; p_{9,11} = 0.67.$$

A request can change its type after being serviced at a node, which allows us changing its route (train reversal). For odd passenger trains (type 3), the probability p_y to change its route at node y, i.e. change type 3 to 4 is as follows: at node 1 – $p_1 = 0.625$; at node 18 – $p_{18} = 0.2$; at node 31 – $p_{31} = 0.33$. Requests with type 1 (odd freight trains) become type 2 at nodes 2, 6, 28, 32 and 40 with probability $p_2 = p_6 = p_{28} = p_{32} = p_{40} = 1$, and type 2 orders become type 1 at node 36 with probability $p_{36} = 1$.

6 Computational Experiment

We performed two computational experiments with different model parameters, first of all, with various volumes of incoming train flows. Tables 3 and 4 present the averaged results of the numerical study obtained from ten runs of the simulation model. The virtual simulation time for each launch was 28 days.

In Tables 3 and 4: P_{loss} is a loss probability, T_{lock} is the average total time (in hours) of blocking one request in all nodes, i.e. total idle time for one train; T_H is average time (in hours) of a train passing along the trunk, excluding freight work; T_S is average sojourn time of a request at the QN (in hours); k_i is an average number of busy channels; t_i is average sojourn time of a request at a node i (in minutes); r_i is an average number of requests received per day at a node i; Tl_i is total blocking time (in minutes) in node i.

Experiment 1. Table 3 shows the results of simulation for the volume of train flows planned for 2025, which is given by matrices 1.

To verify the model, we compared data on train flows (Fig. 2) and the number of requests received by the nodes (Table 3). In the model, the choice of the train that makes the U-turn occurs randomly. Therefore, the number of requests received by the node may differ from the planned value by more than 6%. Thus,

Table 3. The results of simulation 1

Input request			1217.33	P_{loss}	0	T_H (h)	8.54
Input group			1632.00	T_{lock} (h)	1.29	T_S (h)	42.51
	Node 1	**Node 2**	**Node 3**	**Node 4**	**Node 5**	**Node 6**	**Node 7**
k_i	0.34	0.47	0.12	0.13	0.42	0.65	0.44
t_i	14.46	743.19	8.00	8.00	12.55	435.49	31.80
r_i	38.90	0.90	21.71	22.64	48.58	2.13	21.38
Tl_i	1092.33	0	0	0	3894.00	0	0
	Node 8	**Node 9**	**Node 10**	**Node 11**	**Node 12**	**Node 13**	**Node 14**
k_i	0.48	0.40	0.23	0.16	0.37	0.49	0.45
t_i	31.81	12.58	12.99	13.00	12.52	32.00	32.00
r_i	21.45	47.48	25.98	17.43	43.39	23.35	20.08
Tl_i	0	3652.33	0	0	3513.33	10.00	4.00
	Node 15	**Node 16**	**Node 17**	**Node 18**	**Node 19**	**Node 20**	**Node 21**
k_i	0.43	0.50	0.45	0.55	0.64	0.65	0.69
t_i	14.65	32.00	31.99	18.19	43.48	43.47	22.95
r_i	43.44	23.26	20.21	44.44	23.96	21.44	45.38
Tl_i	6045.00	5.00	0	10148.00	24.00	15.50	15831.67
	Node 22	**Node 23**	**Node 24**	**Node 25**	**Node 26**	**Node 27**	**Node 28**
k_i	0.64	0.65	0.71	0.64	0.55	0.87	0.88
t_i	43.49	43.46	25.64	44.46	44.42	28.97	1452.62
r_i	23.90	21.48	45.37	24.80	20.57	47.05	0.87
Tl_i	27.00	9.00	16560.67	11.00	0	18244.67	0
	Node 29	**Node 30**	**Node 31**	**Node 32**	**Node 33**	**Node 34**	**Node 35**
k_i	0.67	0.67	0.74	0.87	0.27	0.22	0.40
t_i	44.81	44.82	21.52	1461.17	13.00	13.00	11.42
r_i	25.32	21.71	51.31	0.86	29.18	24.74	50.50
Tl_i	0	0	16627.00	0	0	1.50	2708.67
	Node 36	**Node 37**	**Node 38**	**Node 39**	**Node 40**	**Node 41**	
k_i	0.31	0.09	0.08	0.46	0.90	0.42	
t_i	69.40	4.01	4.00	10.96	261.97	752.91	
r_i	6.36	32.12	27.65	70.15	4.95	0.80	
Tl_i	210.00	6.33	0	439.00	37.00	0	

we considered the average relative deviation of the number of trains at the RDYs (nodes 1, 5, 9, 12, 15, 18, 21, 24, 27, 29, 31, 35, 36, and 39) from the planned value for all stations. It does not exceed 3.01%.

Analyzing Table 3, we can see the following. The highest channel occupancy rate ($K_{node} = k_i/n_i$, where n_i is the number of channels in a node i) is observed at nodes 28 and 32 (cargo yards at Arshaant and Tolgoit stations). The longest durations of channel blocking (Tl_i) are at nodes 21, 24, 27, and 31 (Tunkh, Mandal, Arshaant and Tolgoit stations). At the same time, $K_{node} \geq 0.55$ in neighboring nodes 19, 20, 22, 23, 25, 26, 29, and 30 (lines between these stations), i.e., the capacity of these nodes is not enough.

The model does not provide for managing the movement of requests. Therefore, during the computational experiment, a collision occurred in four runs of the simulation model. We got a situation when oncoming flows of requests overflow neighboring nodes and block each other. We did not include the results of these launches in Table 3, but highlighted the nodes where the collision occurred: 15, 21, and 24 (Buunhara, Tunkh, and Mandal stations).

The most high-usage nodes are the cargo yards at Arshaant and Tolgoit stations. However, they receive train flows with low intensity (1 train per day), so the loading of cargo yards does not affect significantly on the operation of the trunk as a whole. The lines between Tunkh, Mandal, Arshaant, and Tolgoit stations are bottlenecks since the running time between these stations is the longest - more than 40 min. Therefore, we observe a significant waiting time for train departures (Tl_i) at these stations, which exceeds 9 h per day.

One of the promising ways to increase the capacity of the Ulaanbaatar railway at a relatively small financial cost is to increase the number of tracks at the sidings between stations. This decision can allow two packages of trains to pass along a single-track line at once in different directions. Let us check this proposition by simulation.

Experiment 2. Table 4 presents the simulation results with the following changes in the model: the size of the served group at nodes 8, 11, 14, 17, 20, 23, and 30 is increased to two; the values of the elements in matrices 1 are increased by 20%.

Compared with Experiment 1, we obtained the following results. The average number of busy channels at nodes 8 and 20 decreased on average by 2.6%, in other cases it increased by 17.5% on average. The blocking time of the channels (Tl_i) for nodes that describe stations has increased by 32% on average. The loss probability remained equal to zero. During the experiment, one collision was observed at node 15 (Baruunhara station).

The increase in the number of tracks at the sidings from two to three made it possible to increase the total capacity of the entire section by more than 20%, from 59 to 71.4 trains per day. However, the growth in train traffic leads to an increase in the waiting time for train departures at stations by an average of 32%. Note that the enlarged waiting time is not critical for the total running time of a single train in accordance with the operating regulations. Thus, the proposition is admissible in the given circumstances.

Table 4. The results of simulation 2

Input request			1514.50	P_{loss}	0	T_H (h)	9.00
Input group			1999.25	T_{lock} (h)	1.48	T_S (h)	42.82
	Node 1	**Node 2**	**Node 3**	**Node 4**	**Node 5**	**Node 6**	**Node 7**
k_i	0.43	0.53	0.15	0.15	0.56	0.69	0.56
t_i	14.80	712.30	8.00	8.00	14.31	435.38	31.81
r_i	46.98	1.06	26.09	27.21	57.87	2.29	26.24
Tl_i	1509.00	0	0	0	7262.75	0	0
	Node 8	**Node 9**	**Node 10**	**Node 11**	**Node 12**	**Node 13**	**Node 14**
k_i	0.46	0.50	0.27	0.21	0.50	0.63	0.47
t_i	31.83	13.37	13.00	13.00	13.93	31.98	32.00
r_i	24.05	57.58	30.17	23.60	53.75	30.19	23.57
Tl_i	0	4792.00	0	0	5475.00	10.00	5.33
	Node 15	**Node 16**	**Node 17**	**Node 18**	**Node 19**	**Node 20**	**Node 21**
k_i	0.57	0.62	0.48	0.77	0.79	0.64	0.83
t_i	16.81	32.00	31.99	23.48	43.44	43.43	25.64
r_i	53.74	30.22	23.54	55.25	31.78	24.99	56.77
Tl_i	8843.00	5.33	0	16404.25	0	3.00	18944.00
	Node 22	**Node 23**	**Node 24**	**Node 25**	**Node 26**	**Node 27**	**Node 28**
k_i	0.77	0.66	0.89	0.77	0.68	1.14	0.89
t_i	43.45	43.47	27.78	44.44	44.46	35.43	1440.07
r_i	31.84	24.96	56.80	31.91	24.94	58.59	0.88
Tl_i	7.33	0.00	20672.00	9.00	0	25148.50	0
	Node 29	**Node 30**	**Node 31**	**Node 32**	**Node 33**	**Node 34**	**Node 35**
k_i	0.75	0.70	1.06	0.93	0.34	0.27	0.51
t_i	44.82	44.81	25.43	1466.75	13.00	13.01	12.21
r_i	32.85	25.76	63.83	0.91	37.21	30.03	60.95
Tl_i	3.00	0	25426.00	0	0	11.50	4401.00
	Node 36	**Node 37**	**Node 38**	**Node 39**	**Node 40**	**Node 41**	
k_i	0.48	0.12	0.10	0.56	1.4	0.46	
t_i	70.66	4.03	4.00	10.91	263.08	752.52	
r_i	9.79	40.71	33.49	87.54	7.71	0.88	
Tl_i	642.75	37.00	0	730.00	49.25	0	

Further increase in the volume of train traffic will be difficult to provide due to the small capacity of the stations. In particular, Baruunkharaa has 4 tracks, Tunkh and Mandal have 5 tracks each. Collisions occurred precisely at the nodes of the QN, which correspond to these stations. Therefore, it is possible to significantly increase the overall capacity of the Ulaanbaatar railway section only if a complete reconstruction is carried out with the building of a double-track trunk line.

7 Discussion and Conclusions

The article presents a methodology for modeling the operation of a single-track section of the railway network, the distinctive property of which is a package (group) train schedule. The models constructed on its basis have the form of queuing networks with incoming BMAP flows, several types of requests and their package services at nodes, which makes it possible to take into account the stochastic nature of train arrivals and their non-deterministic running along the railway trunk. This mathematical apparatus also allows us to consider a batch arrival of trains from two or more directions, a non-linear hierarchical structure of the system with several possible routes, and different capacities of stations and lines between them.

Package train traffic is typical for some Asian countries, including Mongolia, through whose territory important transport corridors connecting the central regions of Russia with China pass. Therefore, based on the proposed methodology, the model of the Ulaanbaatar railway section was constructed and studied. As a result, we determined its admissible capacity, "bottlenecks" in the structure, and technical solutions, which can help increase the volume of train traffic in this section by more than 20%.

Further research can be related to improving the accuracy and adequacy of the model. This can be achieved by considering the train as a group of requests (cars), the number of which may change when running along the railway network. Another possible direction is to consider the interaction of railway and other modes of transport, for example, road transport, in the transit of goods and passengers. Besides, it is also interesting to study the transient processes occurring in the railway network operation, as was done for the freight railway station and the transport hub [17,18].

References

1. Vishnevsky, V., Semenova, O.: Polling systems and their application to telecommunication networks. Mathematics **9**(2), 117 (2021)
2. Bushkova, T., Danilyuk, E., Moiseeva, S., Pavlova, E.: Resource queueing system with dual requests and their parallel service. In: Vishnevskiy, V., Samouylov, K. (eds.) DCCN 2019. CCIS, vol. 1141. Springer, Cham (2019). https://doi.org/10.1007/978-3-030-36625-4_29
3. Bushkova, T., Moiseeva, S., Moiseev, A., et al.: Using infinite-server resource queue with splitting of requests for modeling two-channel data transmission. Methodol. Comput. Appl. Probab. **24**, 1753–1772 (2022). https://doi.org/10.1007/s11009-021-09890-6
4. Dudin, A.N., Klimenok, V.I., Vishnevsky, V.M.: The Theory of Queuing Systems with Correlated Flows. Springer, Cham (2020). https://doi.org/10.1007/978-3-030-32072-0
5. Vuuren, van M. : Performance analysis of manufacturing systems: queueing approximations and algorithms. Technische Universiteit Eindhoven, Eindhoven (2007). https://doi.org/10.6100/IR625074

6. Pyrgidis, C.: Railway transportation systems. Design, Construction and Operation. 1st edn. CRC Press, New York (2016)
7. Wilson, N., Fourie, C.J., Delmistro, R.: Mathematical and simulation techniques for modelling urban train networks. South Afr. J. Ind. Eng. **27**, 109–119 (2016). https://doi.org/10.7166/xx-x-1364
8. Kozan, E., Higgens, A.: Modeling train delays in urban networks. Transp. Sci. **32**, 346–357 (1998). https://doi.org/10.1287/trsc.32.4.346
9. Huisman, T., Boucherie, R.: Running times on railway sections with heterogeneous train traffic. Transp. Res. Part B: Methodol. **35**, 271–292 (2001). https://doi.org/10.1016/S0191-2615(99)00051-X
10. Huisman, T., Boucherie, R., Van Dijk, N.: A solvable queueing network model for railway networks and its validation and applications for the Netherlands. Eur. J. Oper. Res. **142**, 30–51 (2002). https://doi.org/10.1016/S0377-2217(01)00269-7
11. Weik, N., Nießen, N.: Quantifying the effects of running time variability on the capacity of rail corridors. J. Rail Transp. Plan. Manag. **15**, 100203 (2020). https://doi.org/10.1016/j.jrtpm.2020.100203
12. Marinov, M., Viegas, J.: A mesoscopic simulation modelling methodology for analyzing and evaluating freight train operations in a rail network. Simul. Model. Pract. Theory **19**(1), 516–539 (2011). https://doi.org/10.1016/j.simpat.2010.08.009
13. Marinov, M., Viegas, J.: A simulation modelling methodology for evaluating flat-shunted yard operations. Simul. Model. Pract. Theory **17**(6), 1106–1129 (2009). https://doi.org/10.1016/j.simpat.2009.04.001
14. Lyubchenko, A., Bartosh, S., Kopytov, et al.: anylogic-based discrete event simulation model of railway junction. In: Silhavy, R., Senkerik, R. (eds.) CSOC 2017. AISC, vol. 574, pp. 141–149. Springer, Cham (2017). https://doi.org/10.1007/978-3-319-57264-2_14
15. Zharkov, M., Lempert, A., Pavidis, M.: Simulation of railway marshalling yards based on four-phase queuing systems. In: Dudin, A., Nazarov, A. (eds.) ITMM 2020. CCIS, vol. 1391. Springer, Cham (2020). https://doi.org/10.1007/978-3-030-72247-0_11
16. Bychkov, I., Kazakov, A., Lempert, A., Zharkov, M.: Modeling of railway stations based on queuing networks. Appl. Sci. **11**(5), 2425 (2021). https://doi.org/10.3390/app11052425
17. Zharkov, M., Kazakov, A., Lempert, A.: On the application of queuing theory in the analysis of transients in the operation of a freight railway station. In: Dudin, A., Nazarov, (eds.) ITMM 2021. CCIS, vol. 1605, pp. 266–277 (2022) https://doi.org/10.1007/978-3-031-09331-9_21
18. Zharkov, M., Kazakov, A., Lempert, A.: Transient process modeling in micrologistic transport systems. IOP Conf. Series: Earth Environ. Sci. 2021 **629**, 012023 (2021). https://doi.org/10.1088/1755-1315/629/1/012023

Asymptotic Analysis of a Multiserver Retrial Queue with Disasters in the Service Block

Natalya Meloshnikova(✉), Ekaterina Fedorova, and Danil Plaksin

National Research Tomsk State University, Lenina Avenue, 36, Tomsk, Russia
meloshnikovana@gmail.com

Abstract. The paper studies a multiserver retrial queuing system with negative calls. The arrival processes of "positive" and "negative" calls are Poisson. Positive call's service time is exponential distributed. Unserved calls go to an orbit, where they wait for random time distributed exponentially. Then they turn up to the service block according to the random multiple access protocol. Disasters are caused by negative calls arrivals. When a negative call comes, it resets all servers. All servicing positive calls leave the system. In the paper, a stationary probability distribution of the number of calls in orbit is found by the method of asymptotic analysis under the condition of a long delay. The results of the numerical analysis are presented.

Keywords: retrial queue · negative customers · orbit · asymptotic analysis · long delay · disasters

1 Introduction

Unmanned aerial vehicles (UAVs), or drones, are unmanned aircraft that are controlled remotely (for example, by a ground station or by another aircraft) or by autonomous software installed on their board or ground control complex [2]. To transmit data from on-board sensors of UAVs to the control point, an UAV has a radio transmitter that provides radio communication with ground-based receiving equipment. UAVs communication can be classified according to data transfer protocols:

- IEEE 802.15.4 data transmission standard using Zigbee, 6LoWPAN protocols;
- Bluetooth;
- WiFi;
- LTE, 3G,4G.

This study was supported by the Tomsk State University Development Programme (Priority-2030).

A. Dudin et al. (Eds.): ITMM 2022, CCIS 1803, pp. 55–67, 2023.
https://doi.org/10.1007/978-3-031-32990-6_5

Of these technologies, 3G and 4G networks have an advantage. Due to the large network coverage, as drones fly over vast distances, Bluetooth and Wi-Fi will be of little use. Also, more bits per second can be transmitted by using the 3G/4G then the Zigbee, 6LoWPAN protocols.

Depending on the data format (text, image, video) and the degree of their compression, the throughput of digital radio data transmission lines from the UAV can be units or hundreds of Mbps. Storing a large amount of data becomes a problem, since data storage equipment provides to increasing of size and weight of UAVs. Thus sometimes UAVs are managed by cloud services [3,4]. In [2] there is a description of drones and cloud services interaction. The cloud service stores and processes information received from drones, and allows companies not to spend a large amount of money on equipment maintenance. But in the case of using external services, there is a threat to the quality of communication and data security. Described communication system can be analyzed via the queuing theory (QT). The data received from the UAV enter in random time, thus it forms an arrival process in queueing system. The cloud service contains resources for incoming information processing or storage, that is called as service block in terms of QT.

In the paper, we consider a mathematical model of UAVs communication as a retrial queueing system. We improve the model by taking into account a negative external impact in networks (e.g. signal loss, technical failures, hacker attacks) as a retrial queueing model with disasters.

Retrial queueing systems [5,6] are new mathematical models of the queueing theory, which are often used to analyze, research and optimize various technical systems: cellular networks, information transmission systems, call centers, cloud computing centers, LANs, etc. [7]. A main feature of such models is the presence of repeated calls to the service block after an unsuccessful attempt to receive the service. There is no a queue in the system, unserved calls go to an orbit (some virtual place), where they perform a random delay. There is a random access protocol for all calls in an orbit.

Queuing models with negative calls (G-systems) were proposed by Gelenbe E. [8,9]. "Negative" calls do not require service, but they affect to a system performance. There are several types of negative effects: negative call destroy calls in a queue, break servers, "reset" a system, etc. As we already said, this feature describes a presence of viruses, hacker attacks,technical problems, etc. Retrial queues with negative calls are studied in [10,11]. In the paper [12], a queuing system with repeated calls and four types of arrivals, positive customers, catastrophes, and two types of negative calls are considered.

The paper is structured as follows. In Sect. 2, a multiserver retrial queueing system with disasters in the service block is described and the problem statement is formulated, system of Kolmogorov equations are written. The Sect. 3 is devoted to the asymptotic analysis method under long delay condition applying for the system of equations solving. In Sect. 4, we demonstrate some numerical experiments that represent the accuracy of the asymptotic results. The Sect. 5 is devoted to some concluding remarks.

2 Model Description

In this paper, we consider a multiserver retrial queueing system (Fig. 1). The arrival process of calls is Poisson with parameter λ, we will call these customers as "positive". Positive customers arrive into a service block (which has N servers), until all servers become busy. The service time of each call is distributed exponentially with parameter μ. If all servers are busy, then the call goes to an orbit where it performs a random delay. The delay duration has an exponential distribution with parameter σ. From the orbit, the call again turns to the service block. If there is a free server, call begins its service, otherwise it returns to the orbit to make next attempt.

Also, there is an arrival process of "negative" calls. It is Poisson with parameter γ. A negative call does not need service. When a negative call enters in the system, it "resets" all servers in the system, so all servicing calls leave the system and the service block becomes free. We will call such model as the retrial queueing system with disasters [12] in the service block.

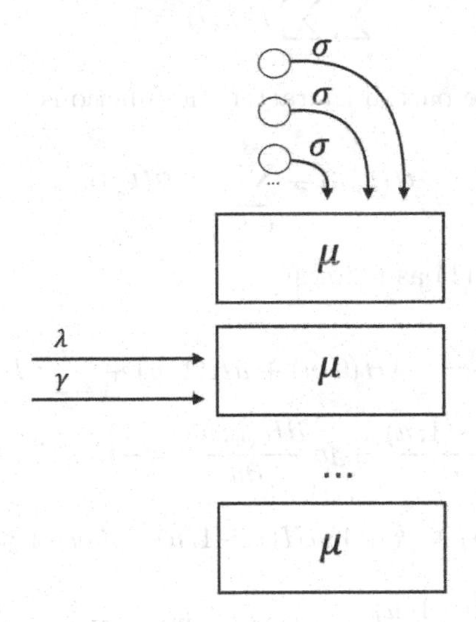

Fig. 1. Multiserver RQ with negative calls

Denote a random process of the number of calls in the orbit by $i(t)$. And process $k(t)$ determines states of the service block as follows:

$$k(t) = \begin{cases} 0, \text{if all servers are free,} \\ 1, \text{if one server is busy,} \\ \dots \\ k, \text{if } k \text{ servers are busy,} \\ \dots \\ K, \text{if all servers are busy.} \end{cases}$$

The problem of finding the stationary probability distribution of the number of calls in the orbit is posed.

Let $P\{k(t) = k, i(i) = i\} = P(k, i)$ be the stationary probabilities that the service block is in state k and there are i calls in the orbit. Obviously, process $\{k(t), i(t)\}$ is Markovian. For probability distribution $P(k, i, t)$, we compose a system of Kolmogorov equations, which in the steady-state has the following form:

$$\begin{cases} -P(0,i)(\lambda + i\sigma) + P(1,i)\mu + \sum\limits_{n=1}^{N} P(n,i)\gamma = 0, \\ -P(k,i)(\lambda + k\mu + i\sigma + \gamma)) + P(k-1,i)\lambda + \\ +P(k-1,i+1)\sigma(i+1) + P(k+1,i)(k+1)\mu = 0, \text{ for } 1 \le k \le K-1, \\ -P(K,i)(\lambda + K\mu + \gamma) + P(K-1,i)\lambda + P(K,i-1)\lambda + \\ +P(K-1,i+1)\sigma(i+1) = 0. \end{cases} \tag{1}$$

To find probabilities $P(k, i)$, it is also necessary to add the normalization condition:

$$\sum_{k=0}^{K} \sum_{i=0}^{\infty} P(k,i) = 1. \tag{2}$$

Let us introduce the partial characteristic functions

$$H(k,u) = \sum_{i=0}^{\infty} e^{jui} P(k,i).$$

We rewrite System (1) as follows

$$\begin{cases} -j\sigma \dfrac{\partial H(0,u)}{\partial u} = -\lambda H(0,u) + \mu H(1,u) + \sum\limits_{n=1}^{K} \gamma H(n,u), \\ j\sigma e^{-ju} \dfrac{\partial H(k-1,u)}{\partial u} - j\sigma \dfrac{\partial H(k,u)}{\partial u} = -(\lambda + k\mu + \gamma)H(k,u) + \\ +\lambda H(k-1,u) + (k+1)\mu H(k+1,u), \; where \, 1 \le k \le K-1, \\ j\sigma e^{-ju} \dfrac{\partial H(K-1,u)}{\partial u} = -(\lambda(1-e^{ju}) + K\mu + \gamma)H(K,u) + \\ +\lambda H(K-1,u). \end{cases}$$

Let's write the system into a matrix form:

$$j\sigma \frac{\partial \mathbf{H}(u)}{\partial u}(\mathbf{A}_0 + e^{-ju}\mathbf{A}_1) = \mathbf{H}(u)(\mathbf{B}_0 + e^{ju}\mathbf{B}_1), \tag{3}$$

$$\mathbf{H}(0)\mathbf{e} = 1, \tag{4}$$

where (4) is the normalization condition, $\mathbf{H}(u) = \{H(0,u), H(1,u), ..., H(k,u), ..., H(K,u)\}$ is row-vector of the partial characteristic functions, matrix $\mathbf{A}_0$, $\mathbf{A}_1$, $\mathbf{B}_0$ and $\mathbf{B}_1$ are following:

$$\mathbf{A}_0 = \begin{pmatrix} -1 & 0 & \dots & 0 & \dots & 0 & 0 \\ 0 & -1 & \dots & 0 & \dots & 0 & 0 \\ \vdots & \vdots & \vdots & \vdots & \vdots & \vdots & \vdots \\ 0 & 0 & \dots & 0 & \dots & 0 & 0 \\ 0 & 0 & \dots & -1 & \dots & 0 & 0 \\ \vdots & \vdots & \vdots & \vdots & \vdots & \vdots & \vdots \\ 0 & 0 & \dots & 0 & \dots & -1 & 0 \\ 0 & 0 & \dots & 0 & \dots & 0 & 0 \end{pmatrix},$$

$$\mathbf{A}_1 = \begin{pmatrix} 0 & 1 & \dots & 0 & \dots & 0 & 0 \\ 0 & 0 & \dots & 0 & \dots & 0 & 0 \\ \vdots & \vdots & \vdots & \vdots & \vdots & \vdots & \vdots \\ 0 & 0 & \dots & 1 & \dots & 0 & 0 \\ 0 & 0 & \dots & 0 & \dots & 0 & 0 \\ \vdots & \vdots & \vdots & \vdots & \vdots & \vdots & \vdots \\ 0 & 0 & \dots & 0 & \dots & 0 & 1 \\ 0 & 0 & \dots & 0 & \dots & 0 & 0 \end{pmatrix},$$

$$\mathbf{B}_0 = \begin{pmatrix} -\lambda & \lambda & \dots & 0 & \dots & 0 \\ \mu+\gamma & -(\lambda+\mu+\gamma) & \dots & 0 & \dots & 0 \\ \gamma & 2\mu & \dots & 0 & \dots & 0 \\ \vdots & \vdots & \vdots & \vdots & \vdots & \vdots \\ \gamma & 0 & \dots & \lambda & \dots & 0 \\ \gamma & 0 & \dots & -(\lambda+k\mu+\gamma) & \dots & 0 \\ \gamma & 0 & \dots & (k+1)\mu & \dots & 0 \\ \vdots & \vdots & \vdots & \vdots & \vdots & \vdots \\ \gamma & 0 & \dots & 0 & \dots & \lambda \\ \gamma & 0 & \dots & 0 & \dots & -(\lambda+N\mu+\gamma) \end{pmatrix},$$

$$\mathbf{B}_1 = \begin{pmatrix} 0 & 0 & \dots & 0 & \dots & 0 \\ 0 & 0 & \dots & 0 & \dots & 0 \\ 0 & 0 & \dots & 0 & \dots & 0 \\ \vdots & \vdots & \vdots & \vdots & \vdots & \vdots \\ 0 & 0 & \dots & 0 & \dots & 0 \\ 0 & 0 & \dots & 0 & \dots & 0 \\ 0 & 0 & \dots & 0 & \dots & 0 \\ \vdots & \vdots & \vdots & \vdots & \vdots & \vdots \\ 0 & 0 & \dots & 0 & \dots & 0 \\ 0 & 0 & \dots & 0 & \dots & \lambda \end{pmatrix}.$$

To solve System (3)–(4), we propose to apply the method of asymptotic analysis under the condition of a long delay of calls in the orbit [13].

3 Asymptotic Analysis

3.1 The First Order Asymptotics

To derive the first order asymptotics, we denote

$$\sigma = \varepsilon, u = \varepsilon w, \mathbf{H}(u) = \mathbf{F}_1(w, \varepsilon).$$

Then Equations (3)–(4) have the form

$$j\frac{\partial \mathbf{F}_1(w,\varepsilon)}{\partial w}(\mathbf{A}_0 + e^{-j\varepsilon w}\mathbf{A}_1) = \mathbf{F}(w,\varepsilon)(\mathbf{B}_0 + e^{j\varepsilon w}\mathbf{B}_1), \tag{5}$$

$$\mathbf{F}_1(0)\mathbf{e} = 1. \tag{6}$$

Let us pass to the limit at $\sigma \to 0$:

$$\begin{cases} j\dfrac{\partial \mathbf{F}_1(w)}{\partial w}(\mathbf{A}_0 + \mathbf{A}_1) = \mathbf{F}_1(w)(\mathbf{B}_0 + \mathbf{B}_1), \\ \mathbf{F}_1(0)\mathbf{e} = 1. \end{cases} \tag{7}$$

Let function $\mathbf{F}_1(w)$ have the form

$$\mathbf{F}_1(w) = \mathbf{R}\phi_1(w). \tag{8}$$

From the first Equation (7), it is obvious that

$$\phi_1(w) = \exp\{jwk_1\},$$

where k_1 is unknown parameter.

Substituting (8) into (7), we obtain a system for vector $\mathbf{R}$ determining:

$$\begin{cases} \mathbf{R}((\mathbf{B}_0 + \mathbf{B}_1) + k_1(\mathbf{A}_0 + \mathbf{A}_1)), \\ \mathbf{R}\mathbf{e} = 1. \end{cases} \tag{9}$$

For k_1 finding, we use a Taylor series $e^{-j\varepsilon w} = 1 - j\varepsilon w + O(\varepsilon^2)$ in Equation (7):

$$j\frac{\partial \mathbf{F}_1(w,\varepsilon)}{\partial w}(\mathbf{A}_0 + \mathbf{A}_1 - (j\varepsilon w)\mathbf{A}_1)\mathbf{e} = \mathbf{F}_1(w,\varepsilon)(\mathbf{B}_0 + \mathbf{B}_1 + (j\varepsilon w)\mathbf{B}_1)\mathbf{e} + O(\varepsilon^2).$$

Given that $(\mathbf{A}_0 + \mathbf{A}_1)\mathbf{e} = 0$, $(\mathbf{B}_0 + \mathbf{B}_1)\mathbf{e} = 0$, we have

$$j\frac{\partial \mathbf{F}_1(w,\varepsilon)}{\partial w}(-(j\varepsilon w)\mathbf{A}_1)\mathbf{e} = \mathbf{F}_1(w,\varepsilon)(j\varepsilon w)\mathbf{B}_1\mathbf{e} + O(\varepsilon^2).$$

Under the limit $\varepsilon \to 0$, the following expression is true

$$\mathbf{R}(\mathbf{B}_1 - k_1\mathbf{A}_1)\mathbf{e} = 0. \tag{10}$$

In this way, we obtain that the first-order asymptotic function has the form $\mathbf{F}_1(w) = \mathbf{R}\exp\{jwk_1\}$, where parameters $\mathbf{R}$ and k_1 are defined by System (9)–(10).

Returning to the substitutions, we obtain the first-order asymptotic characteristic function $h_1(u) = \exp\left\{ju\frac{k_1}{\sigma}\right\}$ of the number of calls in orbit, where value $\frac{k_1}{\sigma}$ is an asymptotic means of the process under study.

3.2 The Second Order Asymptotics

Because of the first-order asymptotics does not conduct to the form probability distribution, we perform the second order asymptotics. First of all we use the following substitution:

$$\mathbf{H}(u) = \exp\left\{\frac{ju}{\sigma}k_1\right\}\mathbf{H}_2(u).$$

From Equation (3), we obtain

$$j\sigma\frac{\partial\mathbf{H}_2(u)}{\partial u}(\mathbf{A}_0 + e^{-ju}\mathbf{A}_1) = \mathbf{H}_2(u)((\mathbf{B}_0 + e^{ju}\mathbf{B}_1) + k_1(\mathbf{A}_0 + e^{-ju}\mathbf{A}_1)), \tag{11}$$

$$\mathbf{H}_2(0)\mathbf{e} = 1. \tag{12}$$

In System (11)–(12), we denote

$$\sigma = \varepsilon^2, \quad u = \varepsilon w, \mathbf{H}_2(u) = \mathbf{F}_2(w,\varepsilon).$$

We have the following equation:

$$\begin{gathered} j\varepsilon\frac{\partial\mathbf{F}_2(w,\varepsilon)}{\partial w}(\mathbf{A}_0 + e^{-j\varepsilon w}\mathbf{A}_1) = \\ = \mathbf{F}_2(w,\varepsilon)((\mathbf{B}_0 + e^{j\varepsilon w}\mathbf{B}_1) + k_1(\mathbf{A}_0 + e^{-j\varepsilon w}\mathbf{A}_1)), \end{gathered} \tag{13}$$

$$\mathbf{F}_2(0,\varepsilon)\mathbf{e} = 1. \tag{14}$$

After passing to the limit (13)-(14) at $\varepsilon \to 0$, we have the system

$$\begin{cases} \mathbf{F}_2(w)((\mathbf{B}_0 + \mathbf{B}_1) + k_1(\mathbf{A}_0 + \mathbf{A}_1)), \\ \mathbf{F}_2(0)\mathbf{e} = 1. \end{cases}$$

solution $\mathbf{F}_2(w)$ which we write in the form of a product

$$\mathbf{F}_2(w) = \mathbf{R}\phi_2(w) = \mathbf{R}\exp\left\{\frac{(jw)^2}{2}k_2\right\},$$

where vector $\mathbf{R}$ is defined by System (9), and the parameter k_2 will be defined below.

Solution $\mathbf{F}_2(w, \varepsilon)$ System (13) can be written in the form of expansion

$$\begin{aligned} \mathbf{F}_2(w, \varepsilon) &= \{\mathbf{R} + j\varepsilon w \mathbf{f}_1(w)\} \phi_2(w) + O(\varepsilon^2) = \\ &= \{\mathbf{R} + j\varepsilon w \mathbf{f}_1(w)\} \exp\left\{\frac{(jw)^2}{2} k_2\right\} + O(\varepsilon^2), \end{aligned} \tag{15}$$

substituting which into System (13), we obtain

$$\begin{gathered} j\varepsilon j^2 w k_2 \mathbf{R}(\mathbf{A}_0 + e^{-j\varepsilon w}\mathbf{A}_1) = \\ \{\mathbf{R} + j\varepsilon w \mathbf{f}_1(w)\}\left((\mathbf{B}_0 + e^{j\varepsilon w}\mathbf{B}_1) + k_1(\mathbf{A}_0 + e^{-j\varepsilon w}\mathbf{A}_1)\right) + O(\varepsilon^2). \end{gathered}$$

Substituting Taylor series $e^{-j\varepsilon w} = 1 - j\varepsilon w + O(\varepsilon^2)$, and taking into account Equality (9), we obtain

$$\mathbf{f}_1(w)((\mathbf{B}_0 + \mathbf{B}_1) + k_1(\mathbf{A}_0 + \mathbf{A}_1)) + \mathbf{R}(\mathbf{B}_1 - k_1\mathbf{A}_1) + k_2\mathbf{R}(\mathbf{A}_0 + \mathbf{A}_1) = \mathbf{0}.$$

from which we obtain the system of equation with respect to $\mathbf{f}_1(w)$.

We write the solution in the form

$$\mathbf{f}_1(w) = G_1(w)\mathbf{R} + \mathbf{f}_1, \tag{16}$$

where $G_1(w)$ is a scalar function, and the vector $\mathbf{f}_1$ is a partial solution of the system.

We get that solution $\mathbf{f}_1$ can be writted as the sum

$$\mathbf{f}_1 = \mathbf{g}_1 + k_2\mathbf{g}_2, \tag{17}$$

where vectors $\mathbf{g}_1$ and $\mathbf{g}_2$ are solutions of systems

$$\mathbf{g}_1((\mathbf{B}_0 + \mathbf{B}_1) + k_1(\mathbf{A}_0 + \mathbf{A}_1)) + \mathbf{R}(\mathbf{A}_0 + \mathbf{A}_1) = \mathbf{0}, \tag{18}$$

$$\mathbf{g}_2((\mathbf{B}_0 + \mathbf{B}_1) + k_1(\mathbf{A}_0 + \mathbf{A}_1)) + \mathbf{R}(\mathbf{B}_1 - k_1\mathbf{A}_1) = \mathbf{0}. \tag{19}$$

To find the value of the quantity k_1 we add all the equations of System (13) and, multiplying this equality on the right by the unit $\mathbf{e}$, we have

$$\begin{aligned} &j\varepsilon \frac{\partial \mathbf{F}_2(w, \varepsilon)}{\partial w}(\mathbf{A}_0 + e^{-j\varepsilon w}\mathbf{A}_1)\mathbf{e} = \\ &= \mathbf{F}_2(w, \varepsilon)((\mathbf{B}_0 + e^{j\varepsilon w}\mathbf{B}_1) + k_1(\mathbf{A}_0 + e^{-j\varepsilon w}\mathbf{A}_1))\mathbf{e}, \end{aligned}$$

Substituting here the expansion in the Taylor series $e^{-j\varepsilon w} = 1 - j\varepsilon w + \frac{(j\varepsilon w)^2}{2} + O(\varepsilon^2)$, and given $(\mathbf{A}_0 + \mathbf{A}_1)\mathbf{e} = 0$, $(\mathbf{B}_0 + \mathbf{B}_1)\mathbf{e} = 0$, we have

$$\begin{aligned} &j\varepsilon \frac{\partial \mathbf{F}_2(w, \varepsilon)}{\partial w}(-(j\varepsilon w)\mathbf{A}_1 + \frac{(j\varepsilon w)^2}{2}\mathbf{A}_1)\mathbf{e} = \\ &= \mathbf{F}_2(w, \varepsilon)(((j\varepsilon w)\mathbf{B}_1 + \frac{(j\varepsilon w)^2}{2}\mathbf{B}_1) + k_1(-(j\varepsilon w)\mathbf{A}_1 + \frac{(j\varepsilon w)^2}{2}\mathbf{A}_1))\mathbf{e}. \end{aligned}$$

Let us substitute Expression (15)

$$-j^2 w\mathbf{R}k_2(-(\varepsilon w)\mathbf{A}_1 + \frac{j(\varepsilon w)^2}{2}\mathbf{A}_1)\mathbf{e} =$$
$$= \{\mathbf{R} + j\varepsilon w\mathbf{f}_1(w)\}\left((jw)(\mathbf{B}_1 - k_1\mathbf{A}_1) - \frac{1}{2}(\varepsilon w^2)(\mathbf{B}_1 + k_1\mathbf{A}_1)\right)\mathbf{e},$$

Carry out some transformations, we obtain the equality

$$-\mathbf{f}_1(w)(\mathbf{B}_1 - k_1\mathbf{A}_1)\mathbf{e} - \frac{1}{2}\mathbf{R}(\mathbf{B}_1 + k_1\mathbf{A}_1)\mathbf{e} + \mathbf{R}k_2\mathbf{A}_1\mathbf{e}, \quad (20)$$

Then, due to equality (17), Equation (20) can be rewritten in the form

$$-(\mathbf{g}_2 + k_2\mathbf{g}_2)(\mathbf{B}_1 - k_1\mathbf{A}_1)\mathbf{e} - \frac{1}{2}\mathbf{R}(\mathbf{B}_1 + k_1\mathbf{A}_1)\mathbf{e} + \mathbf{R}k_2\mathbf{A}_1\mathbf{e},$$

from where we can write the formula for k_2 as follows

$$k_2 = \frac{\mathbf{g}_1(\mathbf{B}_1 - k_1\mathbf{A}_1)\mathbf{e} + \frac{1}{2}\mathbf{R}(\mathbf{B}_1 + k_1\mathbf{A}_1)\mathbf{e}}{-\mathbf{g}_2(\mathbf{B}_1 - k_1\mathbf{A}_1)\mathbf{e} + \mathbf{R}\mathbf{A}_1\mathbf{e}}. \quad (21)$$

As the result of the second order asymptotic analysis, we obtain that asymptotic characteristic function of the number of calls in the orbit has form

$$h_2(u) = \exp\left\{ju\frac{k_1}{\sigma} + \frac{(ju)^2}{2}\frac{k_2}{\sigma}\right\},$$

so it is Gaussian with mean $\frac{k_1}{\sigma}$ and variance $\frac{k_2}{\sigma}$, which are defined by Systems (9)–(10), (18)–(19) and (21).

4 Numerical Analysis

To analyze the range of applicability of the proposed asymptotic method, we numerically compare asymptotic distribution $PA(i)$ and exact distribution $P(i)$ obtained using a numerical algorithm for various values of the system parameters.

Denote the load parameter by $\rho = \frac{\lambda}{K\mu}$. We demonstrate an example for $K = 3$, $\gamma = 0,000001$, and variable values of σ and ρ.

As a measure of the the asymptotic method accuracy, we use the Kolmogorov distance:

$$\Delta = \left|\sum_{n=0}^{i}(PA(n) - P(n))\right|,$$

where $PA(i)$ is an asymptotic probability distribution, $P(n)$ is a corresponding exact distribution.

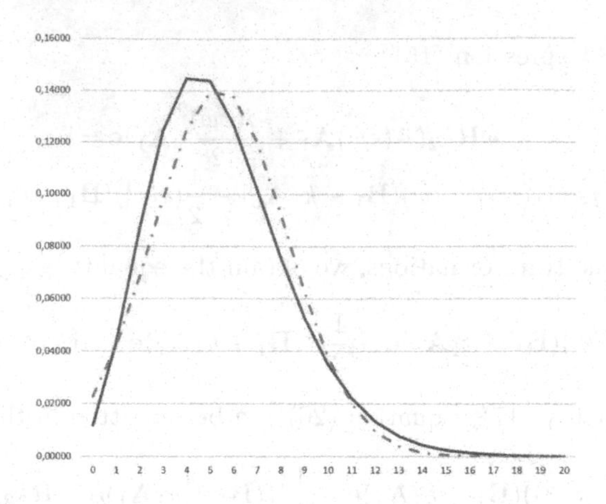

Fig. 2. b) $\rho = 0.3$ and $\sigma = 0.01$

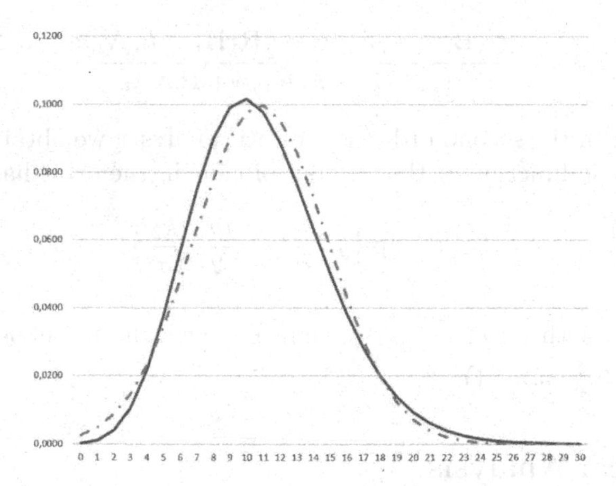

Fig. 3. a) $\rho = 0.3$ and $\sigma = 0.005$

Table 1. Kolmogorov distances for various values of the parameter ρ

	$\sigma = 0.1$	$\sigma = 0.05$	$\sigma = 0.01$	$\sigma = 0.005$
$\rho = 0.3$	0.256	0.184	0.059	0.031
$\rho = 0.5$	0.110	0.054	0.016	0.011
$\rho = 0.7$	0.032	0.032	0.015	0.011

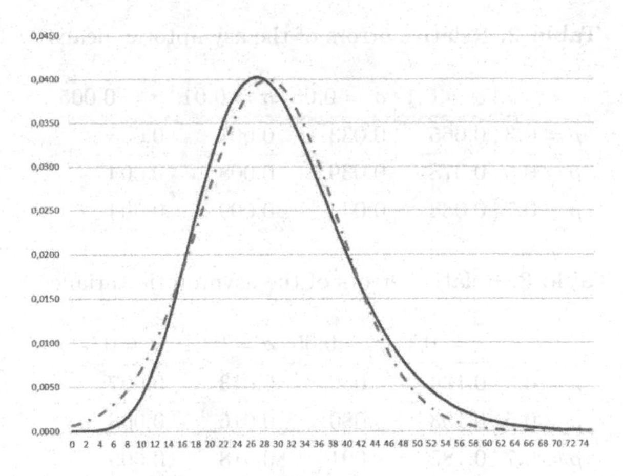

Fig. 4. b) $\rho = 0.7$ and $\sigma = 0.05$

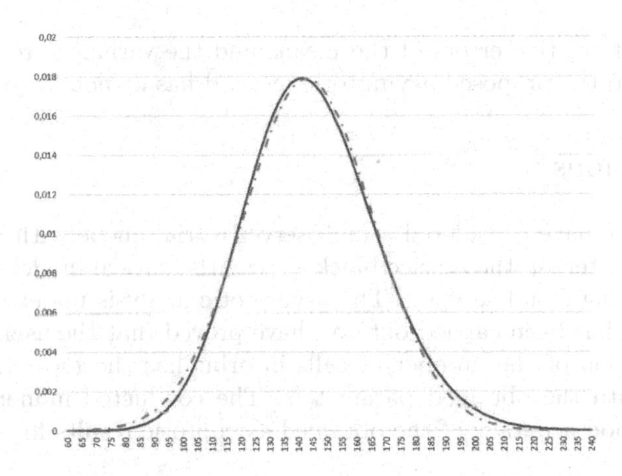

Fig. 5. a) $\rho = 0.7$ and $\sigma = 0.01$

The results of the distributions comparison are presented in Table 1 and Figs. 2, 3, 4, 5.

From Table 1, we conclude that the method accuracy increase with ρ growing and σ decreasing. For other values of the system parameters, the results are same.

We also calculate values of the relative error of the asymptotic means and variance (Tables (2), (3)).

Table 2. Relative errors of the asymptotic means

	$\sigma = 0.1$	$\sigma = 0.05$	$\sigma = 0.01$	$\sigma = 0.005$
$\rho = 0.3$	0.065	0.033	0.007	0.003
$\rho = 0.5$	0.078	0.039	0.008	0.004
$\rho = 0.7$	0.088	0.044	0.009	0.004

Table 3. Relative errors of the asymptotic variance

	$\sigma = 0.1$	$\sigma = 0.05$	$\sigma = 0.01$	$\sigma = 0.005$
$\rho = 0.3$	0.134	0.066	0.013	0.007
$\rho = 0.5$	0.163	0.080	0.016	0.008
$\rho = 0.7$	0.185	0.091	0.018	0.009

We get that relative errors of the means and the variance are less then 2% for $\sigma \leq 0.01$. So the proposed asymptotic method has a enough good accuracy.

5 Conclusions

In the study, we have considered a multiserver retrial queue with negative customers and disasters in the service block as a mathematical model of UAV communications using cloud services. The asymptotic analysis under the condition of a long delay has been carried out. We have proved that the asymptotic characteristic function of the number of calls in orbit has the form of a Gaussian distribution with the obtained parameters. The conducted numerical analysis has shown a good accuracy of the obtained asymptotic results for $\sigma \leq 0.01$.

References

1. Abramov, M.M.: New and promising directions for the application of unmanned aerial vehicles. Tech. Sci. **2**, 227–332 (2022)
2. Poltavskiy, A.V., Zhumbaeva, A.S., Bikkev, R.R.: Multifunctional complexes of unmanned aerial vehicles: development in the weapon system. Reliability Quality Complex Syst. **1**(13), 99–110 (2016)
3. Moeyersons, J., Gevaert, M., Réculé, K.-E., Volckaert, B., Turck, F.D.: UAVs-as-a-d. In: Ahmed, T., Festor, O., Ghamri-Doudanc, Y., Kang, J.M., Schaeffer-Filho, A.E., Lahmadi, A., Madeira, E. (eds.) IFIP/IEEE International Symposium on Integrated Network Management 2021, IM, pp. 926–931. Bordeaux, France (2021)
4. Chen, C., Shi, D., Cui, S., Kang, Y.: Cloud-Based UAV Monitoring and Management Framework. International Conference on Cybernetics, Robotics and Control, CRC, vol. 9999, pp. 61–66. Penang, Malaysia (2018). https://doi.org/10.3390/s16111913
5. Falin, G.I., Templeton, J.G.C.: Retrial queues, 1st edn. Springer, New York, NY (1999)

6. Artalejo, J.R., Gomez-Corral, A.: Retrial Queueing Systems, 1st edn. Springer, Berlin. Heidelberg (2008)
7. Phung-Duc, T.: Retrial Queueing Models: A Survey on Theory and Applications. In: Stochastic Operations Research in Business and Industry, pp. 1–31. World Scientific Publisher, World Scientific Publisher (2017)
8. Gelenbe, E.: Product-form queueing networks with negative and positive customers. J. Appl. Probab. **28**(3), 656–663 (1991)
9. Do, T.V.: Bibliography on G-networks, negative customers and applications. Ann. Oper. Res. **48**, 205–212 (2011)
10. Kirupa, K., Udaya, C.K.: Batch arrival retrial queue with negative customers, multi-optional service and feedback. Commun. Appl. Electron. **2**(4), 14–18 (2015)
11. Fedorova, E.A., Nazarov, A.A., Farkhadov, M.P.: Asymptotic analysis of the RQ-System MMPP/M/1 with negative customers under heavy load condition. Saratov: Izv. Sarat. university New ser. Ser.: Math. Mech. **4**, 534–547 (2016)
12. Shin, Y.W.: Multi-server retrial queue with negative customers and disasters. Queueing Syst. **55**(4), 223–337 (2007)
13. Nazarov, A., Phung-Duc, T., Paul, S.: Slow retrial asymptotics for a single server queue with two-way communication and markov modulated poisson input. J. Syst. Sci. Syst. Eng. **28**(2), 181–193 (2019). https://doi.org/10.1007/s11518-018-5404-6

Refined Limit Theorems for the Critical Continuous-Time Markov Branching Systems

Azam A. Imomov[1,2(✉)] and Misliddin Murtazaev[2]

[1] Karshi State University, Karshi, Uzbekistan
imomov_azam@mail.ru
[2] Romanovskiy Institute of Mathematics, Tashkent, Uzbekistan

Abstract. We consider a critical homogeneous-continuous-time Markov branching system, i.e. the average value of the branching rate is zero. Our basic assumption is that the branching rate generating function of the system regularly varies, in which slowly varying factor varies at infinity with an explicit expression remainder. We essentially rely on the improved version of the Basic Lemma of the critical Markov branching systems theory. First we establish a convergence rate in the Monotone ratio theorem. Subsequently we prove a local-convergence limit theorem on the asymptotic expansion of transition probabilities and their convergence to the invariant measure.

Keywords: Markov branching system · Slowly varying functions with remainder · q-matrix · Transition probabilities · Generating functions · Monotone Ratio theorem · Invariant measure · Convergence rate

1 Introduction and Results

1.1 Preliminaries on Slow Variation

Our arguments in the paper are essentially based on the slow variation, or more general, regular variation conception that was first initiated by the famous Serbian mathematician Jovan Karamata in the early 30s of the last century. Recall that a real-valued, positive and measurable function $L(x)$ is said to be slowly varying (SV) function at infinity (in the sense of Karamata) if $L(\lambda x)/L(x) \to 1$ as $x \to \infty$ for each $\lambda > 0$. Throughout the paper we will use $\mathbf{SV}_\infty$ to denote a class of SV functions at infinity. A function $L(x)$ is said to be SV at zero if $L(1/x) \in \mathbf{SV}_\infty$. Thus the SV property can be defined at any finite point by shifting the origin of the function to this point. There is so-called representation theorem which asserts that if $L(x) \in \mathbf{SV}_\infty$ then $L(x) = c(x) \exp\left\{\int_b^x (\varepsilon(t)/t)dt\right\}$ for some $b > 0$, where $c(x) \to c > 0$ and $\varepsilon(x) \to 0$ as $x \to \infty$. At that, $\varepsilon(x)$ is called the index function of $L(x)$. If $c(x) \equiv const$ then $L(x)$ is said to be normalised SV-function. A function $R(x)$ is called regularly varying at infinity with index δ, if it can be expressed as $R(x) = x^\delta L(x)$ for some $L(x) \in \mathbf{SV}_\infty$.

A. Dudin et al. (Eds.): ITMM 2022, CCIS 1803, pp. 68–79, 2023.
https://doi.org/10.1007/978-3-031-32990-6_6

So that $R(\lambda x)/R(x) \to \lambda^{\delta}$ as $x \to \infty$ for each $\lambda > 0$. For more details on the regular variation conception, we refer to monographs [3] and [17].

In stochastic analysis there are many important issues in which a contribution of the SV conception is extremely useful. Application of Karamata functions in the branching systems theory allows to bypass severe constraints concerning existence of the high-order moments of the infinitesimal characteristics of the system under study. Zolotarev [21] was one of the first who demonstrated the encouraging prospect of application of the SV conception in the theory of Markov branching systems and has obtained principally new results on asymptote of the survival probability of the Markov system under considered.

In this report, delving deeply in a character of the Karamata functions, we learn more subtle properties of the continuous-time Markov branching systems to improve classical limit theorems results.

1.2 Background, Basic Assumptions and Purpose

Models of stochastic branching systems describe the evolution of the population size of the reproductive individuals system. These models most clearly illustrate numerous stochastic phenomena occurring both in nature and in human activity. The simple Galton-Watson model, originally evolved as a family survival model in the second half of the 19th century, today has numerous generalizations and modifications. The integration of various scientific fields has made it possible to find new applications of the branching system models in many fields, such as graph theory, queuing theory, combinatorics, cell biology, molecular biology, etc. Depending on the context, the branching system of one or another model is used to describe an evolution mechanism of individuals; see, e.g., [7,8,12,14,15], [16]. One such model is homogeneous-continuous-time Markov branching systems, which have an obvious influence on the development of the population dynamics theory; see [1,2,6,13].

Let $\mathbb{N}$ be a set of natural numbers and $\mathbb{N}_0 = \{0\} \cup \mathbb{N}$. Consider a population of particles which can die out or produce a random number of particles of the same type. We observe the evolution of the system on a continuous time axis $\mathcal{T}$. The states of the system are determined by the number of particles. The growth of the system occurs according to the following random mechanism. The particles undergo transformations in accordance with the branching rate $\{a_k, k \in \mathbb{N}_0\}$. It means, that each particle in the system has an exponentially distributed random life period with mean $\sum_{k\in\mathbb{N}_0\setminus\{1\}} a_k$ and, at the end of its life, regardless of its history and regardless of the presence of other particles, produces $k \in \mathbb{N}_0\setminus\{1\}$ descendants with probability $-a_k/a_1$. At that the branching rate is subject to the following conditions:

$$a_k \geq 0 \quad for \quad k \in \mathbb{N}_0\setminus\{1\} \qquad and \qquad 0 < a_0 < -a_1 = \sum_{k\in\mathbb{N}_0\setminus\{1\}} a_k < \infty.$$

The stochastic system defined above forms a reducible, homogeneous and continuous-time Markov chain with a state space consisting of two classes:

$\mathcal{S}_0 = \{0\} \cup \mathcal{S}$, where $\mathcal{S} \subset \mathbb{N}$, therein the state $\{0\}$ is absorbing, and $\mathcal{S}$ is the class of possible essential communicating states. Denoting by $Z(t)$ the population size at time $t \in \mathcal{T}$, we have a *homogeneous-continuous-time Markov branching (MB) system* with branching rate $\{a_k, k \in \mathbb{N}_0\}$. The appropriate q-matrix $\mathbb{Q} = \{q_{ij}\}$ of the system $\{Z(t)\}$ is given as follows:

$$q_{ij} = \begin{cases} i a_{j-i+1}\,, & \textit{if} \quad j \geq i \geq 0, \\ i a_0\,, & \textit{if} \quad j = i-1, \\ 0\,, & \textit{otherwise}. \end{cases} \tag{1}$$

The q-matrix fully regulates the further evolution of the MB system; see [4].

Considering transition probabilities

$$P_{ij}(t) := \mathsf{P}\Big\{Z(\tau+t) = j \mid Z(\tau) = i\Big\} \quad \textit{forany} \quad \tau, t \in \mathcal{T},$$

we can see from the q-matrix form (1) that

$$P_{ij}(\varepsilon) = \begin{cases} \delta_{ij} + i a_{j-i+1}\varepsilon + o(\varepsilon)\,, & \textit{if} \quad j \geq i \geq 0, \\ i a_0 \varepsilon + o(\varepsilon)\,, & \textit{if} \quad j = i-1, \\ o(\varepsilon)\,, & \textit{otherwise}, \end{cases} \tag{2}$$

as $\varepsilon \downarrow 0$, where δ_{ij} – Kronecker delta function. Since transition probabilities $P_{ij}(t)$ are i-fold convolution of the distribution $P_{1j}(t)$, i.e.

$$P_{ij}(t) = \sum_{j_1+j_2+\ldots+j_i=j} P_{1j_1}(t) \cdot P_{1j_2}(t) \cdot \ldots \cdot P_{1j_i}(t),$$

in order to study the evolution of MB system, it is sufficient to determine the probabilities $\mathsf{p}_j(t) := P_{1j}(t)$. In turn, the relations (2) show that the probabilities $\mathsf{p}_j(t)$ admit the following local representation (see [18]):

$$\mathsf{p}_j(\varepsilon) = \delta_{1j} + a_j\varepsilon + o(\varepsilon) \qquad \textit{as} \quad \varepsilon \downarrow 0. \tag{3}$$

Now put into consideration the following generating functions (GFs):

$$F(t;s) := \sum_{j \in \mathcal{S}_0} \mathsf{p}_j(t) s^j \qquad \textit{and} \qquad f(s) := \sum_{j \in \mathcal{S}_0} a_j s^j.$$

Then it follows from (3) that

$$F(\tau;s) = s + f(s)\tau + o(\tau) \qquad \textit{as} \quad \tau \downarrow 0 \tag{4}$$

and $F(0;s) = s$ for all $s \in [0,1)$. The formula (4) implies, that the MB system structure is completely defined by given of the branching rate GF $f(s)$.

Let $\sum_{j\in\mathcal{S}} ja_j < \infty$. Then the parameter

$$m := \sum_{j\in\mathcal{S}} ja_j = f'(1-)$$

is the average value of the branching rate in the system, which determines an asymptotic classification of the MB system trajectories. Indeed, we can calculate that the population mean $\sum_{j\in\mathcal{S}} jP_{ij}(t) = ie^{mt}$. It's clear that this tends to zero exponentially if $m < 0$, and goes to infinity at the same rate if $m > 0$ as $t \to \infty$. In this regard, the MB system is divided into the so-called subcritical, critical and supercritical types, depending on $m < 0$, $m = 0$ and $m > 0$ respectively.

In this report we exceptionally consider the critical case, i.e. $m = 0$.

Let $R(t; s) = 1 - F(t; s)$. Sevastyanov [19] proved that if $f'''(1-) < \infty$ then the following asymptotic representation holds:

$$\frac{1}{R(t;s)} - \frac{1}{1-s} = \frac{f''(1-)}{2}t + \mathcal{O}(\ln t) \qquad as \quad t \to \infty \tag{5}$$

for all $s \in [0, 1)$; see [19, p. 72].

Later on Zolotarev [21] has found a principally new result on asymptotic representation for

$$q(t) := R(t; 0) = \mathsf{P}\left\{Z(t) > 0\right\},$$

the survival probability of the MB system at time t. He, refusing the assumption of $f''(1-) < \infty$ and providing that $g(x) = f(1-x)$ is a regularly varying function at zero with index $\gamma \subset (1, 2]$, has proved the following asymptotic relation:

$$\frac{q(t)}{f\left(1 - q(t)\right)} \sim (\gamma - 1)t \qquad as \quad t \to \infty. \tag{6}$$

A more exact asymptotic expansion for $q(t)$ can be obtained from the representation (5) but under the worse condition $f'''(1-) < \infty$.

In 2010 Pakes [14], in connection with the proof of limit theorems, has proved that

$$\frac{1}{R\left(t;s\right)} = U\left(t + V\left(\frac{1}{1-s}\right)\right), \tag{7}$$

where $V(x) = \mathcal{M}(1 - 1/x)$ and $\mathcal{M}(s)$ is in the form

$$\mathcal{M}(s) = \int_0^s \frac{dx}{f(x)} \tag{8}$$

and it generates an invariant measure for the MB system, the function $U(y)$ is the inverse of $V(x)$. Pakes's key assumption is the following representation for the branching rate GF $f(s)$:

$$f(s) = (1 - s)^{1+\nu} \mathcal{L}\left(\frac{1}{1-s}\right) \tag{9}$$

for all $s \in [0, 1)$, where $0 < \nu < 1$ and $\mathcal{L}(x) \in \mathbf{SV}_\infty$. The condition (9) implies that

$$f''(1-) = \lim_{s \uparrow 1} \frac{2f(s)}{(1-s)^2} = \lim_{u \to \infty} 2u^{1-\nu}\mathcal{L}(u) = \infty.$$

It follows from (8) that $\mathcal{M}(0) = 0$, then (7) gives an alternative relation to (6):

$$q(t) = \frac{1}{U(t)}. \tag{10}$$

Though the formulas (6), (7) and (10) certainly improves Sevastyanov's result (5), but it is visually visible that these do not explicitly denote a decreasing rate of $R(t; s)$ and $q(t)$ to zero.

Our purpose in this report is as follows. First we improve the Monotone Ratio theorem from [11], asserting a monotone convergence of $\mathsf{p}_j(t)/\mathsf{p}_1(t)$ to its own limit as $t \to \infty$. Subsequently, we state the local-convergence limit theorem on the asymptotic expansion of transition probabilities $\mathsf{p}_j(t)$ for all $j \in \mathcal{S}$. In doing so, we see that the probabilities $\mathsf{p}_j(t)$ converge to the invariant measure $\{\pi_j\}$ generated by the function $\mathcal{M}(x)$ of the form (8). Along the way, we refine and improve assertion (10). At last, an establishment of the growth rate of the sum $\sum_{k=1}^{n} \pi_k$ is our final conclusion. Thus, our results improve the corresponding results from works [11] and [20], indicating the rate of convergence in these theorems. For our purpose, we require some extra condition, only on the function $\mathcal{L}(x) \in \mathbf{SV}_\infty$ within the condition (9).

1.3 Results

Recently, in [9] obtained results that refine and improve all the listed above statements on this issue. In the cited paper the following extra conditions were imposed on the function $\mathcal{L}(x) \in \mathbf{SV}_\infty$. Denoting $\Lambda(y) := y^\nu \mathcal{L}(1/y)$, we rewrite the assumption (9) as follows:

$$f(1-y) = y\Lambda(y) \qquad [f_\Lambda]$$

for $y \in (0, 1]$. Evidently that the function $\Lambda(y)$ is positive and, it tends to zero and has a ultimately monotone derivative. Then the Lamperti theorem (see [3, p. 401]) implies that $y\Lambda'(y)/\Lambda(y) \to \nu$ as $y \downarrow 0$. Thence it is natural to write

$$\frac{y\Lambda'(y)}{\Lambda(y)} = \nu + \delta(y), \qquad [\Lambda_\delta]$$

where $\delta(y)$ is continuous and $\delta(y) \to 0$ as $y \downarrow 0$. After integrating the relation $[\Lambda_\delta]$, we can see that $L(x)$ is normalised SV-function. Since $\mathcal{L}(x) \in \mathbf{SV}_\infty$, we write

$$\frac{\mathcal{L}(\lambda x)}{\mathcal{L}(x)} = 1 + \omega(x), \tag{11}$$

where $\omega(x) \to 0$ as $x \to \infty$. If in (11) an expression of tail term $\omega(x)$ is given, then $\mathcal{L}(x)$ is said to be *SV with remainder* ω at infinity; see [3, p. 185]. In this case we denote $\mathcal{L}(x) \in \mathbf{SV}_{\infty}(\omega)$. As it has been shown in [9], that the special case when

$$\delta(y) = \Lambda(y)$$

is more interest. In this case, we see that

$$\omega(x) = \mathcal{O}\left(\frac{\mathcal{L}(x)}{x^{\nu}}\right) \qquad as \quad x \to \infty. \tag{12}$$

Throughout the paper $[f_{\nu}]$ and $[\Lambda_{\delta}]$ with (12) we take as *Basic assumptions* for our purpose on our MB system.

So, it has been proved in the work [9] that under Basic assumptions the following relation holds:

$$\mathsf{p}_1(t) = \frac{1}{a_0} q(t) \Lambda\left(q(t)\right), \tag{13}$$

where

$$q(t) = \frac{\mathcal{N}(t)}{(\nu t)^{1/\nu}}\left(1 - \frac{\ln \nu(t)}{\nu^3 t} + o\left(\frac{\ln \nu(t)}{t}\right)\right) \qquad as \quad t \to \infty, \tag{14}$$

herein $\nu(t) = a_0 \nu t + 1$, besides $\mathcal{N}(t)$ is a SV-function, such that

$$\mathcal{N}^{\nu}(t) \cdot \mathcal{L}\left(\frac{(\nu t)^{1/\nu}}{\mathcal{N}(t)}\right) \longrightarrow 1 \qquad as \quad t \to \infty.$$

In [11, Lemma 7] proved the following Monotone Ratio theorem. For all $j \in \mathcal{S}$

$$\frac{\mathsf{p}_j(t)}{\mathsf{p}_1(t)} \uparrow \pi_j < \infty \qquad as \quad t \to \infty, \tag{15}$$

where $\{\pi_j, j \in \mathcal{S}\}$ are an invariant measure for MB system, i.e. it satisfy the following invariant-functional equation:

$$\pi_j = \sum_{i \in \mathcal{S}} \pi_i P_{ij}(t) \qquad for\ any \quad t \in \mathcal{T}. \tag{16}$$

Remind that the discrete-time analogue of statement (15) is already available due to Athreya and Ney [2, p.12, Lemma 2].

Define

$$\pi(t; s) := \sum_{j \in \mathcal{S}} \frac{\mathsf{p}_j(t)}{\mathsf{p}_1(t)} s^j.$$

It follows from (15) that

$$\lim_{t \to \infty} \pi(t; s) = \pi(s) := \sum_{j \in \mathcal{S}} \pi_j s^j.$$

Theorem 1. *Under the Basic assumptions*

$$\pi(t;s) = \pi(s)\left(1 + \frac{\ln \nu(t)}{\nu^2 t} + o\left(\frac{\ln \nu(t)}{t}\right)\right) \qquad as \quad t \to \infty, \tag{17}$$

where $\nu(t) = a_0\nu t + 1$ *and* $\pi(s) = a_0\mathcal{M}(s)$ *and*

$$\mathcal{M}(s) = \int_1^{1/(1-s)} \frac{dx}{x^{1-\nu}\mathcal{L}(x)}. \tag{18}$$

The Theorem 1 shows that $\pi_j = a_0\mu_j$, where $\{\mu_j, j \in \mathcal{S}\}$ are coefficients in the power series expansion $\mathcal{M}(s) = \sum_{j\in\mathcal{S}} \mu_j s^j$. Therefore, they also satisfy the recursive-invariant equation (16). So the two invariant measures $\{\mu_j\}$ and $\{\pi_j\}$ are the same up to a constant factor a_0.

The following local limit theorem follows from Theorem 1 and it shows that transition probabilities $\mathsf{p}_j(t)$ converge to the invariant measure $\{\pi_j\}$, and it also indicates the rate of this convergence.

Theorem 2. *Under the Basic assumptions*

$$\left(\nu t\right)^{1+1/\nu} \cdot \mathsf{p}_j(t) = \pi_j \cdot \left(1 - \frac{\ln \nu(t)}{\nu^3 t} + o\left(\frac{\ln \nu(t)}{t}\right)\right) \qquad as \quad t \to \infty, \tag{19}$$

where $\{\pi_j, j \in \mathcal{S}\}$ *are defined in (15) and* $\nu(t) = a_0\nu t + 1$.

The next theorem is the essence of well-known Slack [20] result on explicit expression for GF of invariant measures, but ours is slightly improving one.

Theorem 3. *Under the Basic assumptions*

$$\pi(s) = \frac{a_0}{\nu\Lambda(1-s)}\Big(1 + \eta(1-s)\Big), \tag{20}$$

where $\eta(x) = \mathcal{O}\big(x^\nu\big)$ *as* $x \downarrow 0$.

According to the power series Tauberian theorem [5, Ch.XIII.5, Th 5], the relation (20) implies that

$$\frac{1}{n^\nu}\sum_{k=1}^{n} \pi_k = \frac{a_0}{\nu\Gamma(1+\nu)}\mathcal{L}_\pi(n)\Big(1 + \rho(n)\Big), \tag{21}$$

where $\Gamma(*)$ is Euler's Gamma function and $\mathcal{L}_\pi(n) := 1/\mathcal{L}(n) \in \mathbf{SV}_\infty$ and the tail term $\rho(n) = \mathcal{O}\big(1/n^\nu\big)$ as $n \to \infty$.

2 Auxiliaries

The section begins by stating and proving the following lemma, describing in essence the same property of the function $L(x) \in \mathbf{SV}_\infty(\omega)$, proved in [10] but with the SV-remainder term $\omega(x) = o\big(L(x)/x^\sigma\big)$.

Lemma 1. *Let $L(x)$ is the normalised SV-function. If $L(x) \in \boldsymbol{SV}_{\infty}(\omega)$ and the SV-remainder term is $\omega(x) = \mathcal{O}\left(L(x)/x^{\sigma}\right)$ with some $\sigma > 0$, then*

$$C_L := \lim_{x \to \infty} L(x) < \infty \tag{22}$$

and

$$L(x) = C_L + \mathcal{O}\left(1\big/x^{\sigma}\right) \qquad as \quad x \to \infty. \tag{23}$$

Proof. We repeat discussions was done in [10]. Then, first we get an evident fact that in the condition of the lemma, $L(x) \in \mathbf{SV}_{\infty}(\omega)$ is normalised SV, for which

$$L(x) = c \cdot \exp \int_{b}^{x} \frac{\varepsilon(t)}{t}\, dt \tag{24}$$

for some $b > 0$, where $c = const$ and $\varepsilon(t) \to 0$ as $t \to \infty$. Therefore, by definition of SV-function with remainder and using the integral mean value theorem, it follows from representation (24) that the index function

$$\varepsilon(x) = \mathcal{O}\left(\frac{L(x)}{x^{\sigma}}\right) \qquad as \quad x \to \infty. \tag{25}$$

Then

$$\int_{b}^{\infty} \frac{|\varepsilon(t)|}{t}\, dt = \int_{b}^{\infty} \mathcal{O}\left(\frac{L(t)}{t^{1+\sigma}}\right) dt < \infty. \tag{26}$$

In the last step, we took into account the fact that integral $\int_{1}^{\infty} t^{-(1+\sigma)} L(t) dt$ converges; see [10, Prop.3]. So combining (24) and (26) we can get to (22). Hence, the SV-remainder term (25) becomes $\varepsilon(x) = \mathcal{O}\left(1/x^{\sigma}\right)$. Therefore

$$\int_{x}^{\infty} \frac{\varepsilon(t)}{t}\, dt = \mathcal{O}\left(\int_{x}^{\infty} \frac{dt}{t^{1+\sigma}}\right) = \mathcal{O}\left(1\big/x^{\sigma}\right) \qquad as \quad x \to \infty.$$

Since

$$C_L - L(x) = C_L \left[1 - \frac{L(x)}{C_L}\right] = C_L \left[1 - \exp\left(-\int_{x}^{\infty} \frac{\varepsilon(t)}{t}\, dt\right)\right],$$

assertion (23) readily follows.

The proof is complete.

The next lemma describes an asymptotic expansion property of integral of the regularly varying function at infinity with remainder in a special case.

Lemma 2. *Let $L(x) \in \boldsymbol{SV}_{\infty}(\omega)$ and it is a locally bounded in (c, ∞) for some $c > 0$. If the SV-remainder term is $\omega(x) = \mathcal{O}\left(L(x)/x^{\sigma}\right)$ with some $\sigma > 0$, then for all $\alpha > -1$*

$$\int_{c}^{t} y^{\alpha} L(y) dy = \frac{1}{\alpha + 1} L(t) t^{\alpha+1} \left(1 + \mathcal{O}\left(1/t^{\beta}\right)\right) \qquad as \quad t \to \infty, \tag{27}$$

where $\beta = \min(\sigma, \alpha + 1)$.

Proof. Making the change of variable $y = ut$, we write

$$I(t) := \int_c^t y^\alpha L(y)dy = L(t)t^{\alpha+1}\left[\int_{c/t}^1 u^\alpha du + \int_{c/t}^1 \left[\frac{L(ut)}{L(t)} - 1\right] u^\alpha du\right]. \quad (28)$$

An evident fact that

$$\int_{c/t}^1 u^\alpha du = \frac{1}{\alpha+1}\frac{1}{t^{\alpha+1}}\left(t^{\alpha+1} - c^{\alpha+1}\right).$$

By the lemma assumption, the expression in brackets of the second integrand in the right-hand side of (28) is the SV-remainder term of the function $L(x) \in \mathbf{SV}_\infty(\omega)$ and by the uniform convergence theorem (see [3, p. 185]) it is $\mathcal{O}\left(L(t)/t^\sigma\right)$ uniformly in $u \in (0,1)$ as $t \to \infty$. In the other hand, by Lemma 1, the SV-function $L(x)$ approaches the finite constant as $t \to \infty$. Then

$$I(t) = \frac{1}{\alpha+1}L(t)t^{\alpha+1}\left[1 + \mathcal{O}\left(\frac{1}{t^{\alpha+1}}\right) + \mathcal{O}\left(\frac{1}{t^\sigma}\right)\right] \quad as \quad t \to \infty.$$

To complete the proof, we choose $\beta = \min(\sigma, \alpha+1)$ at that $\mathcal{O}\left(1/t^\beta\right)$ goes to zero much less rapidly than another one. The assertion (27) follows.

The Lemma 2 is proved. □

We also make essential use of the following lemma, which is an improvement of previous results on the asymptotic expansion of the function $R(t;s)$.

Lemma 3 (Basic Lemma [9]). *Under the Basic assumptions*

$$\frac{1}{\Lambda\big(R(t;s)\big)} - \frac{1}{\Lambda(1-s)} = \nu t + \frac{1}{\nu}\ln \nu(t;s) + o\big(\ln \nu(t;s)\big) \quad (29)$$

as $t \to \infty$, *where* $\nu(t;s) = \Lambda(1-s)\nu t + 1$.

Now let's transform the backward Kolmogorov equation $dF/dt = f(F)$ to the following integral one:

$$\int_0^{F(t;s)} \frac{dx}{f(x)} = t + \mathcal{M}(s), \quad (30)$$

where the function $\mathcal{M}(s)$ is the form of (8). Considering the representation (9) and denoting $1 - x = 1/u$ herewith, we once more transform the left-hand side of (30) and rewrite it as follows:

$$\int_1^{1/R(t;s)} \frac{dx}{x^{1-\nu}\mathcal{L}(x)} = t + \mathcal{M}(s), \quad (31)$$

herein the function $\mathcal{M}(s)$ becomes (18). We use Lemma 1 on the left-hand side of the equality (31) and, as a result obtain the following relation:

$$\int_1^{1/R(t;s)} \frac{dx}{x^{1-\nu}\mathcal{L}(x)} = \frac{1}{\nu}\frac{1}{\Lambda\big(R(t;s)\big)}\left(1 + \mathcal{O}\big(R^\nu(t;s)\big)\right) \quad (32)$$

as $t \to \infty$.

According to Lemma 1, $\mathcal{L}(x) \in \mathbf{SV}_\infty(\omega)$ approaches the finite constant. Consequently, as before, the SV-remainder term becomes $\omega(t) = \mathcal{O}\left(1/t^\nu\right)$ and therefore, the representation (29) implies that $R^\nu(t;s) = \mathcal{O}(1/t)$ as $t \to \infty$ uniformly in $s \in [0,1)$. Thus from (31) and (32)

$$\frac{1}{\nu\Lambda\big(R(t;s)\big)} = \Big(t + \mathcal{M}(s)\Big)\left(1 + \mathcal{O}\left(\frac{1}{t}\right)\right) \qquad as \quad t \to \infty.$$

We can easily transform last equation into the following form:

$$\frac{1}{R(t;s)} = \frac{(\nu t)^{1/\nu}}{\mathcal{N}(t;s)} \cdot \left[1 + \frac{\mathcal{M}(s)}{t}\right]^{1/\nu} \left(1 + \mathcal{O}\left(\frac{1}{t}\right)\right) \qquad as \quad t \to \infty \tag{33}$$

for all $s \in [0,1)$, where $\mathcal{N}(t;s) = \mathcal{L}^{-1/\nu}\left(1/R(t;s)\right)$, for which we can see that

$$\mathcal{N}^\nu(t;s) \cdot \mathcal{L}\left(\frac{(\nu t)^{1/\nu}}{\mathcal{N}(t;s)}\right) \longrightarrow 1 \qquad as \quad t \to \infty.$$

Thus, in the form of (33) we have found an alternative to asymptotic relation (29), depending on the function $\mathcal{M}(s)$.

3 Proof of the Results

Proof of Theorem 1. Recalling $R(t;s) = 1 - F(t;s)$, we write

$$\pi(t;s) = \sum_{j \in \mathcal{S}} \frac{\mathsf{p}_j(t)}{\mathsf{p}_1(t)} s^j = \frac{F(t;s) - F(t;0)}{\mathsf{p}_1(t)} = \left(1 - \frac{R(t;s)}{q(t)}\right) \cdot \frac{q(t)}{\mathsf{p}_1(t)}. \tag{34}$$

Rewrite (13) as follows:

$$\frac{q(t)}{\mathsf{p}_1(t)} = \frac{a_0}{\Lambda\big(q(t)\big)}.$$

It follows from asymptotic relation (29) that

$$\frac{1}{\Lambda\big(q(t)\big)} = \frac{1}{a_0} + \nu t + \frac{1}{\nu}\ln \nu(t) + o\big(\ln \nu(t)\big) \qquad as \quad t \to \infty, \tag{35}$$

where $\nu(t) = a_0 \nu t + 1$. Then

$$\frac{q(t)}{\mathsf{p}_1(t)} = \nu(t) + \frac{a_0}{\nu}\ln \nu(t) + o\big(\ln \nu(t)\big) \qquad as \quad t \to \infty. \tag{36}$$

Since $q(t) = R(t;0)$, asymptotic relation (33) gives

$$\frac{1}{q(t)} = \frac{(\nu t)^{1/\nu}}{\mathcal{N}(t)}\left(1 + \mathcal{O}\left(\frac{1}{t}\right)\right) \qquad as \quad t \to \infty. \tag{37}$$

As it is noted above the function $\mathcal{L}(x) \in \mathbf{SV}_\infty(\omega)$ converges a finite constant, then the function $\mathcal{N}(t;s)$ appeared in (33), has a finite limit, i.e. $\lim_{t\to\infty}\mathcal{N}(t;s) =: C_\mathcal{N} < \infty$. Consequently $\mathcal{N}(t) \sim \mathcal{N}(t;s) \to C_\mathcal{N}$ and relation (33) implies $R(t;s)/q(t) \to 1$ as $t \to \infty$ uniformly in $s \in [0,1)$. Then relation (23) entails

$$\mathcal{L}\left(\frac{1}{R(t;s)}\right) = C_\mathcal{L} + \mathcal{O}\big(q^\nu(t)\big) \qquad as \quad t \to \infty$$

because of $\omega(t) = \mathcal{O}\left(\mathcal{L}(t)/t^\nu\right)$. Thus, since $q^\nu(t) = \mathcal{O}(1/t)$,

$$\frac{\mathcal{N}(t;s)}{\mathcal{N}(t)} = 1 + \mathcal{O}\left(\frac{1}{t}\right) \qquad as \quad t \to \infty. \tag{38}$$

From (33), (37) and (38) we obtain

$$1 - \frac{R(t;s)}{q(t)} = \frac{\mathcal{M}(s)}{\nu t}\left(1 + \mathcal{O}\left(\frac{1}{t}\right)\right) \qquad as \quad t \to \infty. \tag{39}$$

Now combining (34), (36) and (39) it follows

$$\pi(t;s) = \frac{\mathcal{M}(s)}{\nu t}\left[\nu(t) + \frac{a_0}{\nu}\ln\nu(t) + o\big(\ln\nu(t)\big)\right]\left(1 + \mathcal{O}\left(\frac{1}{t}\right)\right)$$

as $t \to \infty$. This implies (17), which completes the proof of the theorem. □

Proof of Theorem 2. By the continuity theorem for GF, (17) implies

$$\frac{\mathsf{p}_j(t)}{\mathsf{p}_1(t)} = \pi_j\left(1 + \frac{\ln\nu(t)}{\nu^2 t} + o\left(\frac{\ln\nu(t)}{t}\right)\right) \qquad as \quad t \to \infty. \tag{40}$$

The combination of relations (13), (14), (35) with equation (40) and after, standard transformations leads us to assertion (19). □

Proof of Theorem 3. It is proved in Theorem 1 that $\pi(s) = a_0\mathcal{M}(s)$, where $\mathcal{M}(s)$ is form of (18). We use Lemma 2 with $\sigma = \nu$ and $\alpha = \nu - 1$ on the right-hand side of (18). Then

$$\pi(s) = \frac{a_0}{\nu(1-s)^\nu\mathcal{L}\left(1/(1-s)\right)}\left(1 + \mathcal{O}(1-s)^\nu\right) \qquad as \quad s \uparrow 1,$$

which is (20).

The proof is complete. □

Author contributions. A. Imomov conceived the relevance of the task and determined the way to solve it. M. Murtazaev, with the support of A. Imomov, implemented the idea of a method for solving the problem, and both wrote this article.

Conflict of Interest. The authors declare that they have no conflict of interest.

References

1. Asmussen, S., Hering, H.: Branching processes. Birkhäuser, Boston (1983)
2. Athreya, K.B., Ney, P.E.: Branching processes. Springer, New York (1972). https://doi.org/10.1007/978-3-642-65371-1
3. Bingham, N.H., Goldie, C.M., Teugels, J.L.: Regular Variation. Univ. Press, Cambridge (1987)
4. Li, J., Cheng, L., Li, L.: Long time behaviour for Markovian branching-immigration systems. Discrete Event Dyn. Syst. **31**(1), 37–57 (2021)
5. Feller, W.: An introduction to probability theory and its applications, v.2. JW & Sons (1971)
6. Harris, T.E.: The Theory of Branching Processes. Springer-Verlag, Berlin (1963)
7. Heathcote C. R. A branching process allowing immigration. Jour. Royal Stat. Soc. **B-27**(1), 138–143 (1965)
8. Imomov, A.A.: On a limit structure of the Galton-Watson branching processes with regularly varying generating functions. Prob. and math. stat. **39**(1), 61–73 (2019)
9. Imomov, A.A., Meyliyev, A.: On application of slowly varying functions with remainder in the theory of Markov Branching Processes with mean one and infinite variance. Ukr. Math. J. **73**(8), 1225–1237 (2022)
10. Imomov A.A., Tukhtaev E.E.: On asymptotic structure of critical Galton-Watson branching processes allowing immigration with infinite variance. Stochastic Models, Publ. (2022). https://doi.org/10.1080/15326349.2022.2033628
11. Imomov A.A.: Limit properties of transition functions of continuous-time Markov branching processes. Int. J Stoch. Analysis **2014**, 409345 (2014). https://doi.org/10.1155/2014/409345
12. Imomov, A.A.: On long-term behavior of continuous-time Markov branching processes allowing immigration. J Sib. Fed. Univ. Math. Phys. **7**(4), 443–454 (2014)
13. Jagers, P.: Branching Progresses with Biological applications. Pitman Press, GB, JW & Sons (1975)
14. Pakes, A.G.: Critical Markov branching process limit theorems allowing infinite variance. Adv. Appl. Prob. **42**, 460–488 (2010)
15. Pakes, A.G.: Revisiting conditional limit theorems for the mortal simple branching process. Bernoulli **5**(6), 969–998 (1999)
16. Pakes, A.G.: Some results for non-supercritical Galton-Watson process with immigration. Math. Biosci. **24**, 71–92 (1975)
17. Seneta, E.: Regularly Varying Functions. Springer, Berlin (1972). https://doi.org/10.1007/BFb0079658
18. Sevastyanov, B.A.: Branching Processes. Nauka, Moscow, Russia (1971)
19. Sevastyanov, B.A.: The theory of Branching stochastic process. Uspekhi Mathematicheskikh Nauk **6**(46), 47–99 (1951)
20. Slack, R.S.: A branching process with mean one and possible infinite variance. Wahrscheinlichkeitstheor. und Verv. Geb. **9**, 139–145 (1968)
21. Zolotarev, V.M.: More exact statements of several theorems in the theory of branching processes. Theory Prob. Appl. **2**, 245–253 (1957)

Retrial Queue MMPP/M/1 with Server Switching

Ksenya Khadzhi-Ogly(✉), Radmir Salimzyanov, and Ekaterina Fedorova

National Research Tomsk State University, Lenina Avenue, 36, Tomsk, Russia
494xad@mail.ru

Abstract. In the paper, we study a retrial queueing system with one server with switching, where a service rate depends on the number of customers in the orbit (it has two values). The arrival process of customers is MMPP, the delay time in the orbit is distributed exponentially. To find the stationary distribution of the number of customers in the orbit, the method of the asymptotic analysis is proposed under the condition of a heavy load. It is proved that the asymptotic characteristic function has the gamma distribution form. The numerical analysis of obtained results is carried out.

Keywords: Retrial queueing system · asymptotic analysis · server switching · MMPP · heavy load

1 Introduction

In many areas of economics, there are systems or tasks with repeated uniform actions. In such system, customers (clients, tasks, information) come at random times, they need service of the same type, they may wait in queues or be lost. Such systems are modelling by queuing systems (QS). Examples of QS can be: various communication systems, loading and unloading complexes (ports, freight stations), currency exchange offices, hospitals, etc. For telecommunication systems (i.e LAN, 4G, FANET, etc.), a new class of QS - queuing systems with repeated calls or retrial queuing systems is usually used. The main difference of such systems is that incoming customers do not leave the system if it found the service device busy. It go to the source of repeated customers (into the orbit) and, after a random delay, try again be serviced. Retrial queueing system have many application examples, i.e. mobile networks, call-centers, different data transmission networks [1–4]. The detailed description of retrial queues is in [5,6].

Because of real information flows have non-Poisson distributions, we consider a retrial queue with MMPP arrivals. Also retrial queuing systems with non-Poisson arrivals are studied in [7–9], etc.

This study was supported by the Tomsk State University Development Programme (Priority-2030).

A. Dudin et al. (Eds.): ITMM 2022, CCIS 1803, pp. 80–91, 2023.
https://doi.org/10.1007/978-3-031-32990-6_7

The models with server switching are not popular. But sometimes in real systems, the intensity of the service changes when the number of waiting tasks grows. Note that the service intensity can became more or less. The most general retrial model of this situation is RQ with the state-dependent service rate [10,11]. But retrial queues with MMPP arrivals and server switching are not studied analytically yet.

The rest of the paper is organized as follows. In Sect. 2, the studied retrial queue is described and the goal of the study is formulated. Section 3 is devoted to applying the asymptotic analysis method under the heavy load condition for the stationary probability distribution of the number of customers in the orbit obtaining. In Sect. 4, we demonstrate some numerical examples and the accuracy of the asymptotic results is estimated. Section 5 is dedicated for some conclusions.

2 Mathematical Model

Let us consider a retrial queueing system (Fig. 1) with one server. Customers arrival to the system according to Markov Modulated Poisson Process (MMPP), which is described by matrices $\mathbf{D_0}$ and $\mathbf{D_1}$ [12,13].

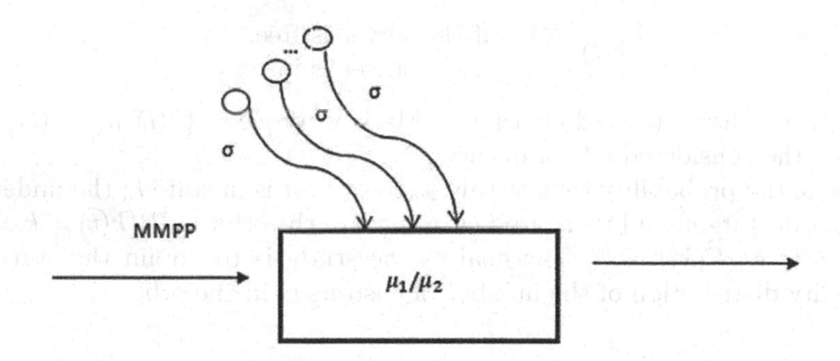

Fig. 1. Retrial Queue MMPP/M/1 with server switching

Let $n(t)$ be underlying Markov chain of the MMPP ($n = 1, 2, \ldots, N$). The generator of $n(t)$ is equal to $\mathbf{Q} = \mathbf{D_0} + \mathbf{D_1}$. $\mathbf{D_1}$ is a diagonal matrix with elements $\rho\lambda_n$, where λ_n are the conditional intensities of the MMPP, ρ is a parameter of the system load (be defined below). Thus, we can write $\mathbf{D_1} = \rho\mathbf{\Lambda}$, where matrix $\mathbf{\Lambda} = \text{diag}\{\lambda_n\}$.

Denote the stationary probability distribution of the states of the process $n(t)$ by vector $\mathbf{r}$, which is defined as follows:

$$\begin{cases} \mathbf{rQ} = \mathbf{0}, \\ \mathbf{re} = 1. \end{cases} \tag{1}$$

Then the rate of the arrival process is calculated as $\lambda = \mathbf{r}\ \rho\mathbf{\Lambda}\mathbf{e}$, where $\mathbf{e}$ is the unit column vector.

We consider the retrial queue with the server switching. The customers are served during a random time distributed according to an exponential law with rates μ_1 or μ_2, where the service parameter depends on the number of customers in the orbit at this moment. Denote a random process of the number of customers in the orbit by $i(t)$. Let I be an some given value (boundary of the number of customers the orbit). While the number of customers $i(t) < I$, the service rate is equal to μ_1. When the number of customers $i(t) \geq I$, the server switches to rate μ_2.

If an incoming customer meets the server busy, it goes to the orbit, where a random delay is occurred. The delay duration has an exponential distribution with rate σ. From the orbit, a customer tries to get service again. If the server is free, the customer occupies it. If it is busy, then the customer returns to the orbit and makes a new delay.

In the paper we suppose that $\mu_1 < \mu_2$. Then the system load is defined as

$$\rho = \lambda/\mu_2,$$

where $\lambda = \mathbf{r}\rho\mathbf{\Lambda}\mathbf{e}$, so $\mathbf{r}\mathbf{\Lambda}\mathbf{e} = \mu_2$ is true for the system parameters.

Introduce process $k(t)$ defining the server states as follows:

$$k(t) = \begin{cases} 0, \text{if the server is free,} \\ 1, \text{if the server is busy.} \end{cases}$$

Thus, we have three-dimensional Markov process $\{k(t), n(t), i(t)\}$ that describes the considered retrial queue.

Denote the probability that at time t, the server is in state k, the underlying process is in state n, and there are i customers in the orbit by $P\{k(t) = k, n(t) = n, i(t) = i\} = P(k, n, i, t)$. The goal of the study is to obtain the stationary probability distribution of the number of customers in the orbit.

Kolmogorov's System of Equations

Let us compose Kolmogorov's differential equations:

for $i < I$

$$\begin{cases} \dfrac{\partial P(0,n,i,t)}{\partial t} = -(\rho\lambda_n + i\sigma - q_{nn})P(0,n,i,t) + \mu_1 P(1,n,i,t) + \\ \quad + \sum\limits_{v \neq n} P(0,v,i,t)q_{vn}, \\ \dfrac{\partial P(1,n,i,t)}{\partial t} = -(\rho\lambda_n + \mu_1 - q_{nn})P(1,n,i,t) + \\ \quad + \rho\lambda_n P(0,n,i,t) + (i+1)\sigma P(0,n,i+1,t) + \\ \quad + \rho\lambda_n P(1,n,i-1,t) + \sum\limits_{v \neq n} P(1,v,i,t)q_{vn}. \end{cases} \tag{2}$$

for $i \geq I$

$$\begin{cases} \dfrac{\partial P(0,n,i,t)}{\partial t} = -(\rho\lambda_n + i\sigma - q_{nn})P(0,n,i,t) + \mu_2 P(1,n,i,t) + \\ \quad + \sum\limits_{v \neq n} P(0,v,i,t)q_{vn}, \\ \dfrac{\partial P(1,n,i,t)}{\partial t} = -(\rho\lambda_n + \mu_2 - q_{nn})P(1,n,i,t) + \\ \quad + \rho\lambda_n P(0,n,i,t) + (i+1)\sigma P(0,n,i+1,t) + \\ \quad + \rho\lambda_n P(1,n,i-1,t) + \sum\limits_{v \neq n} P(1,v,i,t)q_{vn}. \end{cases} \tag{3}$$

Denote $\mathbf{P}(k,i) = \{P(k,1,i), P(k,2,i), \ldots, P(k,N,i)\}$, where $P(k,n,i)$ are stationary probabilities of process $\{k(t), n(t), i(t)\}$. Then in the steady-state, System (2)–(3) has the following matrix form:
for $i < I$

$$\begin{cases} \mathbf{P}(0,i)(\mathbf{Q} - \rho\mathbf{\Lambda} - i\sigma\mathbf{I}) + \mu_1\mathbf{P}(1,i) = \mathbf{0}, \\ \mathbf{P}(1,i)(\mathbf{Q} - \rho\mathbf{\Lambda} - \mu_1\mathbf{I}) + \mathbf{P}(0,i)\rho\mathbf{\Lambda} + \\ \quad + \sigma(i+1)\mathbf{P}(0,i+1) + \mathbf{P}(1,i-1)\rho\mathbf{\Lambda} = \mathbf{0}, \end{cases} \tag{4}$$

for $i \geq I$

$$\begin{cases} \mathbf{P}(0,i)(\mathbf{Q} - \rho\mathbf{\Lambda} - i\sigma\mathbf{I}) + \mu_2\mathbf{P}(1,i) = \mathbf{0}, \\ \mathbf{P}(1,i)(\mathbf{Q} - \rho\mathbf{\Lambda} - \mu_2\mathbf{I}) + \mathbf{P}(0,i)\rho\mathbf{\Lambda} + \\ \quad + \sigma(i+1)\mathbf{P}(0,i+1) + \mathbf{P}(1,i-1)\rho\mathbf{\Lambda} = \mathbf{0}, \end{cases} \tag{5}$$

where $\mathbf{I}$ is the identity matrix.

Let's rewrite System (4)–(5) for the partial characteristic functions:

$$\mathbf{H}(k,u) = \sum_{i=0}^{\infty} e^{jui}\mathbf{P}(k,i),$$

where $j = \sqrt{-1}$ is the imaginary unit.

Thus, we obtain the following equations:

$$\begin{cases} \mathbf{H}(0,u)(\mathbf{Q} - \rho\mathbf{\Lambda}) + j\sigma\dfrac{\partial\mathbf{H}(0,u)}{\partial u} + \mu_1\mathbf{S}_1(u) + \mu_2\mathbf{S}_2(u) = \mathbf{0}, \\ \mathbf{H}(1,u)(\mathbf{Q} - \rho\mathbf{\Lambda}) - \mu_1\mathbf{S}_1(u) - \mu_2\mathbf{S}_2(u) + \mathbf{H}(0,u)\rho\mathbf{\Lambda} - \\ \quad - j\sigma e^{-ju}\dfrac{\partial\mathbf{H}(0,u)}{\partial u} + e^{ju}\mathbf{H}(1,u)\rho\mathbf{\Lambda} = \mathbf{0}. \end{cases} \tag{6}$$

where

$$\mathbf{S}_1(u) = \sum_{i=0}^{I-1}\mathbf{P}(1,i)e^{jui}, \qquad \mathbf{S}_2(u) = \sum_{i=I}^{\infty}\mathbf{P}(1,i)e^{jui}.$$

It's obvious that $\mathbf{S}_1(u) + \mathbf{S}_2(u) = \mathbf{H}(1,u)$.

To solve System (6), the method of asymptotic analysis is proposed under the condition of a heavy load $\rho \to 1$ [14–16].

3 Asymptotic Analysis Under Heavy Load

The asymptotic analysis method is analytical method of solving equations under some limit condition in queueing theory. There are several types of limit conditions, i.e. high intensity of arrival process, heavy load, long delay in orbit, etc. [14–16]. The asymptotic analysis lets obtain an analytical formulas for approximate probability distributions. In the paper, we use the asymptotic analysis method under the heavy load condition $\rho \to 1$ for System (6) solving.

Introduce the following notations:

$$\varepsilon = 1-\rho,\ (\varepsilon \to 0), \quad u = \varepsilon w, \mathbf{H}(0,u) = \varepsilon \mathbf{F}_0(w,\varepsilon), \quad \mathbf{H}(1,u) = \mathbf{F}_1(w,\varepsilon).$$

Also denote

$$\mu_1 \mathbf{S}_1(u) + \mu_2 \mathbf{S}_2(u) = \mu_2 \left(\frac{\mu_1}{\mu_2} \mathbf{S}_1(u) + \mathbf{S_2}(u) \right) = \mu_2 \mathbf{F}_1(w,\varepsilon) - \xi(\varepsilon w).$$

Then System (6) is rewriten as

$$\begin{cases} \varepsilon \mathbf{F}_0(w,\varepsilon)(\mathbf{Q} - (1-\varepsilon)\mathbf{\Lambda}) + j\sigma \dfrac{\partial \mathbf{F}_0(w,\varepsilon)}{\partial w} + \mu_2 \mathbf{F}_1(w,\varepsilon) - \xi(w\varepsilon) = \mathbf{0}, \\ \mathbf{F}_1(w,\varepsilon)(\mathbf{Q} - (1-\varepsilon)(e^{j\varepsilon w} - 1)\mathbf{\Lambda}) - \mu_2 \mathbf{F}_1(w,\varepsilon) + \xi(w\varepsilon) + \\ \varepsilon \mathbf{F}_0(w,\varepsilon)(1-\varepsilon)\mathbf{\Lambda} - -j\sigma e^{-j\varepsilon w} \dfrac{\partial \mathbf{F}_0(w,\varepsilon)}{\partial w} + e^{j\varepsilon w} \mathbf{F}_1(w,\varepsilon)(1-\varepsilon)\mathbf{\Lambda} = \mathbf{0}. \end{cases} \tag{7}$$

The asymptotic method consists of two parts: the asymptotic equations derivation and its solving.

Derivation of Asymptotic Equations

First of all, we write System (7) for $\varepsilon \to 0$. We obtain the following equations:

$$\begin{cases} j\sigma \mathbf{F}_0'(w) = -\mu_2 \mathbf{F}_1(w), \\ \mathbf{F}_1(w)\mathbf{Q} = \mathbf{0}. \end{cases} \tag{8}$$

We use the following expansions of the functions:

$$\begin{cases} \mathbf{F}_0(w,\varepsilon) = \mathbf{F}_0(w) + \varepsilon \mathbf{f}_0(w) + O(\varepsilon^2), \\ \mathbf{F}_1(w,\varepsilon) = \mathbf{F}_1(w) + \varepsilon \mathbf{f}_1(w) + O(\varepsilon^2). \end{cases} \tag{9}$$

where $\mathbf{O}(\varepsilon^2)$ is an infinitesimal value of the order ε^2.

Substituting Expressions (9) in (7) and making some transforms, we obtain the following equations under the limit $\varepsilon \to 0$.

$$\begin{cases} \mathbf{F}_0(w)(\mathbf{Q} - \mathbf{\Lambda}) + j\sigma \mathbf{f}_1'(w) + \mu_2 \mathbf{f}_1(w) = \mathbf{0}, \\ \mathbf{f}_1(w)(\mathbf{Q} - \mu_2 \mathbf{\Lambda}) + \mathbf{F}_0(w)\mathbf{\Lambda} + jw\mathbf{F}_1(w)\mathbf{\Lambda} + j\sigma jw \mathbf{F}_0'(w)\mathbf{\Lambda} + j\sigma \mathbf{f}_0'(w) = \mathbf{0}. \end{cases} \tag{10}$$

Also we add equations of System (7) and multiply it by $\mathbf{e}$ (in this way, we sum up elements of matrices). Taking into account $\mathbf{Qe} = \mathbf{0}$, we obtain:

$$-\mathbf{F}_1(w,\varepsilon)(1-\varepsilon)\mathbf{\Lambda e} + j\sigma e^{-j\varepsilon w}\frac{\partial \mathbf{F}_0(w,\varepsilon)}{\partial w}\mathbf{e} = \mathbf{0}. \tag{11}$$

Substituting Expansions (9), we have two additional asymptotic equations:

$$\begin{cases} \mathbf{F}_1(w)\mathbf{\Lambda e} + j\sigma \mathbf{F}_0'(w)\mathbf{e} = \mathbf{0}, \\ \mathbf{f}_1(w)(\mathbf{f_1}(w)\mathbf{\Lambda e} - \mathbf{F}_1(w)\mathbf{\Lambda e} + j\sigma \mathbf{f}_0'(w)\mathbf{e} - j\sigma jw\mathbf{F}_0'(w)\mathbf{e} = \mathbf{0}. \end{cases} \tag{12}$$

Equations Solving

Under the heavy load, the asymptotic characteristic function of the number of customers in the orbit is defined as

$$h(u) = \mathbf{F}_1\left(\frac{u}{\varepsilon}\right)\mathbf{e} + O(\varepsilon). \tag{13}$$

So we need to find function $\mathbf{F}_1(w)$ from Eqs. (8), (10) and (12). We will obtain it in several stages.

Stage 1. From the second equation of System (8), we can write the equality:

$$\mathbf{F}_1(w) = \mathbf{R}\Phi(w), \tag{14}$$

where $\mathbf{F}_1(w)\mathbf{e} = \Phi(w)\mathbf{Re} = \Phi(w)$.

From the first equation of System (12), we obtain:

$$\mathbf{F}_0'(w) = j\frac{\mu_2}{\sigma}\mathbf{F}_1(w) = j\frac{\mu_2}{\sigma}\mathbf{R}\Phi(w). \tag{15}$$

Stage 2. By summing Eq. (10) and taking into account (15), we obtain the following equation:

$$\{\mathbf{F}_0(w) + \mathbf{f}_1(w)\}\mathbf{Q} = -jw\mathbf{R}\Phi(w)\{\mathbf{\Lambda} - \mu_2\mathbf{I}\}. \tag{16}$$

So we can write

$$\mathbf{F}_0(w) + \mathbf{f}_1(w) = jw\Phi(w)\mathbf{V},$$

where vector $\mathbf{V}$ is a solution of the following equation:

$$\mathbf{VQ} = \mathbf{R}(\mu_2\mathbf{I} - \mathbf{\Lambda}). \tag{17}$$

In order to the solution of System (17) existing, it is necessary that the rank of the extended matrix be equal to the rank of matrix $\mathbf{Q}$. Since det$\mathbf{Q} = 0$, the rank of the extended matrix must also be less than the dimension of the system. Then $(\mu_2\mathbf{R} - \mathbf{R\Lambda})\mathbf{e} = 0$ is true. Then the general solution of (17) can be written as follows

$$\mathbf{V} = C\mathbf{R} + \mathbf{V}_0,$$

where $\mathbf{C} = const$, $\mathbf{V}_0$ is a particular solution, which can be found by some initial condition, for example $\mathbf{Ve} = 0$.

From Eq. (16), we can write:

$$\mathbf{f}_1(w) = jw\Phi(w)\mathbf{V} - \mathbf{F}_0(w). \tag{18}$$

Stage 3. From the first equation of system (10), we get:

$$j\sigma\mathbf{f}_0'(w) = -\mathbf{F}_0(w)(\mathbf{Q} - \mathbf{\Lambda}) - \mu_2\mathbf{f}_1(w).$$

Substitute Expression (18):

$$j\sigma\mathbf{f}_0'(w) = \mathbf{F}_0(w)(\mathbf{\Lambda} - \mathbf{Q} + \mu_2\mathbf{I}) - jw\mu_2\Phi(w)\mathbf{V}. \tag{19}$$

Stage 4. We sum up equations of System (12):

$$\mathbf{f}_1(w)\mathbf{\Lambda}\mathbf{e} + j\sigma\mathbf{f}_0'(w)\mathbf{e} + j\sigma(1 - jw)\mathbf{F}_0'(w)\mathbf{e} = 0. \tag{20}$$

Substituting Formulas (18), (19), we obtain:

$$jw\Phi(w)\mathbf{V}(\mathbf{\Lambda} - \mu_2\mathbf{I})\mathbf{e} + j\sigma(1 - jw)\mathbf{F}_0'(w)\mathbf{e} + \mathbf{F}_0(w)\mu_2\mathbf{e} = 0. \tag{21}$$

Let's differentiate the last expression:

$$\begin{aligned} &jw\left(\Phi(w)\mathbf{V}(\mathbf{\Lambda} - \mu_2\mathbf{I})\mathbf{e} + \Phi'(w)\mathbf{V}(\mathbf{\Lambda} - \mu_2\mathbf{I})\mathbf{e}\right) + \\ &\sigma\mathbf{F}_0'(w)\mathbf{e} + j\sigma(1 - jw)\mathbf{F}_0''(w)\mathbf{e} + \mathbf{F}_0'(w)\mu_2 = 0. \end{aligned} \tag{22}$$

Differentiating Expression (15) and substituting the result into Eq. (22), we obtain:

$$\begin{aligned} &j\Phi(w)(\mathbf{V\Lambda e} - \mu_2\mathbf{Ve} + \mu_2 + \frac{\mu_2^2}{\sigma}) + \\ &+\Phi'(w)[jw(\mathbf{V\Lambda e} - \mu_2\mathbf{Ve} + \mu_2) - \mu_2] = 0. \end{aligned} \tag{23}$$

Let us introduce the notations:

$$\alpha = 1 + \frac{\mu_2^2}{\sigma(\mathbf{V\Lambda e} - \mu_2\mathbf{Ve} + \mu_2)}, \quad \beta = \frac{\mu_2}{\mathbf{V\Lambda e} - \mu_2\mathbf{Ve} + \mu_2}. \tag{24}$$

Then Eq. (23) is rewritten as:

$$\Phi(w)j\alpha = \Phi'(w)[\beta - jw]. \tag{25}$$

Obviously, the solution has the following form

$$\Phi(w) = C\left(1 - \frac{jw}{\beta}\right)^{-\alpha}, \tag{26}$$

where $C = 1$ from the initial condition $\Phi(0) = 1$. Returning to Expression (14), we have:

$$\mathbf{F}_1(w) = \mathbf{R}\left(1 - \frac{jw}{\beta}\right)^{-\alpha}$$

Making the inverse substitutions $w = \dfrac{u}{\varepsilon}$ and $\varepsilon = 1-\rho$, we obtain the following formula for the asymptotic characteristic function

$$h(u) = \left(1 - \frac{ju}{(1-\rho)\beta}\right)^{-\alpha}, \tag{27}$$

that is the characteristic function of the gamma distribution.

4 Numerical Examples

To conduct numerical experiments and analyze the applicability of the asymptotic analysis results, a simulation model was construct using Python programming language.

Let us consider an example for the following system parameters:

$$\mathbf{Q} = \begin{bmatrix} -0.5 & 0.2 & 0.3 \\ 0.1 & -0.3 & 0.2 \\ 0.3 & 0.2 & -0.5 \end{bmatrix}, \quad \mathbf{\Lambda} = \begin{bmatrix} 1.1 & 0 & 0 \\ 0 & 1 & 0 \\ 0 & 0 & 1.15 \end{bmatrix},$$

$$\mu_1 = 0.9, \quad \mu_2 = \lambda/\rho, \quad \sigma = 1, \quad I = 15.$$

In Fig. 2 the comparison of the asymptotic and simulate distributions of the number of customers in the orbit is presented for various values of ρ.

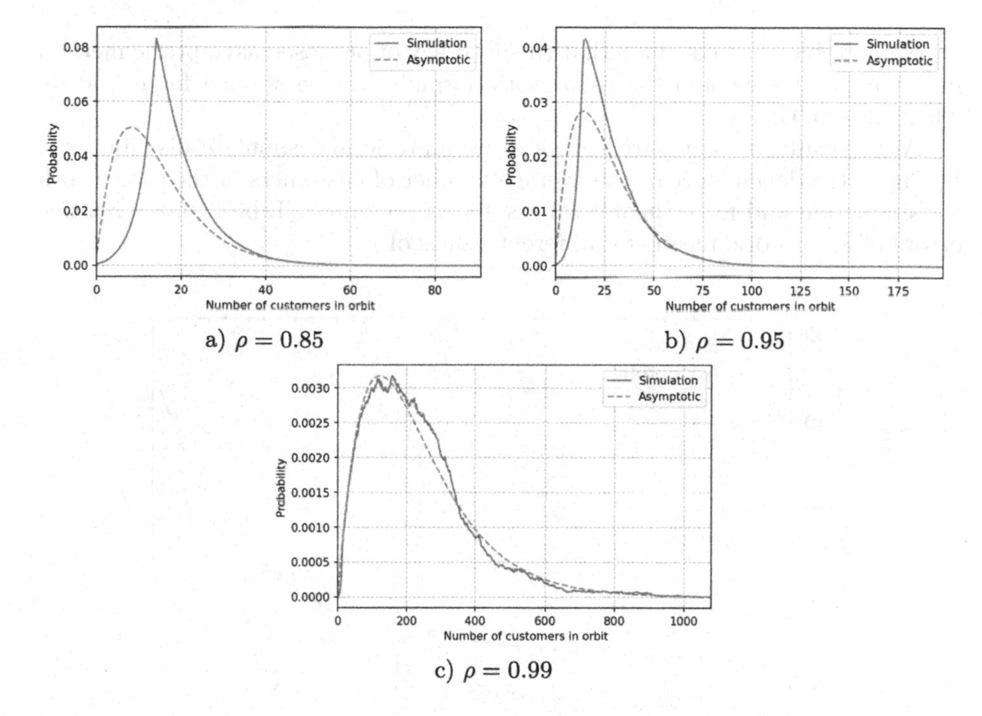

Fig. 2. Probability distributions of the number of customers in the orbit

Note that in case c) $\rho = 0.99$, we obtain non-smooth curve of the simulate probability density. The reason is that the retrial queue operates in close to non stationary mode, so the program works for a long time and the simulate results hardly converges to the distribution. Whereas distributions by analytical formulas (i.e. asymptotic) are always calculated easy and the accuracy of the asymptotic analysis in case of $\rho = 0.99$ is very high.

The applicability of the asymptotic results can be estimated using the Kolmogorov distance:

$$\Delta = \max_i |F_s(i) - F_a(i)|,$$

where F_s is an empirical (simulate) probability distribution function of the number of customers in the orbit, F_a is a corresponding asymptotic probability distribution function.

The results of the distribution comparison are presented in Table 1.

Table 1. Kolmogorov distance for different ρ

ρ	0.75	0.85	0.90	0.95
Δ	0.093	**0.041**	**0.016**	**0.005**

From Table 1, we conclude that the accuracy of proposed asymptotic method increases for $\rho \to 1$ and the asymptotic formula can be applied for $\rho \geq 0.85$, where $\Delta < 0.05$.

Additionally, we compare means of asymptotic and simulate distributions. In Fig. 3, the dependence of the average number of customers in the orbit (both for simulation and for a theoretical results) is presented. Table 2 shows relative errors of asymptotic means for different values of ρ.

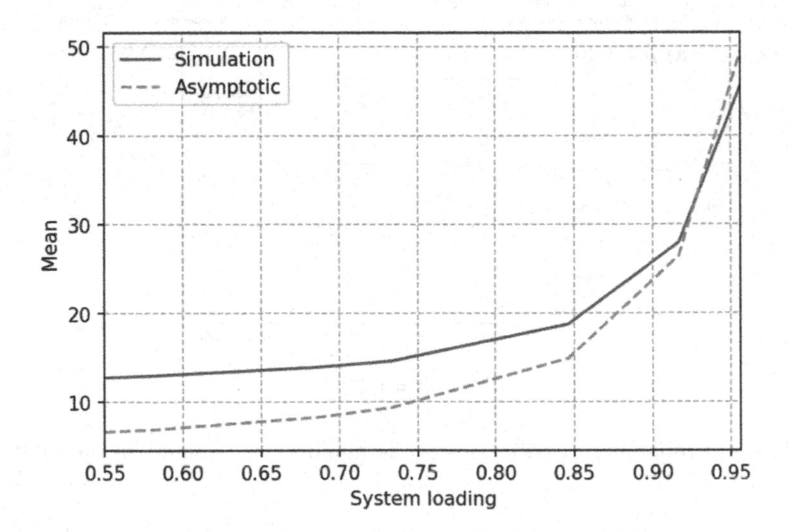

Fig. 3. Asymptotic and simulate means

Table 2. Relative error for different ρ

ρ	0.75	0.85	0.90	0.95
δ	0.361	0.204	**0.056**	**0.006**

Finally, we demonstrate one more interesting example for $\mu_1 > \mu_2$. We take the following values of system parameters:

$$\mathbf{Q} = \begin{bmatrix} -0.5 & 0.2 & 0.3 \\ 0.1 & -0.3 & 0.2 \\ 0.3 & 0.2 & -0.5 \end{bmatrix}, \quad \mathbf{\Lambda} = \begin{bmatrix} 1.1 & 0 & 0 \\ 0 & 1 & 0 \\ 0 & 0 & 1.15 \end{bmatrix},$$

$$\mu_1 = 1.2, \quad \mu_2 = 1, \quad \rho = 0.9, \quad \sigma = 0.1, \quad I = 50.$$

The empirical probability distribution is presented in Fig. 4.

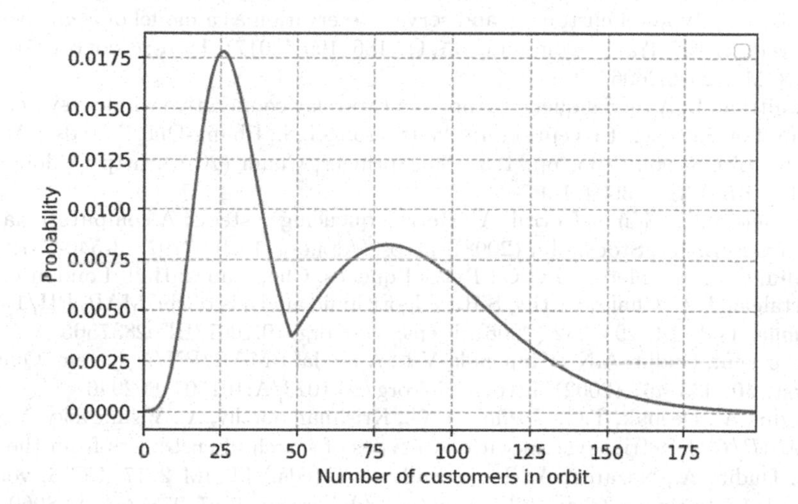

Fig. 4. Asymptotic probability distribution of the number of applications on orbits

The asymptotic results obtained in the paper is applied only for $\mu_1 < \mu_2$ (we define this condition in Sect. 2). So its applicability area does not include the demonstrated case. Thus in future, we plan to develop new method for described case of the retrial queue.

5 Conclusions

In the paper, we have studied a MMPP/M/1 retrial queueing system with switching, where the service rate has two values depending on the number of customers in the orbit. We have proposed the asymptotic analysis method for the obtaining the probability distribution of the number of customers in the orbit. It was proved that the asymptotic characteristic function under the condition of a heavy load has the form of a gamma distribution with the found parameters. The numerical analysis of obtained results shows a good accuracy of the asymptotic method for $\rho \geq 0.9$. Also the asymptotic method applicability area is limited by inequality $\mu_1 < \mu_2$. Absence of analytical formulas for the opposite case is a reason of study development in future.

References

1. Devarajan, K., Senthilkumar, M.: On the retrial-queuing model for strategic access and equilibrium-joining strategies of cognitive users in cognitive-radio networks with energy harvesting. Energies **14**, 2088 (2021)
2. Phung-Duc, T., Kawanishi, K.: Multiserver retrial queue with setup time and its application to data centers. J. Indus. Manage. Optimiz. **15**(1), 15–35 (2019). https://doi.org/10.3934/jimo.2018030
3. Dudin, A.N., Lee, M.H., Dudina, O., Lee, S.K.: Analysis of priority retrial queue with many types of customers and servers reservation as a model of cognitive radio system. IEEE Trans. Commun. **65**(1), 186–199 (2017). https://doi.org/10.1109/TCOMM.2016.2606379
4. Dimitriou, I.: A retrial queue to model a two-relay cooperative wireless system with simultaneous packet reception. In: Wittevrongel, S., Phung-Duc, T. (eds.) ASMTA 2016. LNCS, vol. 9845, pp. 123–139. Springer, Cham (2016). https://doi.org/10.1007/978-3-319-43904-4_9
5. Artalejo, J.R., Gómez-Corral, A.: Retrial queueing systems. A computational approach. Springer, Stockholm (2008). https://doi.org/10.1007/978-3-540-78725-9
6. Falin, G.I., Templeton, J.G.C.: Retrial queues. Chapman & Hall, London (1997)
7. Artalejo, J.R., Chakravarthy, S.R.: Algorithmic analysis of the MAP/PH/1 retrial queue. TOP **14**, 293–332 (2006). https://doi.org/10.1007/BF02837565
8. Breuer, L., Dudin, A.N., Klimenok, V.I.: A retrial BMAP/PN/N system. Queueing Syst. **40**, 433–457 (2002). https://doi.org/10.1023/A:1015041602946
9. Dudin, A., Deepak, T.G., Joshua, V.C., Krishnamoorthy, A., Vishnevsky, V.: On a $BMAP/G/1$ Retrial system with two types of search of customers from the orbit. In: Dudin, A., Nazarov, A., Kirpichnikov, A. (eds.) ITMM 2017. CCIS, vol. 800, pp. 1–12. Springer, Cham (2017). https://doi.org/10.1007/978-3-319-68069-9_1
10. Lv, S., Zhu, L.: Single server repairable queueing system with variable service rate and failure rate. IEEE Access **9**, 1233–1239 (2021). https://doi.org/10.1109/ACCESS.2020.3047815
11. Zhang, X., Wang, J., Ma, Q.: Optimal design for a retrial queueing system with state-dependent service rate. J. Syst. Sci. Complex **30**, 883–900 (2017). https://doi.org/10.1007/s11424-017-5097-9
12. Neuts, M.F.: A Versatile Markovian point process. J. Appl. Probab. **16**(4), 764–779 (1979). https://doi.org/10.2307/3213143
13. Lucantoni, D.M.: New results on the single server queue with a batch Markovian arrival process. Stoch. Model. **7**, 1–46 (1991). https://doi.org/10.1080/15326349108807174
14. Fedorova, E., Nazarov, A., Moiseev, A.: Asymptotic analysis methods for multi-server retrial queueing systems. In: Joshua, V.C., Varadhan, S.R.S., Vishnevsky, V.M. (eds.) Applied Probability and Stochastic Processes. ISFS, pp. 159–177. Springer, Singapore (2020). https://doi.org/10.1007/978-981-15-5951-8_11

15. Danilyuk, E.Y., Fedorova, E.A., Moiseeva, S.P.: Asymptotic analysis of an retrial queueing system M—M—1 with collisions and impatient calls. Autom. Remote. Control. **79**(12), 2136–2146 (2018). https://doi.org/10.1134/S0005117918120044
16. Nazarov, A., Phung-Duc, T., Paul, S.: Slow retrial asymptotics for a single server queue with two-way communication and Markov modulated poisson input. J. Syst. Sci. Syst. Eng. **28**(2), 181–193 (2019). https://doi.org/10.1007/s11518-018-5404-6

Account of Disasters in Analysis of Queueing Systems Modeled by the Quasi-Birth-and-Death-Process

Alexander Dudin(✉)

Belarusian State University, 4 Nezavisimosti Avenue, 220030 Minsk, Belarus
dudin@bsu.by

Abstract. We present a way for the derivation of the generator of a multidimensional continuous-time Markov chain describing the behavior of a queueing model with account of possibility of disaster occurrence under the known generator of the Quasi-Birth-and-Death-Process (QBD) describing behavior of the corresponding queueing model without disasters. The stationary distribution of the Markov chain in the case of the level-independent QBD describing the behavior of a queueing model without disasters is found in the matrix geometric form.

Keywords: Quasi-Birth-and-Death Process · Disaster · Stationary Distribution

1 Introduction

Queueing systems are good mathematical models of versatile real-world systems, networks, processes what makes them attractive subject of study during more than 120 years. The important phenomenon in some real systems described by the single server queues are disasters, i.e., the events arrival of which causes immediate departure of all customers from the system. As the first papers where this phenomenon is taken into account we can mention [1–3]. The queues with disasters are very close to the earlier analysed stochastic clearing systems, see [4–6]. Disasters can be considered as the special case of so called negative customers introduced into consideration by E. Gelenbe and coauthors, see, e.g., [7–9]. Arrival of a negative customer causes removal of one or several usual, positive, customers from the system without service.

The models analysed in [1–3] suppose that the arriving flows of customers and disasters are described by stationary Poisson processes. This essentially restricts their application to the analysis of the modern real-world systems. Flows in these systems have variable instantaneous arrival rate, dependence of successive inter-arrival times and large variance of these times. More general and appropriate for the description of real flows is the model of Markov Arrival Process (MAP), see, e.g., [10–14]. Queueing systems with a single server, the batch MAP flows of

A. Dudin et al. (Eds.): ITMM 2022, CCIS 1803, pp. 92–105, 2023.
https://doi.org/10.1007/978-3-031-32990-6_8

customers and disasters and semi-Markov dependence of service times were analysed in [15–17]. Note that the semi-Markov service process assumed in [15–17] is much more general than the renewal service with an arbitrary or exponential distribution of mutually independent successive service times.

In [15], the analysis was done under the assumption that after disaster occurrence all customers depart from the system, but the server is not broken and is ready to provide service to incoming customers. The analysis included consideration of the embedded at service completion and disaster arrival moments Markov chain, computation of an arbitrary time distribution of the system states with the use of results for Markov renewal processes from [18,19] and computation of the Laplace-Stieltjes transform of sojourn time distribution with the use of the extension of the method of collective marks (method of catastrophes), see, e.g. [20], to the matrix case. In paper [16], it was assumed that recovering of the server during the random time having an arbitrary distribution is required after a disaster arrival. Options of customers loss or accumulation during the recovering time were considered.

The analysis implemented in [15] and [16] exploits, for computation of the stationary distribution of the embedded Markov chain, the use of the vector and matrix generating functions and the property of the analyticity of these functions in the unit circle of the complex plane. The weakness of such approach stems from the necessity to numerically find the roots of some determinant of the matrix function in the unit circle of the complex plane for the computation of the vector of probabilities that the system is empty at an arbitrary service completion epoch and instability of the recursive procedure for computing the other vectors of probabilities of the embedded Markov chain.

Later in [17], the system was analysed with the use of the extension of the well-known approach by M. Neuts, see [25], exploiting the matrix G which describes transition probability of the finite components of the chain during the time until decreasing by one of the value of the denumerable component. This approach provides easier computation of the vector of the stationary probabilities that the system is empty and significantly better stability in the computation of the vectors of the stationary probabilities what is confirmed by the numerical results presented in [17]. As more recent papers where the system with disasters and MAP arrival processes is analysed and useful references are presented, we can cite papers [21–24].

The problem considered in this paper is briefly formulated as follows. Let us have an arbitrary queueing system without disasters, dynamics of which is described by the Quasi-Birth-and-Death-Process (QBD), i.e., the continuous-time multi-dimensional Markov chain having one denumerable component, a finite number of finite components and a block-tridiagonal generator. Let this generator be known. Let now assume that this queueing system is influenced by the flow of disasters described by the MAP. The overwhelming majority of the existing papers devoted to the systems with disasters consider the single server queues where disaster arrival removes, as it was stated above, all customers from the system. In this paper we analyse the QBD, which is suitable for description of behavior of multi-server systems with disasters. For such systems, three scenarios of disaster impact are possible: (i) disaster arrival removes all customers from the servers but does not impact other customers presenting in the system; (ii) disaster arrival removes all customers from the system; (iii) disaster arrival removes all customers that are waiting for processing (stay in the buffer or the orbit) but do not impact on customers receiving service. Analysis presented below is useful for scenario (iii). However, it can be easily adopted, via another definition of the components of the Markov chain, for scenario (ii).

We suggest that arrival of a disaster causes immediate transition of the denumerable component of the QBD into the zero state and does not affect the finite components. It is required to write down the generator of the continuous-time multi-dimensional Markov chain describing this queueing system with account of disasters. This Markov chain does not belong to the class of QBD. In case of the level-independent QBD describing the initial queueing system (system without disasters), the stationary distribution of the states of the constructed Markov chain is computed.

The reminder of the paper is as follows. In Sect. 2, the considered problem is formulated in more detail. In Sect. 3, the solution of the problem is given. Section 4 contains the algorithm for computation of the stationary distribution of the constructed Markov chain for the case when the initial QBD has the level-independent transitions. Section 5 concludes the paper.

2 Problem Formulation

Let us consider some queueing system operation of which is defined by a QBD $\xi_t = \{i_t, n_t\}$, $t \geq 0$. The component i_t admits values from the infinite set $0, 1, \ldots$. The component n_t has a finite state space. Indeed, in description of a concrete queueing system the component n_t can be a finite set of components, say $(n_t^{(1)}, \ldots, n_t^{(J)})$, with the finite state spaces having certain physical sense, e.g., number of customers of different types receiving service, the underlying processes of arrival processes, etc. Usually, it is useful to consider namely multi-dimensional process $n_t = (n_t^{(1)}, \ldots, n_t^{(J)})$ for derivation of the generator of QBD. But for the purposes of our analysis without restriction of generality we consider this set as one finite component n_t. Also, we suppose that the state space of the component n_t does not depend on the value of the components i_t and the cardinality of this space is equal to K. The cases when such a dependence takes place, e.g. when independence takes place only for the values of i_t greater than some threshold, can be considered separately.

Components of the QBD ξ_t are supposed to be enumerated in the direct lexicographic order and the set of the states having the value i of the first component is called as level i.

Let Q be the generator of the QBD ξ_t. It has the following structure:

$$Q = \begin{pmatrix} Q_{0,0} & Q_{0,1} & O & O & O & \cdots \\ Q_{1,0} & Q_{1,1} & Q_{1,2} & O & O & \cdots \\ O & Q_{2,1} & Q_{2,2} & Q_{2,3} & O & \cdots \\ O & O & Q_{3,2} & Q_{3,3} & Q_{3,4} & \cdots \\ \vdots & \vdots & \vdots & \vdots & \vdots & \ddots \end{pmatrix} \quad (1)$$

where the blocks $Q_{i,j}$, $\max\{0, i-1\} \leq j \leq i+1$, contain the intensities of transitions of the chain ξ_t from the states that belong to level i to the states that belong to level j.

As the examples of queueing systems described by such a QBD we can mention the huge variety of various systems (multi-server systems, systems with MAP arrival processes of different kinds of customers, breakdowns, interruptions, systems with service, repair, vacation and other times having so-called phase-type (PH) distribution, see [26], tandem queues, queues operating in a random environment, queues with passive servers, queueing/inventory systems, communication systems with energy harvesting, semi-open queueing networks, etc.).

In particular, such a QBD describes the dynamics of the operation of the cell of a cognitive radio system with servers reservation for the licensed users, see, e.g., [28–30] where the component i_t is the number of low-priority cognitive users in the orbit (in [28,30]) or in the buffer (in [29]). The finite component n_t is the multidimensional one and includes the number of busy servers, number of servers occupied by the cognitive users, the states of the underlying process of arrival of two types of customers. In [30], where the strategy of servers reservation is assumed to be not of a threshold type, but of a hysteresis type, additionally the component, which tracks the state of the admission managing process, is included into the process n_t. In [28], the arrivals are defined by the marked Markov Arrival Process. In [29,30], they are defined by two independent Markov arrival processes.

Let now the behavior of the considered queueing system be influenced by the occurrence of the disasters. We suppose that disaster occurrence causes the immediate transition of the component i_t to the state 0, while other components do not change their values. In application to real-world systems, we can interpret the component i_t as the number of customers waiting for service in the buffer or in the orbit. The component n_t describes service and arrival processes. Disaster can have the meaning of a power supply failure in the buffer or periodic deliberate devastation of a buffer or orbit to avoid congestion of the system.

The process of disasters occurrence is assumed here to be defined by the MAP having the underlying process η_t, $t \geq 0$, with the state space $\{1, \ldots, z\}$ and transition intensities given by the entries of the matrices Z_0 and Z_1. Transitions of the process η_t with intensities given by the entries of the matrix Z_1 lead to a disaster arrival. Transitions of the process η_t with intensities given by the entries of the matrix Z_0 are not accompanied by a disaster arrival. Note that the diagonal components of the matrix Z_0 are negative and define, up to the sign, intensity of departure of the process η_t from the corresponding state.

The matrix $Z_0 + Z_1$ is the generator of the process η_t. The invariant probability vector $\boldsymbol{\psi}$ of this process is the unique solution to the system of equations

$$\boldsymbol{\psi}(Z_0 + Z_1) = \mathbf{0}, \ \boldsymbol{\psi}\mathbf{e} = 1$$

where $\mathbf{e}$ is a column vector of appropriate size consisting of ones, and $\mathbf{0}$ is a row vector of appropriate size consisting of zeroes. The average disaster arrival rate ϕ is given by $\phi = \boldsymbol{\psi} Z_1 \mathbf{e}$.

Disaster arrival during the time when the component i_t is equal to zero is assumed to be ignored. Variants when the disaster occurrence affects not only the component i_t but also the component n_t can be considered analogously to the way presented below.

The dynamics of the system influenced by the disasters is defined by the Markov chain

$$\boldsymbol{\xi}_t = (\xi_t, \eta_t) = \{i_t, n_t, \eta_t\}, \ t \geq 0.$$

Our goal is to present simple formulas showing the relationship between the generators of the initial QBD ξ_t and the continuous-time Markov chain $\boldsymbol{\xi}_t$.

3 The Main Result

Let $\mathcal{Q}$ be the generator of the Markov chain $\boldsymbol{\xi}_t$. As above, we call as level i the set of the states of the Markov chain having the value i of the first component. Cardinality of levels of the Markov chain $\boldsymbol{\xi}_t$ is equal to the cardinality of the corresponding levels of the Markov chain ξ_t multiplied by z.

Theorem 1. The generator $\mathcal{Q}$ of the Markov chain $\boldsymbol{\xi}_t$ has the following form

$$\mathcal{Q} = \begin{pmatrix} \mathcal{Q}_{0,0} & \mathcal{Q}_{0,1} & O & O & O & \cdots \\ \mathcal{Q}_{1,0} & \mathcal{Q}_{1,1} & \mathcal{Q}_{1,2} & O & O & \cdots \\ \mathcal{Q}_{2,0} & \mathcal{Q}_{2,1} & \mathcal{Q}_{2,2} & \mathcal{Q}_{2,3} & O & \cdots \\ \mathcal{Q}_{3,0} & O & \mathcal{Q}_{3,2} & \mathcal{Q}_{3,3} & \mathcal{Q}_{3,4} & \cdots \\ \mathcal{Q}_{4,0} & O & O & \mathcal{Q}_{4,3} & \mathcal{Q}_{4,4} & \cdots \\ \vdots & \vdots & \vdots & \vdots & \vdots & \ddots \end{pmatrix} \tag{2}$$

where the blocks $\mathcal{Q}_{i,j}$ contain intensities of transition of the chain $\boldsymbol{\xi}_t$ from the states that belong to the level i to the states that belong to the level j.

The relations between the blocks $Q_{i,j}$ of generator the Q given by formula (1) and the blocks $\mathcal{Q}_{i,j}$ of the generator $\mathcal{Q}$ given by formula (2) are as follows:

$$\begin{aligned} \mathcal{Q}_{i,i+1} &= Q_{i,i+1} \otimes I_z, \ i \geq 0, \\ \mathcal{Q}_{0,0} &= Q_{0,0} \oplus (Z_0 + Z_1), \\ \mathcal{Q}_{i,i} &= Q_{i,i} \oplus Z_0, \ i \geq 1, \\ \mathcal{Q}_{1,0} &= Q_{1,0} \oplus Z_1, \\ \mathcal{Q}_{i,i-1} &= Q_{i,i-1} \otimes I_z, \ i \geq 2, \\ \mathcal{Q}_{i,0} &= I_K \otimes Z_1, \ i \geq 2. \end{aligned} \tag{3}$$

Here $O_{a\times b}$ denotes a zero matrix of size $a\times b$, I_a denotes the identity matrix of size a, $\otimes$ and $\oplus$ are symbols of Kronecker product and sum of matrices, see [31–33].

Proof 1. Let us analyse all transition rates of the Markov chain $\boldsymbol{\xi}_t$.

The transitions from the level i to the level $i+1$, $i\geq 0$, and from the level i to the level $i-1$, $i\geq 2$, are possible only via the corresponding transitions of the process $\{i_t, n_t\}$. The component η_t does not change at the moments of such transitions (i.e., the probabilities of its transitions are given by the identity matrix I_z). Taking into account the definition of Kronecker product of matrices, we obtain relations $\mathcal{Q}_{i,i+1} = Q_{i,i+1}\otimes I_z$, $i\geq 0$, and $\mathcal{Q}_{i,i-1} = Q_{i,i-1}\otimes I_z$, $i\geq 2$.

The transitions within the level i, $i\geq 1$, are possible only via the corresponding transitions of the process $\{i_t, n_t\}$ within the level i, $i\geq 1$, (without any transitions of the underlying process of disaster arrival) or transitions of the component η_t that do not lead to a disaster occurrence (without any transitions of the QBD). Therefore, $\mathcal{Q}_{i,i} = Q_{i,i}\otimes I_z + I_K\otimes Z_0 = Q_{i,i}\oplus Z_0$, $i\geq 1$.

For level 0, transitions of the component η_t that lead to a disaster occurrence also do not cause the change of the level because it is assumed that disasters are ignored during the stay of the process at level 0. Thus, $\mathcal{Q}_{0,0} = Q_{0,0}\oplus(Z_0+Z_1)$.

The transition from the level i, $i\geq 2$, to level 0 is possible only via a disaster arrival. Therefore, $\mathcal{Q}_{i,0} = I_K\otimes Z_1$, $i\geq 2$.

For level 1, the transition to level 0 is possible due to the both corresponding transition of the chain $\{i_t, n_t\}$ from level 1 to level 0 and a disaster arrival. Therefore, we have the formula $\mathcal{Q}_{1,0} = Q_{1,0}\oplus Z_1$.

Theorem 1 is proved.

Further analysis of the queueing system with disaster can be implement via the use of the algorithmic results presented in [34].

4 Level-Independent *QBD*

The most well-analysed in the literature, see, e.g., [26], is the case of level-independent QBD. In this case, the triblock-diagonal generator Q of the QBD ξ_t has the following quasi-Toeplitz form:

$$Q = \begin{pmatrix} Q_{0,0} & Q_{0,1} & O & O & O & \cdots \\ Q_{1,0} & Q^0 & Q^+ & O & O & \cdots \\ O & Q^- & Q^0 & Q^+ & O & \cdots \\ O & O & Q^- & Q^0 & Q^+ & \cdots \\ \vdots & \vdots & \vdots & \vdots & \vdots & \ddots \end{pmatrix}, \tag{4}$$

i.e., $Q_{i,i} = Q^0$, $Q_{i,i+1} = Q^+$ for all i, $i\geq 1$, and $Q_{i,i-1} = Q^-$ for all i, $i\geq 2$.

For this case, it is well-known that the criterion for existence of the stationary distribution of the states of the QBD is the fulfilment of the inequality

$$\mathbf{y}Q^-\mathbf{e} > \mathbf{y}Q^+\mathbf{e} \tag{5}$$

where the vector $\mathbf{y}$ is the unique solution to the system

$$\mathbf{y}(Q^0 + Q^+ + Q^-) = \mathbf{0}, \ \ \mathbf{ye} = 1.$$

Let condition (3) be fulfilled. Then the following limits called as the stationary (invariant) probabilities of the states of the QBD exist:

$$\pi(i,n) = \lim_{t\to\infty} P\{i_t = i, n_t = n\}, \ i \geq 0.$$

Denote by $\boldsymbol{\pi}_i$ the row vector of the stationary probabilities of the states of the QBD that belong to level i enumerated in the lexicographic order.

It is known, see, e.g., [26], that the vectors $\boldsymbol{\pi}_i, \ i \geq 0$, can be computed as follows.

The vectors $\boldsymbol{\pi}_i, \ i \geq 2$, are expressed via the vector $\boldsymbol{\pi}_1$ as:

$$\boldsymbol{\pi}_i = \boldsymbol{\pi}_1 \mathcal{R}^{i-1}, \ i \geq 1, \tag{6}$$

where the matrix $\mathcal{R}$ is the minimal non-negative solution of the non-linear matrix equation

$$\mathcal{R}^2 Q^- + \mathcal{R} Q^0 + Q^+ = O. \tag{7}$$

The vector $(\boldsymbol{\pi}_0, \boldsymbol{\pi}_1)$ is the unique solution to the system

$$(\boldsymbol{\pi}_0, \boldsymbol{\pi}_1) \begin{pmatrix} Q_{0,0} & Q_{0,1} \\ Q_{1,0} & Q^0 + \mathcal{R} Q^- \end{pmatrix} = \mathbf{0}, \tag{8}$$

$$\boldsymbol{\pi}_0 \mathbf{e} + \boldsymbol{\pi}_1 (I - \mathcal{R})^{-1} \mathbf{e} = 1. \tag{9}$$

Equation (4) can be solved by the method of successive iterations, see, e.g., [26], [27].

The formal easy proof of the presented way for computation of the vectors $\boldsymbol{\pi}_i, \ i \geq 0$, can be implemented via the substitution of the expression (4) into equilibrium equations

$$(\boldsymbol{\pi}_0, \boldsymbol{\pi}_1, \boldsymbol{\pi}_2, \ldots) Q = \mathbf{0}, \ (\boldsymbol{\pi}_0, \boldsymbol{\pi}_1, \boldsymbol{\pi}_2, \ldots) \mathbf{e} = 1 \tag{10}$$

and verification that the equilibrium equations turn into the identities conditional that the matrix $\mathcal{R}$ is the solution of non-linear matrix Eq. (5). More rigorous and detailed proof, including the proof of existence, under fulfillment of ergodicity condition (3), of the minimal non-negative solution of Eq. (5), can be found in [26].

5 Consideration of the System Described by the Level-Independent QBD that is Influenced by the Disasters

Assume that the initial queueing system is described by the level-independent QBD with the generator of form (4). Let now this system be influenced by the MAP flow of disasters as it is described above.

Then, the following statement is true.

Corollary. In the case of the level-independent QBD with the generator of form (4), the generator $\mathcal{Q}$ of the Markov chain $\boldsymbol{\xi}_t$ has the following form

$$\mathcal{Q} = \begin{pmatrix} \mathcal{Q}_{0,0} & \mathcal{Q}_{0,1} & O & O & O & \dots \\ \mathcal{Q}_{1,0} & \mathcal{Q}^0 & \mathcal{Q}^+ & O & O & \dots \\ \mathcal{Q}^* & \mathcal{Q}^- & \mathcal{Q}^0 & \mathcal{Q}^+ & O & \dots \\ \mathcal{Q}^* & O & \mathcal{Q}^- & \mathcal{Q}^0 & \mathcal{Q}^+ & \dots \\ \mathcal{Q}^* & O & O & \mathcal{Q}^- & \mathcal{Q}^0 & \dots \\ \vdots & \vdots & \vdots & \vdots & \vdots & \ddots \end{pmatrix} \tag{11}$$

where the blocks $\mathcal{Q}_{0,0}$, $\mathcal{Q}_{0,1}$, $\mathcal{Q}_{1,0}$ have the same form as in Theorem 1 and the rest of the non-zero blocks are defined by

$$\mathcal{Q}^0 = \mathcal{Q}_{i,i},\ i \geq 1,\ \mathcal{Q}^- = \mathcal{Q}_{i,i-1},\ \mathcal{Q}^* = \mathcal{Q}_{i,0},\ i \geq 2,\ \mathcal{Q}^+ = \mathcal{Q}_{i,i+1},\ i \geq 1,$$

where the formulas for the matrices $\mathcal{Q}_{i,i}$, $\mathcal{Q}_{i,i-1}$, $\mathcal{Q}_{i,0}$, $\mathcal{Q}_{i,i+1}$ are also given in Theorem 1.

Consider the problem of computation of the stationary distribution of the Markov chain $\boldsymbol{\xi}_t$. We suggest that the average disaster arrival rate ϕ is positive. It is possible to prove that under this suggestion the stationary distribution of Markov chain $\boldsymbol{\xi}_t$ exists for any values of the system parameters.

Denote

$$\pi(i, n, \eta) = \lim_{t \to \infty} P\{i_t = i, n_t = n, \eta_t = \eta\},\ i \geq 0,$$

and let $\boldsymbol{\pi}_i$ be the row vectors of the stationary probabilities $\pi(i, n, \eta)$ of the states of the Markov chain $\boldsymbol{\xi}_t$ that belong to level i enumerated in the lexicographic order.

Theorem 2. The vectors $\boldsymbol{\pi}_i$, $i \geq 2$, can be computed as follows.

$$\boldsymbol{\pi}_i = \boldsymbol{\pi}_1 \mathcal{F}^{i-1},\ i \geq 2, \tag{12}$$

where the matrix $\mathcal{F}$ is defined by

$$\mathcal{F} = -\mathcal{Q}^+(\mathcal{Q}^0 + \mathcal{Q}^+\mathcal{G})^{-1}, \tag{13}$$

the matrix $\mathcal{G}$ is the solution of the non-linear matrix equation

$$\mathcal{Q}^- + \mathcal{Q}^0\mathcal{G} + \mathcal{Q}^+\mathcal{G}^2 = O. \tag{14}$$

The vector $(\boldsymbol{\pi}_0, \boldsymbol{\pi}_1)$ is computed as the unique solution to the system

$$(\boldsymbol{\pi}_0, \boldsymbol{\pi}_1) \begin{pmatrix} \mathcal{Q}_{0,0} & \mathcal{Q}_{0,1} \\ \mathcal{Q}^- + (I - \mathcal{F})^{-1}(I_K \otimes Z_1) & \mathcal{Q}^0 + \mathcal{F}\mathcal{Q}^- \end{pmatrix} = \mathbf{0}, \tag{15}$$

$$\boldsymbol{\pi}_0 \mathbf{e} + \boldsymbol{\pi}_1 (I - \mathcal{F})^{-1} \mathbf{e} = 1. \tag{16}$$

Remark 1. Instead of computation of the matrix $\mathcal{F}$ via the use of the matrix $\mathcal{G}$ that is the solution of equation (12), it can be computed directly as the solution of equation

$$\mathcal{F}^2\mathcal{Q}^- + \mathcal{F}\mathcal{Q}^0 + \mathcal{Q}^+ = O \tag{17}$$

of the same form as (5).

Note that in (5) the sum $(Q^0 + Q^+ + Q^-)$ of the matrices Q^0, Q^+, and Q^- is the generator while the sum $(\mathcal{Q}^0 + \mathcal{Q}^+ + \mathcal{Q}^-)$ of the matrices $\mathcal{Q}^0, \mathcal{Q}^+$, and $\mathcal{Q}^-$ is the sub-generator with strong domination of the diagonal entry in at least one row. The generator is the matrix $(\mathcal{Q}^0 + \mathcal{Q}^+ + \mathcal{Q}^- + \mathcal{Q}^*)$.

Note also that the matrix $\mathcal{G}$, which is the solution of non-linear matrix Eq. (12), here is not stochastic, but sub-stochastic one. This fact should be taken in mind when the iterative procedure for computation of this matrix is constructed.

Proof. The easy formal proof also can be implemented by the substitution of the vectors $\boldsymbol{\pi}_i$, $i \geq 0$, defined by formulas given above into the equilibrium equations

$$(\boldsymbol{\pi}_0, \boldsymbol{\pi}_1, \boldsymbol{\pi}_2, \dots)\mathcal{Q} = \mathbf{0}, \ (\boldsymbol{\pi}_0, \boldsymbol{\pi}_1, \boldsymbol{\pi}_2, \dots)\mathbf{e} = 1 \tag{18}$$

and verification that these equations turn into the identities.

Here we present another proof based on the derivation of another than (16) system of equations for the vector $(\boldsymbol{\pi}_0, \boldsymbol{\pi}_1, \boldsymbol{\pi}_2, \dots)$. To this end, we use the notion of the censored Markov chain, for more details see, e.g., [35–37]. This notion was effectively used in [38] for derivation of the alternative system of equations for the stationary probability vectors of asymptotically quasi-Toeplitz Markov chains, see also [13]. The notion was also already used in [17] for the analysis of the $BMAP/SM/1$ type queues with the MAP flow of disasters, embedded Markov chain for which has more general than (9) upper-Hessenberg block structure of one-step transition probability matrix. The absence of non-zero blocks above the first up-diagonal in (9) allows to simplify the result of [17] and obtain matrix-geometric form (10) of the stationary distribution for the Markov chain $\boldsymbol{\xi}_t$.

To derive the alternative system of equations via the use of the notion of the censored Markov chain, we need first obtain two matrices, denotes by $\mathcal{G}$ and $\mathcal{H}$.

The entries of the matrix $\mathcal{G}$ define the transition probabilities of the components $\{n_t, \eta_t\}$ of the Markov chain $\boldsymbol{\xi}_t$ during the time until the component i_t first time reaches the value i starting from the value $i+1$, $i \geq 1$. In the case of the level-dependent QBD, the matrix G depends on the value of i (i.e., instead of $\mathcal{G}$ we consider the matrix $\mathcal{G}_i$), and the matrices $\mathcal{G}_i$ satisfy the system of equations

$$\mathcal{Q}_{i+1,i} + \mathcal{Q}_{i,i}\mathcal{G}_i + \mathcal{Q}_{i+1,i+2}\mathcal{G}_{i+1}\mathcal{G}_i = O, \ i \geq 1.$$

In the considered in this section case of the level-independent process, this system turns to the equation of form (12).

The entries of the matrix $\mathcal{H}$ define the transition probabilities of the components $\{n_t, \eta_t\}$ of the Markov chain $\boldsymbol{\xi}_t$ during the time until the component i_t

reaches value 0 starting from the value i, $i \geq 1$, without visiting the level $i-1$. In case of the level-dependent QBD, the matrix $\mathcal{H}$ depends on the value of i (i.e., instead of $\mathcal{H}$ we consider the matrix $\mathcal{H}_i$), and the matrices $\mathcal{H}_i$ satisfy the system of equations

$$\mathcal{Q}_{i,0} + (\mathcal{Q}_{i,i} + \mathcal{Q}_{i,i+1}(\mathcal{G}_{i+1} + I))\mathcal{H}_i = O, \ i \geq 2.$$

In the considered in this section case of the level-independent process, from this system we obtain the following formula for calculation of the matrix $\mathcal{H}$:

$$\mathcal{H} = -(\mathcal{Q}^0 + \mathcal{Q}^+(\mathcal{G} + I))^{-1}\mathcal{Q}^*.$$

It is easy to check that the matrix $\mathcal{G} + \mathcal{H}$ is the stochastic one, i.e., it is the matrix with non-negative entries and $(\mathcal{G} + \mathcal{H})\mathbf{e} = \mathbf{e}$.

Having obtained the ways to compute the matrices $\mathcal{G}$ and $\mathcal{H}$, we are able to derive the alternative system of equations for the probability vectors $\boldsymbol{\pi}_i$, $i \geq 0$.

Let us fix an arbitrary integer i, $i \geq 2$, and construct the Markov chain which is censored with respect to the Markov chain $\boldsymbol{\xi}_t$ with the censoring level i. The trajectories of the censored Markov chain are obtained from the trajectories of the original Markov chain $\boldsymbol{\xi}_t$ by means of removal the parts of the trajectories of the original Markov chain when the value of the first component of the chain i_t exceeds the value i.

Let us denote by $\bar{\mathcal{Q}}_{l,i}$ the matrices of transition intensities of the censored Markov chain with the censoring level i. It is easy to understand that the values of $\bar{\mathcal{Q}}_{l,i}$ are different from zero only for $l = i-1$ and $l = i$ and that

$$\bar{\mathcal{Q}}_{i-1,i} = \mathcal{Q}_{i-1,i} = \mathcal{Q}^+, \ \bar{\mathcal{Q}}_{i,i} = \mathcal{Q}^0 + \mathcal{Q}^+\mathcal{G}. \tag{19}$$

Correspondingly, we obtain the following equations for the vectors of stationary probabilities of the censored Markov chain that coincide, up to the normalizing multiplier, with the respective vectors of stationary probabilities of the initial Markov chain:

$$\boldsymbol{\pi}_i\bar{\mathcal{Q}}_{i,i} + \boldsymbol{\pi}_{i-1}\bar{\mathcal{Q}}_{i-1,i} = \mathbf{0}. \tag{20}$$

Such equations are valid for the arbitrarily fixed integer i, $i \geq 2$. Therefore, (18) is the alternative system that was the goal of our construction.

Taking in mind formulas (17), we obtain that

$$\boldsymbol{\pi}_i = \boldsymbol{\pi}_{i-1}(-\mathcal{Q}^+(\mathcal{Q}^0 + \mathcal{Q}^+\mathcal{G})^{-1}), \ i \geq 2,$$

from which, by introducing the notation $\mathcal{F}$, we obtain relation (10). The inverse matrix here exists because the inverted matrix is the irreducible sub-generator.

Equation (13) for still unknown vector $(\boldsymbol{\pi}_0, \boldsymbol{\pi}_1)$ is obtained from equilibrium Eq. (18) taking into account formula (10). Equation (14). is obtained from the normalization condition.

Theorem 2 is proved.

Remark 2. If one pursues the way of the proof of relation (10) with the matrix $\mathcal{F}$ defined as solution of Eq. (15) by means of the direct substitution of (10) to Eq. (16), then he/she is obliged to prove that Eq. (15) has a solution and this solution is such that vectors $\boldsymbol{\pi}_i$, $i \geq 0$, define the stationary distribution of the considered Markov chain. Proving the Theorem 2 via the derivation of the alternative system of Eqs. (18), we give the explicit form (11) of the matrix $\mathcal{F}$ obtained via the pure probabilistic considerations and do not have any need to prove its existence.

Remark 3. Analysis of level dependent case (when generator $\mathcal{Q}$ has general form (2)) can be found in [34]. The stationary probability vectors $\boldsymbol{\pi}_i$ are offered to be computed recursively as $\boldsymbol{\pi}_{i+1} = \boldsymbol{\pi}_i \mathcal{R}_i$ where the sequence of matrices $\mathcal{R}_i$ satisfies the derived set of equations. However, the question of the proof of such matrices existence is not discussed.

Having solved the problem of computation of the stationary probability vectors $\boldsymbol{\pi}_i$, $i \geq 0$, it is possible to compute various performance measures of the queueing system and solve optimization problems.

Several formulas for computation of performance measures are as follows.

The average number L of customers in the system is calculated as

$$L = \boldsymbol{\pi}_1 (I - \mathcal{F})^{-1} \mathbf{e}.$$

The probability P_{ignore} that an arbitrary disaster will be ignored (because the system is empty at its occurrence moment) is calculated as

$$P_{ignore} = \frac{1}{\phi} \boldsymbol{\pi}_0 (I_{K_1} \otimes Z_1) \mathbf{e}.$$

Let λ be the customers arrival rate in the original queueing system described by the QBD.

The probability P_{loss} that an arbitrary customer will be lost in this system (due to the disaster occurrence) is calculated as

$$P_{loss} = \frac{1}{\lambda} \sum_{i=1}^{\infty} i \boldsymbol{\pi}_i (I_K \otimes Z_1) \mathbf{e} = \frac{1}{\lambda} \boldsymbol{\pi}_1 (I - \mathcal{F})^{-1} (I_K \otimes Z_1) \mathbf{e}.$$

The average number L_{arr} of customers in this queueing system at a moment of a new customer arrival is calculated as

$$L_{arr} = \boldsymbol{\pi}_1 (I - \mathcal{F})^{-1} \frac{(\mathcal{Q}^+ \otimes I_z)}{\lambda} \mathbf{e}.$$

6 Conclusion

In this paper, the following problem is solved. Having an arbitrary queueing system, dynamics of which is described by a QBD, it is necessary to write down the infinitesimal generator of the multidimensional stochastic process describing the behavior of the similar queueing system which is influenced by the disasters occurring according to the MAP.

An arrival of a disaster causes immediate removal from the system of all customers waiting for processing in the system (their number is defined by the denumerable first component of the QBD) but does not change the state of the finite components of the QBD.

The generator of the multidimensional Markov chain describing the behavior of the queueing system with disasters does not have a triblock-diagonal structure and has an additional non-zero block column corresponding to state 0 of the denumerable component. Explicit expressions for all blocks of the generator of the multidimensional Markov chain via the blocks of the generator of QBD and the matrices defining transitions of the underlying process of the MAP describing arrival of disasters are presented.

In more detail, the case when the initial QBD is level independent (i.e., its generator is the quasi-Toeplitz matrix) is considered in more detail. The stationary distribution of the states of the multidimensional Markov chain describing the behavior of the queueing system with disaster is obtained in the matrix geometric form. Formulas for calculation of several performance measures of such a system are presented.

Various generalizations of the considered problem are possible. (i) Disaster occurrence leads to the temporal break in the work of the system during which the value of the denumerable component is frozen in state 0 or changes with another dynamics of the finite components than in the original QBD. This generalization can be done with the use of the results from [39]. (ii) Disaster occurrence leads to some jumps in the state of the finite component n_t or its sub-components. E.g., in application to the models of cognitive radio systems like considered in [28–30] the arrival of disaster can cause removal from the system of all or some part of the licensed and (or) cognitive users receiving service, transition of the process that controls admission of cognitive users, presence of effect of disasters arriving to the empty system. (iii) Disaster occurrence leads to the complete emptying the system, etc.

References

1. Bayer, N., Boxma, O.J.: Wiener-Hopf analysis of an $M/G/1$ queue with negative customers and of a related class of random walks. Queueing Syst. **23**(1), 301–316 (1996)
2. Chen, A., Renshaw, E.: The $M/M/1$ queue with mass exodus and mass arrivals when empty. J. Appl. Probab. **34**, 192–207 (1997)
3. Jain, G., Sigman, K.: A Pollaczeck-Khinchine formula for $M/G/1$ queues with disasters. J. Appl. Probab. **33**, 1191–1200 (1996)
4. Serfozo, R., Stidham, S.: Semi-stationary clearing processes. Stoch. Proc. Appl. **6**, 165–178 (1978)
5. Stidham, S.: Stochastic clearing systems. Stoch. Proc. Appl. **2**, 85–113 (1974)
6. Sigman, K.: Stationary Marked Point Processes: An Intuitive Approach. Chapman & Hall, London (1995)

7. Gelenbe, E.: Random neural networks with positive and negative signals and product form solution. Neural Comput. **1**, 502–510 (1989)
8. Gelenbe, E.: Product form networks with negative and positive customers. J. Appl. Probab. **28**, 655–663 (1991)
9. Gelenbe, E., Glynn, P., Sigman, K.: Queues with negative arrivals. J. Appl. Probab. **28**, 245–250 (1991)
10. Chakravarthy, S.R.: The batch Markovian arrival process: a review and future work. Adv. Probability Theory Stochastic Processes **1**, 21–49 (2001)
11. Chakravarthy, S.R.: Introduction to Matrix-Analytic Methods in Queues 1: Ana lytical and Simulation Approach Basics. ISTE Ltd, London and John Wiley and Sons, New York (2022)
12. Chakravarthy, S.R.: Introduction to Matrix-Analytic Methods in Queues 2: An alytical and Simulation Approach Queues and Simulation. ISTE Ltd, London and John Wiley and Sons, New York (2022)
13. The Theory of Queuing Systems with Correlated Flows. Springer, Cham (2020). https://doi.org/10.1007/978-3-030-32072-0_6
14. Lucantoni, D.M.: New results on the single server queue with a batch Markovian arrival process. Commun. Stat. Stoch. Models **7**(1), 1–46 (1991)
15. Dudin, A.N., Nishimura, S.: A $BMAP/SM/1$ queueing system with Markovian arrival input of disasters. J. Appl. Probab. **36**(3), 868–881 (1999)
16. Dudin, A.N., Karolik, A.V.: $BMAP/SM/1$ queue with Markovian input of disasters and non-instantaneous recovery. Perform. Eval. **45**(1), 19–32 (2001)
17. Dudin, A., Semenova, O.: A stable algorithm for stationary distribution calculation for a $BMAP/SM/1$ queueing system with Markovian arrival input of disasters. J. Appl. Probab. **41**(2), 547–556 (2004)
18. Cinlar, E.: Markov renewal theory. Adv. Appl. Probab. **1**(2), 123–187 (1969)
19. Cinlar, E.: Introduction to stochastic processes. Courier Corporation (2013)
20. van Dantzig, D.: Chaines de Markof dans les ensembles abstraits et applications aux processus avec regions absorbantes et au probleme des boucles, Ann. de l'Inst. H. Poincare 14(facs. 3), 145–199 (1955)
21. Chakravarthy, S.R.: A disaster queue with Markovian arrivals and impatient customers. Appl. Math. Comput. **214**(1), 48–59 (2009)
22. Chakravarthy, S.R.: A catastrophic queueing model with delayed action. Appl. Math. Model. **46**, 631–649 (2017)
23. Kumar, N., Gupta, U.C.: Analysis of $BMAP/MSP/1$ queue with MAP generated negative customers and disasters. Commun. Stat.-Theory Methods, 1–27 (2021)
24. Kuki, A., Bérczes, T., Sztrik, J.: Analyzing the effect of catastrophic breakdowns with retrial queues in a two-way communication system. In: Dudin, A., Nazarov, A., Moiseev, A. (eds) ITMM 2021. CCIS, vol. 1605, pp. 144–156. Springer, Cham (2022). https://doi.org/10.1007/978-3-031-09331-9_12
25. Neuts, M.F.: Structured Stochastic Matrices of $M/G/1$ Type and Their Applications. Marcel Dekker, New York (1989)
26. Neuts, M.F.: Matrix-Geometric Solutions in Stochastic Models. The Johns Hopkins University Press, Baltimore (1981)
27. Ye, Q.: High accuracy algorithms for solving nonlinear matrix equations in queueing models, pp. 401–415. Advances in Algorithmic Methods for Stochastic Models. Notable Publications, New Jersey (2000)
28. Sun, B., Lee, M.H., Dudin, S.A., Dudin, A.N.: Analysis of multiserver queueing system with opportunistic occupation and reservation of servers. Math. Probl. Eng. **2014**(178108), 1–13 (2014)

29. Sun, B., Lee, M.H., Dudin, S.A., Dudin, A.N.: $MAP + MAP/M_2/N/\infty$ Queueing System with Absolute Priority and Reservation of Servers. Math. Probl. Eng. **2014**(813150), 1–15 (2014)
30. Dudin, A., Dudin, S., Manzo, R., Raritá, L.: Analysis of multi-server priority queueing system with hysteresis strategy of server reservation and retrials. Mathematics 10(20)(3747) (2022)
31. Graham, A.: Kronecker Products and Matrix Calculus with Applications. Ellis Horwood, Cichester (1981)
32. Horn, R.A., Johnson, C.R.: Topics in Matrix Analysis. Cambridge University Press, Cambridge, UK (1991)
33. Zhang, H., Ding, F.: On the Kronecker products and their applications. J. Appl. Math. 2013(296185) (2013)
34. Baumann, H., Sandmann, W.: Steady state analysis of level dependent quasi-birth-and-death processes with catastrophes. Comput. Oper. Res. **39**(2), 413–423 (2012)
35. Kemeny, J.G., Snell, J.L., Knapp, A.W.: Denumerable Markov chains, vol. 40. Springer Science & Business Media (2012)
36. Zhao, Y.Q., Liu, D.: The censored Markov chain and the best augmentation. J. Appl. Probab. **33**(3), 623–629 (1996)
37. Zhao, Y.Q.: Censoring technique in studying block-structured Markov chains. Advances in algorithmic methods for stochastic models, pp. 417–433 (2000)
38. Klimenok, V.I., Dudin, A.N.: Multi-dimensional asymptotically quasi-Toeplitz Markov chains and their application in queueing theory. Queueing Syst. **54**, 245–259 (2006)
39. Dudin, A.N., Kim, C.S., Klimenok, V.I.: Markov chains with hybrid repeated rows - upper-Hessenberg quasi-Toeplitz structure of block transition probability matrix. J. Appl. Probab. **45**(1), 211–225 (2008)

Retrial Queue with Two-Way Communication and Collisions

Anatoly Nazarov and Olga Lizyura(✉)

Institute of Applied Mathematics and Computer Science,
National Research Tomsk State University,
36 Lenina ave., Tomsk 634050, Russia
oliztsu@mail.ru

Abstract. In this paper, we consider retrial queue with two types of calls: incoming and outgoing. If the server is busy upon arrival, the incoming call provokes collisions. If two incoming calls cause collision, both of them join the orbit. On the other hand, if an outgoing call is in service upon arrival, incoming call preempts outgoing. In this case, the outgoing call is lost.

Keywords: Retrial queue · Two-way communication · Incoming calls · Outgoing calls · Collisions

1 Introduction

Retrial queues with two-way communication arose as mathematical models of blended call centers. In such systems, the server fills the idle time serving outgoing calls [1,2]. This behavior allows to increase the efficiency of the system. Outgoing calls can model not only calls but any kind of additional work.

The retrial behavior is also important property of telephone communication systems. When the arriving call finds the server busy, it is not lost, but joins the virtual pool named orbit and retries to receive service after some random delay. The impact of retrials on modeling call centers is considered in papers [4,7].

In communication services, collisions is common problem. The calls, which try to receive service in the same time can interrupt each other. Random access protocols with interrupting both calls upon collision are considered in papers [3,5].

We propose investigation of retrial queue with two-way communication and collisions using asymptotic-diffusion method. Previously, we have provided similar study for the model with renewal input and without collisions [6].

This study was supported by the Tomsk State University Development Programme (Priority-2030).

A. Dudin et al. (Eds.): ITMM 2022, CCIS 1803, pp. 106–115, 2023.
https://doi.org/10.1007/978-3-031-32990-6_9

2 Model Description

We consider two-way communication retrial queue with collisions (Fig. 1).

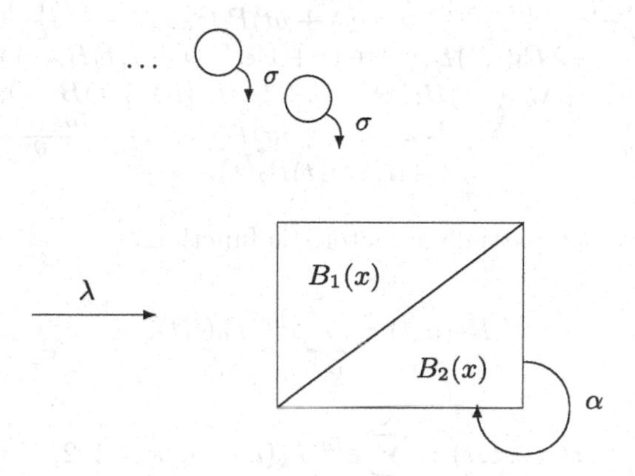

Fig. 1. Retrial queue with collisions and two-way communication

The input is Poisson process with rate λ. Service times of incoming calls follow general distribution with distribution function $B_1(x)$. If the server is busy serving incoming call upon arrival, both calls join the orbit and make a random delay before retrial. The rate of retrials for each call in the orbit is σ.

The server also makes outgoing calls itself when idle. The rate of outgoing calls is α. Service times of outgoing calls are defined by distribution function $B_2(x)$. In case when the outgoing call is in service upon arrival, fresh incoming call starts service and the outgoing call is lost.

We introduce the following notations:

- $i(t)$ is the number of calls in the orbit at instant t;
- $k(t)$ is the state of the server at instant t: 0, if the server is idle, 1, if the server is busy with incoming call, 2, if the server is busy with outgoing call;
- $z(t)$ is the residual service time at instant t.

The problem is to obtain the steady state distribution of the number of calls in the orbit.

3 Kolmogorov System

We denote probability distribution

$$P_k(i,t) = \mathbb{P}\{i(t) = i,\ k(t) = k,\ z(t) < z\}.$$

This distribution satisfy the Kolmogorov system of equations

$$\begin{aligned}
&\frac{\partial P_0(i,t)}{\partial t} = -(\lambda + i\sigma + \alpha)P_0(i,t) + \frac{\partial P_1(i,0,t)}{\partial z} + \\
&+\frac{\partial P_2(i,0,t)}{\partial z} + \lambda P_1(i-2,t) + (i-1)\sigma P_1(i-1,t), \\
&\frac{\partial P_1(i,z,t)}{\partial t} - \frac{\partial P_1(i,z,t)}{\partial z} = -(\lambda + i\sigma)P_1(i,z,t) - \frac{\partial P_1(i,0,t)}{\partial z} + \\
&+\lambda P_0(i,t)B_1(z) + (i+1)\sigma P_0(i+1,t)B_1(z) + \\
&+\lambda P_2(i,t)B_1(z) + (i+1)\sigma P_2(i+1,t)B_1(z), \\
&\frac{\partial P_2(i,z,t)}{\partial t} - \frac{\partial P_2(i,z,t)}{\partial z} = -(\lambda + i\sigma)P_2(i,z,t) - \frac{\partial P_2(i,0,t)}{\partial z} + \\
&+\alpha P_0(i,t)B_2(z).
\end{aligned} \quad (1)$$

Then we introduce partial characteristic functions

$$H_0(u,t) = \sum_{i=0}^{\infty} e^{jui} P_0(i,t),$$

$$H_k(u,z,t) = \sum_{i=0}^{\infty} e^{jui} P_k(i,z,t), \ k = 1,2,$$

and rewrite system (1) as follows:

$$\begin{aligned}
&\frac{\partial H_0(u,t)}{\partial t} = -(\lambda + \alpha)H_0(u,t) + j\sigma\frac{\partial H_0(u,t)}{\partial u} + \frac{\partial H_1(u,0,t)}{\partial z} + \frac{\partial H_2(u,0,t)}{\partial z} + \\
&+\lambda e^{2ju}H_1(u,t) - j\sigma e^{ju}\frac{\partial H_1(u,t)}{\partial u}, \\
&\frac{\partial H_1(u,z,t)}{\partial t} - \frac{\partial H_1(u,z,t)}{\partial z} = -\lambda H_1(u,z,t) + j\sigma\frac{\partial H_1(u,z,t)}{\partial u} - \frac{\partial H_1(u,0,t)}{\partial z} + \\
&+\lambda H_0(u,t)B_1(z) - j\sigma e^{-ju}\frac{\partial H_0(u,t)}{\partial u}B_1(z) + \\
&+\lambda H_2(u,t)B_1(z) - j\sigma e^{-ju}\frac{\partial H_2(u,t)}{\partial u}B_1(z), \\
&\frac{\partial H_2(u,z,t)}{\partial t} - \frac{\partial H_2(u,z,t)}{\partial z} = -\lambda H_2(u,z,t) + j\sigma\frac{\partial H_2(u,z,t)}{\partial u} - \frac{\partial H_2(u,0,t)}{\partial z} + \\
&+\alpha H_0(u,t)B_2(z).
\end{aligned} \quad (2)$$

We also derive additional equation by summing up system (2) and taking the limit by $z \to \infty$:

$$\begin{aligned}
\frac{\partial H(u,t)}{\partial t} = (e^{ju}-1)\Big\{ & j\sigma e^{-ju}\frac{\partial H_0(u,t)}{\partial u} + \lambda(e^{ju}+1)H_1(u,t) - \\
& -j\sigma\frac{\partial H_1(u,t)}{\partial u} + j\sigma e^{-ju}\frac{\partial H_2(u,t)}{\partial u}\Big\}.
\end{aligned} \quad (3)$$

4 Asymptotic-Diffusion Method

In system (2) and Eq. (3), we introduce the notations

$$\sigma = \varepsilon,\ u = w\varepsilon,\ \tau = t\varepsilon,\ H_0(u,t) = F_0(w,\tau,\varepsilon),\ H_k(u,z,t) = F_k(w,z,\tau,\varepsilon), k = 1,2$$

and derive system

$$
\begin{aligned}
&\varepsilon\frac{\partial F_0(w,\tau,\varepsilon)}{\partial \tau} = -(\lambda+\alpha)F_0(w,\tau,\varepsilon) + j\frac{\partial F_0(w,\tau,\varepsilon)}{\partial w} + \frac{\partial F_1(w,0,\tau,\varepsilon)}{\partial z}+\\
&\quad+\frac{\partial F_2(w,0,\tau,\varepsilon)}{\partial z} + \lambda e^{2jw\varepsilon}F_1(w,\tau,\varepsilon) - je^{jw\varepsilon}\frac{\partial F_1(w,\tau,\varepsilon)}{\partial w},\\
&\varepsilon\frac{\partial F_1(w,z,\tau,\varepsilon)}{\partial \tau} - \frac{\partial F_1(w,z,\tau,\varepsilon)}{\partial z} = -\lambda F_1(w,z,\tau,\varepsilon) + j\frac{\partial F_1(w,z,\tau,\varepsilon)}{\partial w} -\\
&\quad-\frac{\partial F_1(w,0,\tau,\varepsilon)}{\partial z} + \lambda F_0(w,\tau,\varepsilon)B_1(z) -\\
&-je^{-jw\varepsilon}\frac{\partial F_0(w,\tau,\varepsilon)}{\partial w}B_1(z) + \lambda F_2(w,\tau,\varepsilon)B_1(z) - je^{-jw\varepsilon}\frac{\partial F_2(w,\tau,\varepsilon)}{\partial w}B_1(z),\\
&\varepsilon\frac{\partial F_2(w,z,\tau,\varepsilon)}{\partial \tau} - \frac{\partial F_2(w,z,\tau,\varepsilon)}{\partial z} = -\lambda F_2(w,z,\tau,\varepsilon) + j\frac{\partial F_2(w,z,\tau,\varepsilon)}{\partial w} -\\
&\quad-\frac{\partial F_2(w,0,\tau,\varepsilon)}{\partial z} + \alpha F_0(w,\tau,\varepsilon)B_2(z),\\
&\varepsilon\frac{\partial F(w,\tau,\varepsilon)}{\partial \tau} = (e^{jw\varepsilon}-1)\Big\{je^{-jw\varepsilon}\frac{\partial F_0(w,\tau,\varepsilon)}{\partial w} + \lambda(e^{jw\varepsilon}+1)F_1(w,\tau,\varepsilon) -\\
&\quad-j\frac{\partial F_1(w,\tau,\varepsilon)}{\partial w} + je^{-jw\varepsilon}\frac{\partial F_2(w,\tau,\varepsilon)}{\partial w}\Big\}.
\end{aligned}
\tag{4}
$$

Solving system (4) in the limit by $\varepsilon \to 0$, we prove Theorem 1.

Theorem 1. *Asymptotic characteristic functions* $F_0(w,\tau) = \lim_{\varepsilon\to 0} F_0(w,\tau,\varepsilon)$, $F_k(w,z,\tau) = \lim_{\varepsilon\to 0} F_k(w,z,\tau,\varepsilon)$, $k=1,2$, *can be presented as*

$$F_0(w,\tau) = r_0 e^{jwx(\tau)},$$

$$F_k(w,z,\tau) = r_k(z)e^{jwx(\tau)},$$

where r_0 *and* $r_k = \lim_{z\to\infty} r_k(z)$ *is the distribution of the server state, which is given by*

$$
\begin{aligned}
r_0(x) &= \frac{1}{(2-B_1^*(\lambda+x))\left(1+\frac{\alpha}{\lambda+x}(1-B_2^*(\lambda+x))\right)},\\
r_1(x) &= \frac{1-B_1^*(\lambda+x)}{2-B_1^*(\lambda+x)},\\
r_2(x) &= \frac{\frac{\alpha}{\lambda+x}(1-B_2^*(\lambda+x))}{(2-B_1^*(\lambda+x))\left(1+\frac{\alpha}{\lambda+x}(1-B_2^*(\lambda+x))\right)}.
\end{aligned}
\tag{5}
$$

$x(\tau)$ *satisfies differential equation*

$$x'(\tau) = -x(\tau)r_0 + (2\lambda + x(\tau))r_1 - x(\tau)r_2. \tag{6}$$

Proof. We derive the system for functions $F_0(w,\tau)$ and $F_k(w,z,\tau)$, considering system (4) in the limit by $\varepsilon \to 0$. We present functions $F_0(w,\tau)$ and $F_k(w,z,\tau)$ in the form $F_0(w,\tau) = r_0e^{jwx(\tau)}$ and $F_k(w,z,\tau) = \lim_{\varepsilon\to 0} F_k(w,z,\tau,\varepsilon)$, $k=1,2$, respectively. After this, we obtain system

$$
\begin{aligned}
-(\lambda+\alpha+x)r_0 + r_1'(0) + r_2'(0) + (\lambda+x)r_1 = 0,\\
r_1'(z) - r_1'(0) - (\lambda+x)r_1(z) + (\lambda+x)r_0B_1(z) + (\lambda+x)r_2B_1(z) = 0,\\
r_2'(z) - r_2'(0) - (\lambda+x)r_2(z) + \alpha r_0 B_2(z) = 0,
\end{aligned}
\tag{7}
$$

and equation

$$x'(\tau) = -x(\tau)r_0 + (2\lambda + x(\tau))r_1 - x(\tau)r_2.$$

Solving the system with normalization condition $\sum_{k=0}^{2} r_k = 1$, we obtain formalae (5). The last equation coincides with (6). Theorem is proved.

Having (6), we denote function

$$a(x) = -xr_0(x) + (2\lambda + x)r_1(x) - xr_2(x), \tag{8}$$

which will serve as a drift coefficient of additional diffusion process approximating the number of calls in the orbit.

For the next step of analysis, we introduce the following notations in system (2) and equation (3):

$$H_0(u,t) = e^{j\frac{u}{\sigma}x(\sigma t)}H_0^{(2)}(u,t),\ H_k(u,z,t) = e^{j\frac{u}{\sigma}x(\sigma t)}H_k^{(2)}(u,z,t), k = 1,2,$$

which yields

$$\begin{gathered}
\frac{\partial H_0^{(2)}(u,t)}{\partial t} + jua(x)H_0^{(2)}(u,t) = -(\lambda+\alpha+x)H_0^{(2)}(u,t)+ \\
+j\sigma\frac{\partial H_0^{(2)}(u,t)}{\partial u} + \frac{\partial H_1^{(2)}(u,0,t)}{\partial z} + \frac{\partial H_2^{(2)}(u,0,t)}{\partial z} + \\
+e^{ju}(\lambda e^{ju}+x)H_1^{(2)}(u,t) - j\sigma e^{ju}\frac{\partial H_1^{(2)}(u,t)}{\partial u}, \\
\frac{\partial H_1^{(2)}(u,z,t)}{\partial t} + jua(x)H_1^{(2)}(u,z,t) = \frac{\partial H_1^{(2)}(u,z,t)}{\partial z} - (\lambda+x)H_1^{(2)}(u,z,t)+ \\
+j\sigma\frac{\partial H_1^{(2)}(u,z,t)}{\partial u} - \frac{\partial H_1^{(2)}(u,0,t)}{\partial z} + (\lambda+e^{-ju}x)H_0^{(2)}(u,t)B_1(z)- \\
-j\sigma e^{-ju}\frac{\partial H_0^{(2)}(u,t)}{\partial u}B_1(z) + (\lambda+e^{-ju}x)H_2^{(2)}(u,t)B_1(z)- \\
-j\sigma e^{-ju}\frac{\partial H_2^{(2)}(u,t)}{\partial u}B_1(z), \\
\frac{\partial H_2^{(2)}(u,z,t)}{\partial t} + jua(x)H_2^{(2)}(u,z,t) = \frac{\partial H_2^{(2)}(u,z,t)}{\partial z} - \\
-(\lambda+x)H_2^{(2)}(u,z,t) + j\sigma\frac{\partial H_2^{(2)}(u,z,t)}{\partial u} - \frac{\partial H_2^{(2)}(u,0,t)}{\partial z} + \alpha H_0^{(2)}(u,t)B_2(z), \\
\frac{\partial H^{(2)}(u,t)}{\partial t} + jua(x)H^{(2)}(u,t) = (e^{ju}-1)\Big\{j\sigma e^{-ju}\frac{\partial H_0^{(2)}(u,t)}{\partial u} - \\
-xe^{-ju}H_0^{(2)}(u,t) + (\lambda(e^{ju}+1)+x)H_1^{(2)}(u,t) - j\sigma\frac{\partial H_1^{(2)}(u,t)}{\partial u} + \\
+j\sigma e^{-ju}\frac{\partial H_2(u,t)}{\partial u} - xe^{-ju}H_2^{(2)}(u,t)\Big\}.
\end{gathered}$$

In the obtained system, we present σ as $\sigma = \varepsilon^2$ and denote

$$u = w\varepsilon,\ \tau = t\varepsilon^2,\ H_0^{(2)}(u,t) = F_0^{(2)}(w,\tau,\varepsilon),$$

$$H_k^{(2)}(u,z,t) = F_k^{(2)}(w,z,\tau,\varepsilon),\ k = 1,2.$$

Then we obtain system

$$
\begin{aligned}
&\varepsilon^2\frac{\partial F_0^{(2)}(w,\tau,\varepsilon)}{\partial\tau}+jw\varepsilon a(x)F_0^{(2)}(w,\tau,\varepsilon)=-(\lambda+\alpha+x)F_0^{(2)}(w,\tau,\varepsilon)+\\
&\quad+j\varepsilon\frac{\partial F_0^{(2)}(w,\tau,\varepsilon)}{\partial w}+\frac{\partial F_1^{(2)}(w,0,\tau,\varepsilon)}{\partial z}+\frac{\partial F_2^{(2)}(w,0,\tau,\varepsilon)}{\partial z}+\\
&\quad+e^{jw\varepsilon}(\lambda e^{jw\varepsilon}+x)F_1^{(2)}(w,\tau,\varepsilon)-j\varepsilon e^{jw\varepsilon}\frac{\partial F_1^{(2)}(w,\tau,\varepsilon)}{\partial w},\\
&\varepsilon^2\frac{\partial F_1^{(2)}(w,z,\tau,\varepsilon)}{\partial\tau}+jw\varepsilon a(x)F_1^{(2)}(w,z,\tau,\varepsilon)=\frac{\partial F_1^{(2)}(w,z,\tau,\varepsilon)}{\partial z}-\\
&\quad-(\lambda+x)F_1^{(2)}(w,z,\tau,\varepsilon)+j\varepsilon\frac{\partial F_1^{(2)}(w,z,\tau,\varepsilon)}{\partial w}-\frac{\partial F_1^{(2)}(w,0,\tau,\varepsilon)}{\partial z}+\\
&\quad+(\lambda+e^{-jw\varepsilon}x)F_0^{(2)}(w,\tau,\varepsilon)B_1(z)-j\varepsilon e^{-jw\varepsilon}\frac{\partial F_0^{(2)}(w,\tau,\varepsilon)}{\partial w}B_1(z)+\\
&\quad+(\lambda+e^{-jw\varepsilon}x)F_2^{(2)}(w,\tau,\varepsilon)B_1(z)-j\varepsilon e^{-jw\varepsilon}\frac{\partial F_2^{(2)}(w,\tau,\varepsilon)}{\partial w}B_1(z),\\
&\varepsilon^2\frac{\partial F_2^{(2)}(w,z,\tau,\varepsilon)}{\partial\tau}+jw\varepsilon a(x)F_2^{(2)}(w,z,\tau,\varepsilon)=\frac{\partial F_2^{(2)}(w,z,\tau,\varepsilon)}{\partial z}-\\
&\quad-(\lambda+x)F_2^{(2)}(w,z,\tau,\varepsilon)+j\varepsilon\frac{\partial F_2^{(2)}(w,z,\tau,\varepsilon)}{\partial w}-\frac{\partial F_2^{(2)}(w,0,\tau,\varepsilon)}{\partial z}+\\
&\quad+\alpha F_0^{(2)}(w,\tau,\varepsilon)B_2(z),
\end{aligned}\tag{9}
$$

and equation

$$
\begin{aligned}
&\varepsilon^2\frac{\partial F^{(2)}(w,\tau,\varepsilon)}{\partial\tau}+jw\varepsilon a(x)F^{(2)}(w,\tau,\varepsilon)=(e^{jw\varepsilon}-1)\Big\{j\varepsilon e^{-jw\varepsilon}\frac{\partial F_0^{(2)}(w,\tau,\varepsilon)}{\partial w}-\\
&-xe^{-jw\varepsilon}F_0^{(2)}(w,\tau,\varepsilon)+(\lambda(e^{jw\varepsilon}+1)+x)F_1^{(2)}(w,\tau,\varepsilon)-j\varepsilon\frac{\partial F_1^{(2)}(w,\tau,\varepsilon)}{\partial w}+\\
&\quad+j\varepsilon e^{-jw\varepsilon}\frac{\partial F_2^{(2)}(w,\tau,\varepsilon)}{\partial w}-xe^{-jw\varepsilon}F_2^{(2)}(w,\tau,\varepsilon)\Big\}.
\end{aligned}\tag{10}
$$

Theorem 2. *Asymptotic characteristic functions* $F_0^{(2)}(w,\tau,\varepsilon)$ *and* $F_k^{(2)}(w,z,\tau,\varepsilon)$ *can be presented as*

$$F_0^{(2)}(w,\tau,\varepsilon)=\Phi(w,\tau)\{r_0+jw\varepsilon f_0\}+O(\varepsilon^2),\tag{11}$$

$$F_k(w,z,\tau,\varepsilon)=\Phi(w,\tau)\{r_k(z)+jw\varepsilon f_k(z)\}+O(\varepsilon^2),k=1,2,\tag{12}$$

where

$$f_0=Cr_0+g_0-\frac{\partial\Phi(w,\tau)/\partial w}{w\Phi(w,\tau)}\varphi_0,\tag{13}$$

$$f_k(z)=Cr_k(z)+g_k(z)-\frac{\partial\Phi(w,\tau)/\partial w}{w\Phi(w,\tau)}\varphi_k(z),k=1,2.\tag{14}$$

Here r_k are determined in Theorem 1, g_0 and $g_k = \lim_{z\to\infty} g_k(z)$ are additional coefficients that are determined by

$$\begin{gathered}
g_0(x) = \\
= \frac{a(x)}{\lambda+x} \frac{r_2(x)+\alpha r_0(x)B_2^{*\prime}(\lambda+x))(1-B_2^*(\lambda+x))}{(2-B_1^*(\lambda+x))\left(1+\frac{\alpha}{\lambda+x}(1-B_2^*(\lambda+x))\right)} + \\
+ \frac{a(x)}{\lambda+x} \frac{r_1(x)}{(2-B_1^*(\lambda+x))\left(1+\frac{\alpha}{\lambda+x}(1-B_2^*(\lambda+x))\right)} + \\
+ \left(a(x)B_1^{*\prime}(\lambda+x) + \tfrac{x}{\lambda+x}(1-B_1^*(\lambda+x))\right) \times \\
\times \frac{r_0(x)+r_2(x)}{(2-B_1^*(\lambda+x))\left(1+\frac{\alpha}{\lambda+x}(1-B_2^*(\lambda+x))\right)} + \\
+ \frac{a(x)}{\lambda+x} \frac{r_2(x)+\alpha r_0(x)B_2^{*\prime}(\lambda+x))}{(2-B_1^*(\lambda+x))\left(1+\frac{\alpha}{\lambda+x}(1-B_2^*(\lambda+x))\right)},
\end{gathered} \tag{15}$$

$$\begin{gathered}
g_2(x) = \tfrac{\alpha}{\lambda+x} g_0(x)(1-B_2^*(\lambda+x)) - \tfrac{a(x)}{\lambda+x}\left(r_2(x) + \alpha r_0(x)B_2^{*\prime}(\lambda+x)\right), \\
g_1(x) = (g_0(x)+g_2(x))(1-B_1^*(\lambda+x)) - \\
- \tfrac{a(x)}{\lambda+x} r_1(x) - (r_0(x)+r_2(x))\left(a(x)B_1^{*\prime}(\lambda+x) - \tfrac{x}{\lambda+x}(1-B_1^*(\lambda+x))\right),
\end{gathered}$$

where $B_k^*(s)$ is the LST of $B_k(x)$. φ_0 and $\varphi_k(z)$ are determined as

$$\varphi_0(x) = r_0'(x), \ \varphi_k(z) = \frac{\partial r_k(z,x)}{\partial x}.$$

Asymptotic characteristic function $\Phi(w,\tau)$ satisfy the equation

$$\frac{\partial \Phi(w,\tau)}{\partial \tau} = a'(x)w\frac{\partial \Phi(w,\tau)}{\partial w} + b(x)\frac{(jw)^2}{2}\Phi(w,\tau), \tag{16}$$

where $a(x)$ is defined by (8) and $b(x)$ is as follows:

$$\begin{gathered}
b(x) = a(x) + 2[-xg_0(x) + (2\lambda+x)g_1(x) - \\
-xg_2(x) + xr_0(x) + \lambda r_1(x) + xr_2(x)].
\end{gathered} \tag{17}$$

Proof. The process of proving Theorem 2 is given in paper [6]. Here we just highlight the key steps of analysis.

First, we use decompositions to the Tailor series up to ε^2 in system (9). Substituting (11) and (12) into equations, we consider the system in the limit by $\varepsilon \to 0$:

$$\begin{gathered}
a(x)r_0 = -(\lambda+\alpha+x)f_0 + r_0\tfrac{\partial\Phi(w,\tau)/\partial w}{w\Phi(w,\tau)} + f_1'(0) + f_2'(0) + \\
+(2\lambda+x)r_1 + (\lambda+x)f_1 - r_1\tfrac{\partial\Phi(w,\tau)/\partial w}{w\Phi(w,\tau)}, \\
a(x)r_1(z) = f_1'(z) - f_1(z)(\lambda+x) + r_1(z)\tfrac{\partial\Phi(w,\tau)/\partial w}{w\Phi(w,\tau)} - f_1'(0) - xr_0B_1(z) + \\
+(\lambda+x)f_0B_1(z) - r_0\tfrac{\partial\Phi(w,\tau)/\partial w}{w\Phi(w,\tau)}B_1(z) - xr_2B_1(z) + (\lambda+x)f_2B_1(z) - \\
-r_2\tfrac{\partial\Phi(w,\tau)/\partial w}{w\Phi(w,\tau)}B_1(z), \\
a(x)r_2(z) = f_2'(z) - (\lambda+x)f_2(z) + r_2(z)\tfrac{\partial\Phi(w,\tau)/\partial w}{w\Phi(w,\tau)} - \\
-f_2'(0) + \alpha f_0B_2(z).
\end{gathered}$$

After that, we use decompositions (13) and (14), which yields two systems of equations:

$$\begin{aligned} &-(\lambda+\alpha+x)g_0+g_1'(0)+g_2'(0)+(\lambda+x)g_1= \\ &=a(x)r_0-(2\lambda+x)r_1, \\ &g_1'(z)-g_1'(0)-(\lambda+x)g_1(z)+(\lambda+x)g_0B_1(z)+(\lambda+x)g_2B_1(z)= \\ &=a(x)r_1(z)+xr_0B_1(z)+xr_2B_1(z), \\ &g_2'(z)-g_2'(0)-(\lambda+x)g_2(z)+\alpha g_0B_2(z)= \\ &=a(x)r_2(z), \end{aligned} \tag{18}$$

and

$$\begin{aligned} &-(\lambda+\alpha+x)\varphi_0+\varphi_1'(0)+\varphi_2'(0)+(\lambda+x)\varphi_1= \\ &=r_0-r_1, \\ &\varphi_1'(z)-\varphi_1'(0)-(\lambda+x)\varphi_1(z)+(\lambda+x)\varphi_0B_1(z)+(\lambda+x)\varphi_2B_1(z)= \\ &=r_1(z)-r_0B_1(z)-r_2B_1(z), \\ &\varphi_2'(z)-\varphi_2'(0)-(\lambda+x)\varphi_2(z)+\alpha\varphi_0B_2(z)=r_2(z). \end{aligned} \tag{19}$$

As we can see, system (19) can be obtained by differentiating of system (7) by x. Thus, we conclude that

$$\varphi_0(x)=r_0'(x),\ \varphi_k(z)=\frac{\partial r_k(z,x)}{\partial x}.$$

Solving system (18), we obtain (15).

Finally, we consider the same transformations in Eq. (10) making decompositions in Taylor series up to ε^3, which yields (16). Theorem is proved.

The final step of asymptotic-diffusion method is building an additional diffusion process approximating the number of calls in the orbit of retrial queue. We apply the inverse Fourier transform to the (16) and obtain the Fokker-Planck equation for the probability density of some diffusion process $y(\tau)$. The stochastic differential equation of the process is given by

$$dy(\tau)=a'(x)yd\tau+\sqrt{b(x)}dw(\tau), \tag{20}$$

where $w(\tau)$ is a Wiener process.

Then we introduce additional process $z(\tau)$

$$z(\tau)=x(\tau)+\varepsilon y(\tau), \tag{21}$$

where $x(\tau)$ is the solution of (6) and $\varepsilon=\sqrt{\sigma}$. The stochastic differential equation for process $z(\tau)$ has the following form:

$$dz(\tau)=d(x(\tau)+\varepsilon y(\tau))=(a(x)+\varepsilon ya'(x))d\tau+\varepsilon\sqrt{b(x)}dw(\tau). \tag{22}$$

We use decompositions

$$a(z)=a(x+\varepsilon y)=a(x)+\varepsilon ya'(x)+o(\varepsilon^2),$$

$$\varepsilon\sqrt{b(z)} = \varepsilon\sqrt{b(x+\varepsilon y)} = \varepsilon\sqrt{b(x)+o(\varepsilon)} = \varepsilon\sqrt{b(x)} + o(\varepsilon^2),$$

to transform equation (22) in the following form up to $o(\varepsilon^2)$:

$$dz(\tau) = a(z)d\tau + \sqrt{\sigma b(z)}dw(\tau). \tag{23}$$

Stationary probability density $S(z)$ of $z(\tau)$ is the solution of Fokker-Planck equation

$$-\frac{\partial}{\partial z}\{a(z)S(z)\} + \frac{1}{2}\frac{\partial^2}{\partial z^2}\{\sigma b(z)S(z)\} = 0. \tag{24}$$

Solving the equation, we obtain

$$S(z) = \frac{C}{b(z)}\exp\left\{\frac{2}{\sigma}\int\limits_0^z \frac{a(x)}{b(x)}dx\right\}, \tag{25}$$

where C is an integration constant.

Discrete approximate distribution of the number of calls in the orbit is determined by formula

$$P(i) = \frac{S(\sigma i)}{\sum\limits_{n=0}^{\infty} S(\sigma n)}. \tag{26}$$

5 Numerical Example

Consider the retrial queue with the following parameters. The rate of arrivals is $\lambda = 0.2$. Service times are defined by gamma distributions with shape and rate parameter equal to 1 for incoming calls and equal to 0.5 for outgoing calls. Mean service time is equal to 1 for both types of calls. The rate of outgoing calls is $\alpha = 1$ and retrial rate is $\sigma = 0.01$.

For parameters defined above, we show the approximation (26) in Fig. 2.

6 Conclusion

We have considered retrial queue with two-way communication and collisions. Using asymptotic-diffusion method, we have derived the approximation for probability distribution of the number of calls in the orbit.

References

1. Artalejo, J.R., Phung-Duc, T.: Single server retrial queues with two way communication. Appl. Math. Model. **37**(4), 1811–1822 (2013)
2. Artalejo, J.R., Phung-Duc, T.: Markovian retrial queues with two way communication. J. Ind. Manag. Optim. **8**(4), 781–806 (2012)
3. Choi, B.D., Shin, Y.W., Ahn, W.C.: Retrial queues with collision arising from unslotted CSMA/CD protocol. Queue. Syst. **11**, 335–356 (1992)

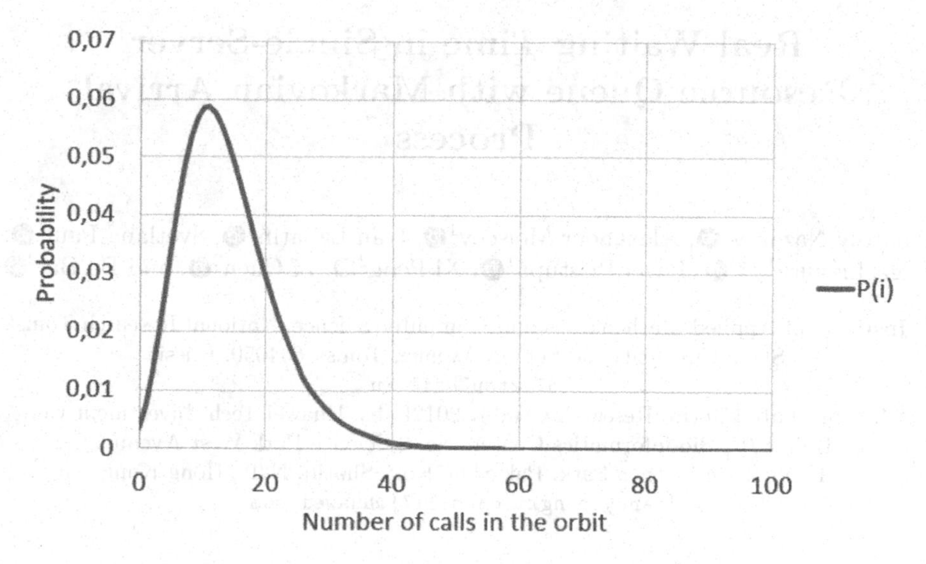

Fig. 2. Graph of the approximation $P(i)$ of the probability distribution of the number of calls in the orbit

4. Hu, K., Allon, G., Bassamboo, A.: Understanding customer retrials in call centers: preferences for service quality and service speed. Manuf. Serv. Oper. Manag. **24**(2), 1002–1020 (2022)
5. Lakaour, L., Aïssani, D., Adel-Aissanou, K., Barkaoui, K.: M/M/1 retrial queue with collisions and transmission errors. Methodol. Comput. Appl. Probab. **21**, 1395–1406 (2019)
6. Nazarov, A., Phung-Duc, T., Paul, S., Lizyura, O.: Diffusion limit for single-server retrial queues with renewal input and outgoing calls. Mathematics **10**(6), 948 (2022)
7. Pustova, S.V.: Investigation of call centers as retrial queuing systems. Cybern. Syst. Anal. **46**, 494–499 (2010)

Real Waiting Time in Single-Server Resource Queue with Markovian Arrival Process

Anatoly Nazarov[1], Alexander Moiseev[1], Ivan Lapatin[1], Svetlana Paul[1], Olga Lizyura[1(✉)], Pavel Pristupa[1], Xi Peng[2], Li Chen[2], and Bo Bai[2]

[1] Institute of Applied Mathematics and Computer Science, National Research Tomsk State University, 36 Lenina Avenue, Tomsk 634050, Russia
oliztsu@mail.ru

[2] Theory Lab, Central Research Center, 2012Labs, Huawei Tech. Investment Co., Ltd., 8/F, Bio-informatics Center, No. 2 Science Park West Avenue, Hong Kong Science Park, Pak Shek Kok, Shatin, N.T., Hong Kong
{pancy.pengxi,chen.li7}@huawei.com

Abstract. We consider single-server queue, which models a telecommunication node. Packets arrive in the system according to the Markovian arrival process (MAP). The packets bring some amount of information, which is stored in a continuous-type unlimited buffer before it will be transmitted out of the system. We consider the system with fluid manner of service, which means that the server takes the information from the buffer with constant speed. Our aim is to derive the probability distribution of the real waiting time in the system.

Keywords: Queueing system · Markovian arrival process · Resource flow · Real waiting time

1 Introduction

The study of telecommunication systems using stochastic modeling methods is often associated with obtaining the probability distribution of the number of customers in the system. An equally important problem is the study of the waiting time for a customer in the system. Most of the research on this topic deals with the so-called virtual waiting time. The distribution of such value is based on the assumption that a customer can arrive to the system at any instant of time, which is valid for systems with Poisson input flows.

However, in the case of correlated incoming flows, we faced the fact that the distribution of the virtual waiting time does not coincide with the distribution of the waiting time obtained during the statistical analysis of the corresponding real system. This is due to the fact that in flows, the structure of which is different from the Poisson flow, there are periods of time in which the request cannot be received due to the correlation between interarrival times in the flow.

A. Dudin et al. (Eds.): ITMM 2022, CCIS 1803, pp. 116–125, 2023.
https://doi.org/10.1007/978-3-031-32990-6_10

In paper [2], Cox and Isham studied the behavior of the virtual waiting time in the system with Poisson input and arbitrary distribution of the service times. In this case, there is no problem with real waiting time since it coincide with the virtual one. They also studied some generalizations of the model. Rubin [6] considered the model of communication network in form of queueing system with several Poisson inputs. He took into account the distribution of the waiting time in the system. As we can see, if the input is Poisson, no matter if it is single or multiple, the study of the virtual waiting time is enough to estimate the delay of packets in the system.

Let us consider the studies of queueing systems with non-Poisson inputs. Paper [1] is dedicated to the analysis of the multiserver queueing system with infinite buffer, MAP input and phase-type distribution of the service times. The authors consider the mean waiting time in the system and present the analysis of the relation of the waiting time to the system load and the correlation coefficient of the flow. In paper [5], the author proposes the model of queue with phase-type renewal input and arbitrary distribution of service times. The study is focused on the virtual waiting time in such system, which is also quite complicated problem. However, the author does not consider the difference between virtual and real waiting time. The similar results are described in [3], where the author present the modification of Pollaczek-Khinchin formula for the virtual waiting time in the model with BMAP input and arbitrary distribution of the service times.

The technique of obtaining the distribution of the real waiting time are described in papers [4,7]. The authors use the joint distribution of the virtual waiting time and the state of correlated input process to derive the distribution of the real waiting time.

In this paper, we have considered the model with MAP flow of packets, which fed to the system with infinite buffer and arbitrary distribution of the packet length. The service is assumed to be the transmission of the packet through the communication node with constant rate. We consider the joint process of the input flow state and the amount of information in the system to derive the virtual waiting time and, consequently, the distribution of the real waiting time. Due to the computational complexity of the numerical algorithm, we also propose the approximation of cumulative distribution function of the real waiting time. Numerical examples shows the high accuracy of the approximation.

2 Model Description

In queueing theory, the generally accepted term is virtual waiting time, which means the waiting time $W(t)$ of the request arrived at instant t. Here, the specified condition for the arrival of a request at the time t, at which it most likely will not arrive, is of fundamental importance.

We will call the real waiting time of the request the length of interval from the real moment of arrival to the moment of the beginning of its service.

In case of Poisson input flow, the virtual and real waiting times have the same probability distribution. For other classes of input flows, the probability distributions of virtual and real waiting times can be significantly different.

We consider the queuing system with infinite buffer. The input is MAP flow of packets, which is defined as follows. We have homogeneous Markov process $k(t)$ with finite set of states $\{1, 2, \ldots, K\}$ and continuous time t. Thus, we set the conditional rates λ_k, $k, n = 1, 2, \ldots, K$. We also set probabilities $p_{kn}(0)$ and $p_{kn}(1)$ of that at the instant of transmission of Markov chain from the state k to the state n there are zero or one packet arrives in the flow, respectively. We denote the matrices $\mathbf{D}_0 = [d_{kn}(0)]$ and $\mathbf{D}_1 = [d_{kn}(1)]$, the elements of which are defined as follows

$$d_{kn}(0) = \begin{cases} -\lambda_k, & k = n, \\ \lambda_k p_{kn}(0), & k \neq n, \end{cases}$$

$$d_{kn}(1) = \lambda_k p_{kn}(1), \ k, n = 1, 2, \ldots, K.$$

The distribution of the amount of information in one packet is arbitrary and defined by cumulative distribution function $B(x)$.

We assume that the service in the system has the meaning of transmission the information through the communication node. Therefore, we denote v as the constant rate of transmission.

In order to find the probability distribution of the real waiting time in the system with incoming MAP and arbitrary distribution of the packet lengths, we need to obtain the vector characteristic function of the virtual waiting time. In turn, the virtual waiting time is related to the amount of information in the buffer by distribution. Therefore, we start by obtaining the Kolmogorov system of equations for this random process.

3 Amount of Information in the Buffer

Consider two-dimensional process $\{k(t), S(t)\}$, where $k(t)$ is the state of underlying Markov chain of MAP and $S(t)$ is the amount of information in the buffer at instant t. Denoting the probability distribution of two-dimensional process $\{k(t), S(t)\}$ as $P_k(s, t)$, we derive the system of equalities

$$P_k(s - v\Delta t, t + \Delta t) = P_k(s, t)(1 - \lambda_k \Delta t) + \sum_{n=1}^{K} P_n(s, t)\lambda_n p_{nk}(0)\Delta t$$

$$+ \sum_{n=1}^{K} \int_0^s P_n(s - x, t)\lambda_n p_{nk}(1)\Delta t + o(\Delta t), \ k = 1, 2, \ldots, K.$$

We divide the equations by Δt and take the limit by $\Delta t \to 0$:

$$\frac{\partial P_k(s,t)}{\partial t} - v\frac{\partial P_k(s,t)}{\partial s} = -\lambda_k P_k(s,t) + \sum_{n=1}^{K} P_n(s,t)\lambda_n p_{nk}(0)$$

$$+ \sum_{n=1}^{K} \int_0^s P_n(s - x, t)\lambda_n p_{nk}(1) dB(x), \ k = 1, 2, \ldots, K. \quad (1)$$

After that, we transform system of Eq. (1) to the system for the steady state distribution of two-dimensional process $\{k(t), S(t)\}$

$$vP_k'(s) - \lambda_k P_k(s) + \sum_{n=1}^{K} P_n(s)\lambda_n p_{nk}(0)$$

$$+\sum_{n=1}^{K}\int_0^s P_n(s-x)\lambda_n p_{nk}(1)dB(x) = 0,\ k = 1, 2, \ldots, K. \tag{2}$$

We denote $\mathbf{P}(s)$ as a vector of probabilities $P_k(s)$ and write the system (2) in the matrix form

$$v\mathbf{P}'(s) + \mathbf{P}(s)\mathbf{D}_0 + \int_0^s \mathbf{P}(s-x)dB(x)\mathbf{D}_1 = \mathbf{0}. \tag{3}$$

Using Laplace-Stieltjes transform (LST) in (3), we obtain

$$\mathbf{S}(\alpha)\{v\alpha\mathbf{I} + \mathbf{D}_0 + \beta(\alpha)\mathbf{D}_1\} = v\alpha\mathbf{V}, \tag{4}$$

where

$$\mathbf{S}(\alpha) = \int_0^\infty e^{-\alpha s}d\mathbf{P}(s),\ \beta(\alpha) = \int_0^\infty e^{-\alpha s}dB(x),$$

$\mathbf{I}$ is the identity matrix of corresponding dimension, $\mathbf{V}$ is the vector of probabilities

$$\int_0^{+0} e^{-\alpha s}d\mathbf{P}(s) = P_k(+0) = V_k.$$

4 Virtual Waiting Time

Since the virtual waiting time $W(t)$ and the amount of information in the buffer $S(t)$ are related by formula

$$W(t) = \frac{S(t)}{v}, \tag{5}$$

we denote vector LST of the virtual waiting time as

$$\mathbf{W}(\alpha) = \mathbf{S}\left(\frac{\alpha}{v}\right) = \mathbb{E}\{e^{-\alpha W(t)}I(k(t) = k)\}, \tag{6}$$

where $I(k(t) = k)$ is the indicator function. Hence, function $\mathbf{W}(\alpha)$ is the solution of matrix equation

$$\mathbf{W}(\alpha)\left\{\alpha\mathbf{I} + \mathbf{D}_0 + \beta\left(\frac{\alpha}{v}\right)\mathbf{D}_1\right\} = \alpha\mathbf{V}. \tag{7}$$

Further, we seek vector $\mathbf{V}$ denoting

$$\mathbf{G}(\alpha) = \alpha\mathbf{I} + \mathbf{D}_0 + \beta\left(\frac{\alpha}{v}\right),$$

and transforming Eq. (7) as follows:

$$\mathbf{W}(\alpha)\mathbf{G}(\alpha) = \alpha\mathbf{V}. \tag{8}$$

We will seek the roots α_k, $k = 1, 2, \ldots, K-1$, of the equation $\det(\mathbf{G}(\alpha)) = 0$ and non-trivial solutions of the system $\mathbf{G}(\alpha_k)\mathbf{x}_k = 0$. Obviously, the root $\alpha_0 = 0$ of the equation $\det(\mathbf{G}(\alpha)) = 0$ always exists. Thus, the number of non-trivial solutions of the system $\mathbf{G}(\alpha_k)\mathbf{x}_k = 0$ is equal to $(K-1)$.

Substituting $\alpha = \alpha_k$ into Eq. (8) and multiplying by vector $\mathbf{x}_k$ for each $k = 1, 2, \ldots, K-1$ and taking into account that $\mathbf{G}(\alpha_k)\mathbf{x}_k = 0$, we obtain equalities

$$\mathbf{V}\mathbf{x}_k = 0, k = 1, 2, \ldots, K-1. \tag{9}$$

After that, we consider Eq. (7) and multiply it by vector of ones $\mathbf{e}$:

$$\mathbf{W}(\alpha)\left\{\alpha\mathbf{I} + \mathbf{D}_0 + \beta\left(\frac{\alpha}{v}\right)\mathbf{D}_1\right\}\mathbf{e} = \alpha\mathbf{V}\mathbf{e}.$$

Differnetiating the last equation by α, we obtain

$$\mathbf{W}'(\alpha)\left\{\alpha\mathbf{I} + \mathbf{D}_0 + \beta\left(\frac{\alpha}{v}\right)\mathbf{D}_1\right\}\mathbf{e} + \mathbf{W}(\alpha)\left\{\mathbf{I} + \frac{1}{v}\beta'\left(\frac{\alpha}{v}\right)\mathbf{D}_1\right\}\mathbf{e} = \alpha\mathbf{V}\mathbf{e}. \tag{10}$$

Setting $\alpha = 0$, we can write

$$\mathbf{V}\mathbf{e} = \mathbf{r}\left\{\mathbf{I} + \frac{b_1}{v}\mathbf{D}_1\right\}\mathbf{e} = 1 - \frac{b_1}{v}\mathbf{r}\mathbf{D}_1\mathbf{e},$$

where $\mathbf{r}$ is the vector of steady state distribution of the MAP states, b_1 is the mean packet length. Finally, considering the last equation together with (9), we derive the system of linear algebraic equations for the elements of vector $\mathbf{V}$

$$\mathbf{V}\mathbf{x}_k = 0, k = 1, 2, \ldots, K-1,$$

$$\mathbf{V}\mathbf{e} = 1 - \frac{b_1}{v}\mathbf{r}\mathbf{D}_1\mathbf{e}. \tag{11}$$

From Eq. (7), we obtain the scalar LST of virtual waiting time

$$\mathbb{E}e^{-\alpha W(t)} = \mathbf{W}(\alpha)\mathbf{e} = \alpha\mathbf{V}\left\{\alpha\mathbf{I} + \mathbf{D}_0 + \beta\left(\frac{\alpha}{v}\right)\mathbf{D}_1\right\}^{-1}\mathbf{e}. \tag{12}$$

5 Real Waiting Time

Having the scalar LST for the virtual waiting time, we can derive the LST for the real waiting time using the formula

$$RWT(\alpha) = \mathbf{W}(\alpha)\frac{\mathbf{D}_1\mathbf{e}}{\mathbf{r}\mathbf{D}_1\mathbf{e}}. \tag{13}$$

Formula (13) is derived in [7]. Denoting distribution function $F(x)$ of the real waiting time, we can write the following relations

$$L = F(+0) = \lim_{x\to 0} F(x) = \lim_{\alpha\to\infty} \alpha \int_0^\infty e^{-\alpha x}F(x)dx = \mathbf{V}\frac{\mathbf{D}_1\mathbf{e}}{\mathbf{r}\mathbf{D}_1\mathbf{e}}. \tag{14}$$

Here L is the value of distribution function $F(x)$ in zero point. We also derive the relation between vector $\mathbf{V}$ and the system load ρ

$$\mathbf{V}\mathbf{e} = 1 - \frac{b_1}{v}\mathbf{r}\mathbf{D}_1\mathbf{e} = 1 - \rho,\ \rho = 1 - \mathbf{V}\mathbf{e}. \tag{15}$$

We transform the LST of the real waiting time to the characteristic function as follows:

$$H(u) = \mathbf{W}(-ju)\frac{\mathbf{D}_1\mathbf{e}}{\mathbf{r}\mathbf{D}_1\mathbf{e}}, \tag{16}$$

where $j = \sqrt{-1}$. Having the characteristic function, we obtain the cumulative distribution function of the real waiting time using the inverse Fourier transform

$$F(x) = \frac{1}{2\pi}\int_{-\infty}^{\infty} \frac{1-e^{-jux}}{ju}H(u)du. \tag{17}$$

Since the distribution function $F(x)$ has the gap at zero point, the computation of the inverse Fourier transform can be difficult. Therefore, we present characteristic function in the following form:

$$h(u) = \frac{H(u) - L}{1 - L},$$

and apply the inverse Fourier transform to function $h(u)$

$$f(x) = \frac{1}{2\pi}\int_{-\infty}^{\infty} \frac{1-e^{-jux}}{ju}h(u)du.$$

Then the distribution function of the real waiting time in the system is given by

$$F(x) = L + (1-L)f(x). \tag{18}$$

We also propose an arrpoximation of the distribution function $F(x)$ in the form of

$$\tilde{F}(x) = 1 - (1-L)e^{-\gamma x}. \tag{19}$$

Here L is the value of the function in zero point and it is given by the formula (14). In order to evaluate parameter γ, we use the method of moments.

In order to obtain the first raw vector moment, we consider equations (7) and (10) differentiating them by α:

$$\mathbf{W}'(\alpha)\left\{\alpha\mathbf{I}+\mathbf{D}_0+\beta\left(\frac{\alpha}{v}\right)\mathbf{D}_1\right\}+\mathbf{W}(\alpha)\left\{\mathbf{I}+\frac{1}{v}\beta'\left(\frac{\alpha}{v}\right)\mathbf{D}_1\right\}=\alpha\mathbf{V},$$

$$\mathbf{W}''(\alpha)\left\{\alpha\mathbf{I}+\mathbf{D}_0+\beta\left(\frac{\alpha}{v}\right)\mathbf{D}_1\right\}\mathbf{e}+2\mathbf{W}'(\alpha)\left\{\mathbf{I}+\frac{1}{v}\beta'\left(\frac{\alpha}{v}\right)\mathbf{D}_1\right\}\mathbf{e}$$

$$+\frac{1}{v^2}\mathbf{W}(\alpha)\beta''\left(\frac{\alpha}{v}\right)\mathbf{D}_1\mathbf{e}=0.$$

Then we set $\alpha=0$ and obtain the system of equations

$$\mathbf{m}_1(\mathbf{D}_0+\mathbf{D}_1)=\mathbf{r}\left\{\mathbf{I}-\frac{b_1}{v}\mathbf{D}_1\right\}-\mathbf{V},$$

$$\mathbf{m}_1\left\{\mathbf{I}-\frac{b_1}{v}\mathbf{D}_1\right\}\mathbf{e}=\frac{b_2}{2v^2}\mathbf{r}\mathbf{D}_1\mathbf{e}, \tag{20}$$

where $\mathbf{m}_1$ is the first raw vector moment of the virtual waiting time, b_1 and b_2 are the first and second raw moments of a packet length.

From formula (16), we obtain the formula for the mean real waiting time

$$m_{real}=-jH'(0)=-\mathbf{W}'(0)\frac{\mathbf{D}_1\mathbf{e}}{\mathbf{r}\mathbf{D}_1\mathbf{e}}=\mathbf{m}_1\frac{\mathbf{D}_1\mathbf{e}}{\mathbf{r}\mathbf{D}_1\mathbf{e}}. \tag{21}$$

Finally, we estimate the parameter γ of the proposed approximation using formula

$$\gamma=\frac{1-L}{m_{real}}. \tag{22}$$

6 Numerical Example

To illustrate the possibility of using the proposed approximation (19) of the distribution function of the real waiting time in the $MAP/GI/1/\infty$ system, consider a numerical example with the following parameters of the incoming flow

$$\mathbf{D}_0=\begin{bmatrix}-8.5 & 0.16 & 0.1 & 0\\ 0.56 & -1.3 & 0 & 0\\ 0.8 & 0 & -0.8 & 0\\ 0.18 & 0 & 0 & -0.9\end{bmatrix},\ \mathbf{D}_1=\begin{bmatrix}8 & 0.04 & 0.1 & 0.1\\ 0.14 & 0.6 & 0 & 0\\ 0 & 0 & 0 & 0\\ 0.72 & 0 & 0 & 0\end{bmatrix}.$$

The amount of information in the packets is Gamma distributed with shape parameter equal to 2 and scale parameter equal to 8. We consider the system for different rates of transmission $v=4.317, 2.59, 1.85, 1.439$, i.e. for different system loads $\rho=0.3, 0.5, 0.7, 0.9$.

For each system load, we obtain the theoretical distribution function Theor_WT of the real waiting time using formula (18) and its approximation

Approx_WT using formula (19). We also build the empirical distribution function Sim_WT of the real waiting time using simulation. VWT line depicts the distribution function of the virtual waiting time. Figures 1, 2, 3 and 4 depict the comparison between these four functions.

We note that the distribution function of the virtual waiting time signifacntly differs with simulation results. The gap between virtual and real waiting times is maximal in zero point. This is the reason why analysis of the real waiting time is important for systems with non-Poisson arrival processes.

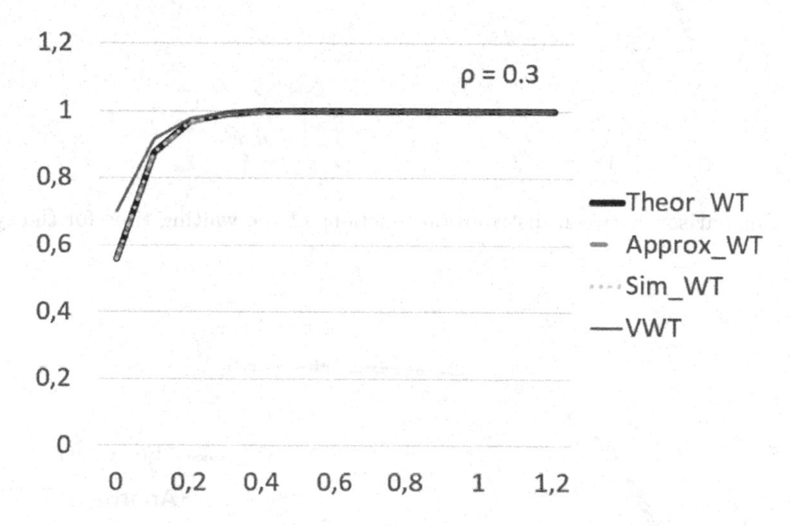

Fig. 1. Comparison between distribution functions of the waiting time for the system load 0.3

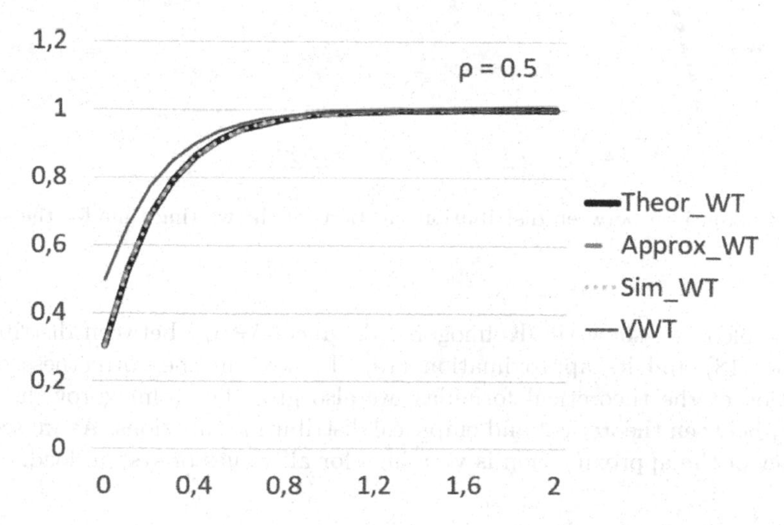

Fig. 2. Comparison between distribution functions of the waiting time for the system load 0.5

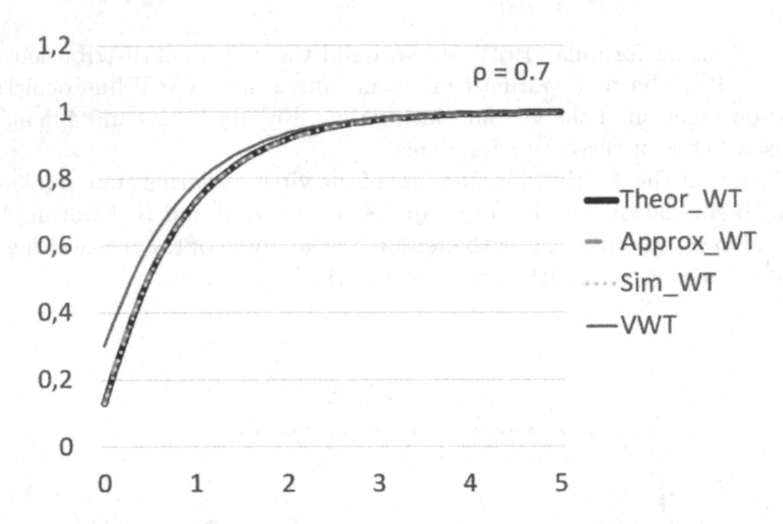

Fig. 3. Comparison between distribution functions of the waiting time for the system load 0.7

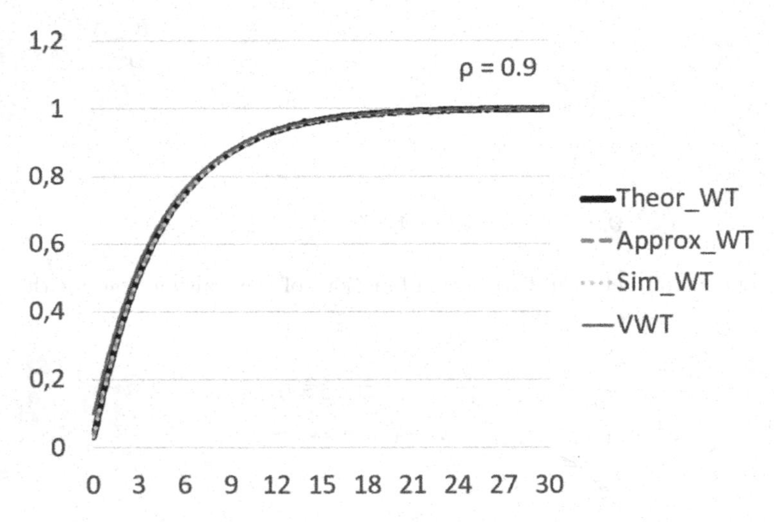

Fig. 4. Comparison between distribution functions of the waiting time for the system load 0.9

In Table 1, we show the Kolmogorov distance Δ_{T_App} between distribution function (18) and its approximation (19). To confirm the correctness of the derivation of the theoretical formulas, we also give the Kolmogorov distances Δ_{T_Sim} between theoretical and empirical distribution functions. As we see, the accuracy of the approximation is very high for all values of system load.

Table 1. Kolmogorov distances Δ_{T_App} and Δ_{T_Sim} for different values of system load

ρ	0.3	0.5	0.7	0.9
Δ_{T_App}	0.0018	0.0077	0.0078	0.0042
Δ_{T_Sim}	0.0006	0.0009	0.0018	0.0072

7 Conclusion

We have considered the real waiting time in the queue with MAP input and arbitrary distribution of packet length. For the system, we have obtained the LST of the virtual waiting time and the cumulative distribution function of the real waiting time. We have also proposed an approximation of the distribution function of the real waiting time and provided the results of numerical experiments. It is shown that the difference between virtual and real waiting times is greater when the system load is low. The results of experiments proof the accuracy of proposed approximation. Besides this it has simple structure, which allows reducing computational costs of obtaining the distribution of the real waiting time in the system. Our results will be helpful in analysis of the real commication systems with non-Poisson input processes where the real waiting time defines the conditions for the provision of quality of service.

References

1. Al-Begain, K., Dudin, A., Klimenok, V., Dudin, S.: Generalized survivability analysis of systems with propagated failures. Comput. Math. Appl. **64**(12), 3777–3791 (2012)
2. Cox, D., Isham, V.: The virtual waiting-time and related processes. Adv. Appl. Probab. **18**(2), 558–573 (1986)
3. Lucantoni, D.M.: New results on the single server queue with a batch markovian arrival process. Commun. Stat. Stoch. Models **7**(1), 1–46 (1991)
4. Ozawa, T.: Sojourn time distributions in the queue defined by a general QBD process. Queueing Syst. **53**(4), 203–211 (2006)
5. Ramaswami, V.: The N/G/1 queue and its detailed analysis. Adv. Appl. Probab. **12**(1), 222–261 (1980)
6. Rubin, I.: An approximate time-delay analysis for packet-switching communication networks. IEEE Trans. Commun. **24**(2), 210–222 (1976)
7. Vishnevskii, V.M., Dudin, A.N.: Queueing systems with correlated arrival flows and their applications to modeling telecommunication networks. Autom. Remote Control **78**(8), 1361–1403 (2017). https://doi.org/10.1134/S000511791708001X

Controlled Queueing Systems G/G/1 with Time Shift

V. N. Tarasov(✉) and N. F. Bakhareva

Volga State University of Telecommunications and Informatics, Samara, Russia
veniamin_tarasov@mail.ru

Abstract. The article presents the results of research on queueing systems (QS) described by density functions shifted to the right along the time axis. It is obvious that the shift of the distribution law to the right increases the mathematical expectation of the described random variable and thereby reduces its coefficient of variation. From the queuing theory, it is known that the average waiting time for applications in the queue is directly proportional to the squares of the coefficients of variation of the intervals between arrivals and the service time. Then, in systems with a time shift, the average waiting time in comparison with conventional classical systems should decrease many times over. The numerical characteristics and, consequently, the parameters of the laws of distributions that form the QS with a time shift become functionally dependent on the magnitude of the shift parameter. The average waiting time as the main characteristic of the QS and other characteristics of the queueing system, which are derivatives of the average waiting time in such systems, become functions of the shift parameter. The shift parameter becomes a control parameter, and a queueing system with a time shift becomes a controlled queuing system. The adequacy of the presented numerical-analytical models of systems with a time shift is confirmed, on the one hand, by a computational experiment in Mathcad and simulation in the GPSS World system, on the other hand.

Keywords: shifted distribution laws · system with time shift · Laplace transform · spectral solution · QS characteristics

1 Introduction

Let us consider single-channel queuing systems (QS), for which the laws of distribution of intervals between arrivals $a(t)$ and service time $b(t)$ are given by shifted density functions of the form with the shift parameter $t_0 > 0$

$$a(t) = \begin{cases} a(t - t_0), & t \geq t_0 \\ 0, & 0 \leq t < t_0 \end{cases}, \quad b(t) = \begin{cases} b(t - t_0), & t \geq t_0 \\ 0, & 0 \leq t < t_0 \end{cases} \tag{1}$$

from the class of functions transformable according to Laplace. This limitation is related to the method of spectral decomposition used in the work, which uses

A. Dudin et al. (Eds.): ITMM 2022, CCIS 1803, pp. 126–137, 2023.
https://doi.org/10.1007/978-3-031-32990-6_11

Laplace transforms. The time shift operation transforms classical QS into non-Markovian G/G/1 type systems due to the change in the coefficients of variation of random variables. An example of a shifted exponential function is shown in Fig. 1.

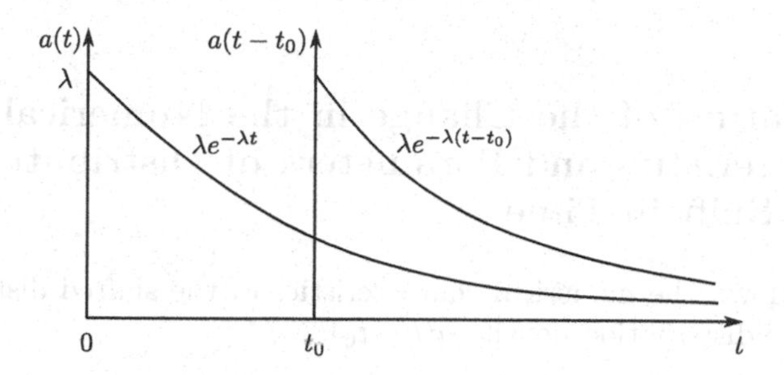

Fig. 1. An example of $a(t)$ function.

As the main method in the analysis of QS G/G/1, we will use the spectral solution of the Lindley integral equation, one of the forms of which with respect to the distribution function of the waiting time $W(y)$ is given in the form [1]:

$$W(y) = \int_0^y W(y-u)dC(u), \quad y \geq 0,$$

where $C(u)$ are the distribution functions of a random variable – the difference between the service time of a claim and the time between arrivals of two neighboring claims. The spectral solution, in addition to the theory of queuing, is also used in many other areas of scientific research [1–3].

When solving the Lindley integral equation by the spectral method, the waiting time distribution function is determined, and then the average waiting time of requests in the queue $\bar{W}$ is determined through the Laplace transform of the waiting time density function. This is achieved by constructing in the complex s-plane a fractional-rational function with expansion components $\alpha(s)$ and $\beta(s)$:

$$A^*(-s) \cdot B^*(s) - 1 = \alpha(s)/\beta(s), \tag{2}$$

where $A^*(s)$, $B^*(s)$ Laplace transforms of the corresponding density functions (1).

Based on this method, the author obtained spectral solutions with the derivation of calculation formulas for the average waiting time for sixteen QS with a time shift, formed by exponential (M), Erlang (E_2), hyperexponential (H_2), hyper-Erlang (HE_2) distributions of the second order. From these four most commonly used distribution laws, sixteen different QS were formed. These results have been published in [4–7] and others. Both in domestic and foreign scientific literature, the authors did not find studies on such QS. For the first time such a study of the author was published in [4], further development of this topic is presented in [5–7] and other works. The works [8,9] can be most closely related to this topic. In the proposed work, methods for approximating distribution laws using the method of moments are used; these approaches are described in more detail in [10–13], and similar studies in queuing theory have recently been carried out in [14–17].

2 The Nature of the Change in the Numerical Characteristics and Parameters of Distributions with a Shift in Time

Let us write down the numerical characteristics of the shifted distributions for the case of the distribution density $a(t - t_0)$.

$$\text{M: } \bar{\tau}_\lambda = 1/\lambda + t_0,\ c_\lambda = 1/(1 + \lambda t_0);$$

$$\text{E}_2:\ \bar{\tau}_\lambda = 2/\lambda + t_0,\ c_\lambda = \sqrt{2}/(2 + \lambda t_0);$$

$$\text{H}_2:\ \bar{\tau}_\lambda = p\lambda_1^{-1} + (1-p)\lambda_2^{-1} + t_0,\ c_\lambda^2 = \frac{[(1-p^2)\lambda_1^2 - 2\lambda_1\lambda_2 p(1-p) + p(2-p)\lambda_2^2]}{[t_0\lambda_1\lambda_2 + (1-p)\lambda_1 + p\lambda_2]^2};$$

$$\text{HE}_2:\ \bar{\tau}_\lambda = 2p\lambda_1^{-1} + 2(1-p)\lambda_2^{-1} + t_0,\ c_\lambda^2 = \frac{\lambda_1^2 - 2p\lambda_2(\lambda_1 - \lambda_2) + p(1-2p)(\lambda_1 - \lambda_2)^2}{2[t_0\lambda_1\lambda_2 - p(\lambda_1 - \lambda_2) + \lambda_1]^2}.$$

Here, $\bar{\tau}_\lambda$ is the average interval between neighboring requests, c_λ is the coefficient of variation of the interval between neighboring requests.

In the case of the distribution $b(t - t_0)$ in these expressions, the symbol λ will change to μ.

Changing the parameters of distributions under shift.

$$\text{M: } \lambda = 1/(\bar{\tau}_\lambda - t_0);$$

$$\text{E}_2:\ \lambda = 2/(\bar{\tau}_\lambda - t_0);$$

$$\text{H}_2:\ \lambda_1 = \frac{2p}{\bar{\tau}_\lambda - t_0},\ \lambda_2 = \frac{2(1-p)}{\bar{\tau}_\lambda - t_0},\ p = \frac{1}{2} \pm \sqrt{\frac{1}{4} - \frac{(\bar{\tau}_\lambda - t_0)^2}{2[(\bar{\tau}_\lambda - t_0)^2 + c_\lambda^2\bar{\tau}_\lambda^2]}};$$

$$\text{HE}_2:\ \lambda_1 = \frac{4p}{\bar{\tau}_\lambda - t_0},\ \lambda_2 = \frac{4(1-p)}{\bar{\tau}_\lambda - t_0},\ p = \frac{1}{2} \pm \sqrt{\frac{1}{4} - \frac{3(\bar{\tau}_\lambda - t_0)^2}{8[(\bar{\tau}_\lambda - t_0)^2 + c_\lambda^2\bar{\tau}_\lambda^2]}}.$$

As can be seen from the above expressions, the numerical characteristics and parameters of the distribution laws that form the QS with a time shift, and hence the average waiting time for applications in the queue, become functionally dependent on the value of the shift parameter $t_0 > 0$. The shift parameter can thus act as a control parameter, and a queueing system with a time shift can act as a controlled queuing system. When deriving results for all QS with a time shift, the following theorem was used.

3 Theorem on the Spectral Expansion

Consider the class of density functions $f(t)$ that are Laplace-transformable, i.e., for which there is a function . Consider the class of density functions $f(t)$ that are Laplace-transformable, i.e., for which there is a function $F^*(s) = \int_0^\infty e^{-st} f(t)dt \equiv L[f(t)]$. The distribution laws most frequently used in queuing theory, such as exponential, hyperexponential, Erlang and hyper-Erlang, are included in this class. From the theory of Laplace transformations, the delay property is known: for any, the equality

$$L[f(t - t_0)] = e^{-st_0} \cdot F^*(s), \tag{3}$$

where $\text{Re}(s) > 0$.

The generalization of the author's results [4–7] and others with input distributions transformed according to Laplace allows us to formulate the following theorem, which is necessary for further presentation of the results.

Theorem 1. *Function $A^*(-s) \cdot B^*(s) - 1 = \alpha(s)/\beta(s)$ for all QS with delay is invariant to the shift operation, i.e. it does not depend on the shift operation of the distribution law.*

Proof. To do this, consider a QS with a time shift, formed by two density functions (1) with Laplace transforms $A^*(s)$, $B^*(s)$ respectively.

Let us write a function $A^*(-s) \cdot B^*(s) - 1 = \alpha(s)/\beta(s)$ for a QS with a time shift, taking into account property (3):

$$\alpha(s)/\beta(s) = e^{t_0 s} A^*(-s) e^{-t_0 s} B^*(s) - 1 = A^*(-s) B^*(s) - 1. \tag{4}$$

It follows from (4) that the spectral expansion (2) does not depend on the shift of the distribution law and coincides in form with the expansion for the classical system. Consequently, the solutions for the average waiting time for both systems with a time shift and conventional systems will have the same form, but due to the change in the parameters of distributions during a shift, the latter must be redefined.

The theorem is proved.

From (4) it also follows that functions (1) must have exactly the same delay time t_0. Because the parameters of distributions (1) and their numerical characteristics will be functions of the shift parameter, then all the probabilistic-temporal characteristics of the QS will depend on the value of the shift parameter. Consequently, the shift parameter becomes a control parameter for regulating the probabilistic-temporal characteristics of the QS: the average waiting time, the average queue length, the average residence time, and the average number of requests in the QS.

4 Examples of Systems with a Time Shift

To simplify the presentation of the results, we consider examples of systems formed by the Erlang distribution of the second order and the exponential distribution. The choice of such systems is because the GPSS Word simulation system has high-quality generators of such random sequences and we can compare the results of analytical and simulation modeling.

4.1 E$_2$/M/1 System with Time Shift

Consider a system formed by the density functions $a(t) = \lambda^2(t - t_0)e^{-\lambda(t-t_0)}$ and $b(t) = \mu e^{-\mu(t-t_0)}$. In the classics, such a system corresponds to QS E$_2$/M/1. According to the theorem, for the system under consideration the spectral expansion takes the same form as for the system E$_2$/M/1:

$$A^*(-s)B^*(s) - 1 = \frac{\alpha(s)}{\beta(s)} = \left(\frac{\lambda}{\lambda - s}\right)^2 \frac{\mu}{\mu + s} - 1 =$$

$$= -\frac{s[s^2 + (\mu - 2\lambda)s + \lambda(\lambda - 2\mu)]}{(\lambda - s)^2(\mu + s)} = -\frac{s(s + s_1)(s - s_2)}{(\lambda - s)^2(\mu + s)},$$

where $-s_1 = -(\mu - 2\lambda)/2 - \sqrt{\mu(\mu + 4\lambda)}/2$, $s_2 = (2\lambda - \mu)/2 + \sqrt{\mu(\mu + 4\lambda)}/2$ zeros of the expansion numerator. Expansion components $\alpha(s) = \frac{s(s+s_1)}{s+\mu}$, $\beta(s) = -\frac{(\lambda-s)^2}{s-s_2}$. Omitting intermediate calculations in the spectral decomposition method, we write the Laplace transform of the waiting time density function $W^*(s) = \frac{s_1(s+\mu)}{\mu(s+s_1)}$ [5]. Hence the average waiting time

$$\bar{W} = 1/s_1 - 1/\mu. \tag{5}$$

For the practical application of this calculation formula in the expression for s_1, the distribution parameters E$_2$ and M must be taken considering the shift operation. The algorithm for calculating the average waiting time includes the following steps. We set the input parameters for the calculation: the load factor $\rho = \bar{\tau}_\mu/\bar{\tau}_\lambda$ — as the ratio of two average values of time intervals, the coefficients of variations of these random intervals c_λ, c_μ and the shift parameter t_0. From the moment equations

$$\bar{\tau}_\lambda = 2/\lambda + t_0, \quad c_\lambda = \sqrt{2}/(2 + \lambda t_0), \quad \bar{\tau}_\mu = 1/\mu + t_0, \quad c_\mu = 1/(1 + \mu t_0)$$

we determine the unknown parameters of the distributions λ, μ: $\lambda = \frac{2}{\bar{\tau}_\lambda - t_0}$, $\mu = \frac{1}{\bar{\tau}_\mu - t_0}$.

Whence it follows that a constraint is imposed on the shift parameter $0 < t_0 < \bar{\tau}_\mu$. Then the coefficients of variations c_λ, c_μ for this system will vary in the following ranges: $0 < c_\lambda < 1/\sqrt{2}$, $0 < c_\mu < 1$. The nature of this change is easy to trace in Table 1. For other systems, the calculation algorithms will be similar.

Thus, as a result of shifting the distribution laws, we get completely different QS, for which, instead of fixed coefficients of variations in the case of classical systems, there are ranges of their variation.

4.2 M/E$_2$/1 System with Time Shift

Consider a system formed by the density functions $a(t) = \lambda e^{-\lambda(t-t_0)}$ and $b(t) = \mu^2(t-t_0)e^{-\mu(t-t_0)}$. This is a dual system in relation to the previous system. The spectral solution for this system takes the form

$$\frac{\alpha(s)}{\beta(s)} = \frac{\lambda}{\lambda - s} \cdot \left(\frac{\mu}{\mu+s}\right)^2 - 1 = \frac{s[s^2 + (2\mu - \lambda)s + \mu(\mu - 2\lambda)]}{(\lambda - s)(\mu + s)^2} = \frac{s(s+s_1)(s+s_2)}{(\lambda - s)(\mu+s)^2}.$$

The expansion numerator in the case of a stable system has two real negative roots, which for convenience we denote by $-s_1$, $-s_2$:

$$-s_1 = -(2\mu - \lambda)/2 + \sqrt{\lambda(\lambda + 4\mu)}/2, \quad -s_2 = -(2\mu - \lambda)/2 - \sqrt{\lambda(\lambda + 4\mu)}/2.$$

Taking into account the rules for constructing the expansion components $\alpha(s)$ and $\beta(s)$, we obtain $\alpha(s) = \frac{s(s+s_1)(s+s_2)}{(\mu+s)^2}$, $\beta(s) = \lambda - s$. Laplace transform of waiting time density function in the M/E$_2$/1 system

$W^*(s) = s \cdot \Phi_+(s) = \frac{(1-\rho)(\mu+s)^2}{(s+s_1)(s+s_2)}$. Hence the average waiting time for requests in the queue

$$\bar{W} = \frac{1}{s_1} + \frac{1}{s_2} - \frac{2}{\mu}. \tag{6}$$

Transformation of formula (6), considering the expressions for s_1 and s_2, inevitably leads to the Polyachek–Khinchin formula for the M/G/1 system

$$\bar{W} = \frac{3\rho}{2\mu(1-\rho)}, \tag{7}$$

where $\rho = \bar{\tau}_\mu / \bar{\tau}_\lambda = 2\lambda/\mu$ load factor [5].

The parameters of the calculation formula are again taken taking into account the shift operation.

4.3 M/M/1 System with Time Shift

Now, as another example, consider a system formed by shifted exponential density functions $a(t) = \lambda e^{-\lambda(t-t_0)}$ and $b(t) = \mu e^{-\mu(t-t_0)}$. For it, we immediately write the finished result for the average waiting time [4]

$$\bar{W} = \frac{\lambda/\mu}{\mu - \lambda}, \tag{8}$$

where the parameters λ and μ are again determined taking into account the shift operation. Moment equations for numerical characteristic $\bar{\tau}_\lambda = 1/\lambda + t_0$, $\bar{\tau}_\mu = 1/\mu + t_0$, $c_\lambda = 1/(1 + \lambda t_0)$, $c_\mu = 1/(1 + \mu t_0)$ make it possible to express the unknown parameters λ and μ in terms of the variable shift parameter t_0 and transform expression (8) to the form

$$\bar{W} = \frac{(\bar{\tau}_\mu - t_0)^2}{(\bar{\tau}_\lambda - \bar{\tau}_\mu)}. \tag{9}$$

For the previous systems, formulas (5) and (6) can also be expressed in terms of the shift parameter, but they will no longer be so compact.

5 Results of Numerical Experiments

As confirmation of the adequacy of the proposed QS models with a time shift, the results of numerical simulation of the average waiting time $\bar{W}$ in Mathcad for the three systems considered above are presented below. In the tables, the average waiting time in the classical system is denoted by $\bar{W}_{\text{cl}}$.

The load factor in all tables is determined by the ratio of average intervals $\rho = \bar{\tau}_\mu / \bar{\tau}_\lambda$. The calculations used a single service time $\bar{\tau}_\mu = 1$.

Table 1. Results of computational experiments for QS $E_2/M/1$ with a shift.

Input parameter				Average waiting time	
ρ	c_λ	c_μ	t_0	$\bar{W}$	$\bar{W}_{\text{cl}}$
0,1	0,706	0,99	0,01	0,029	0.030
	0,700	0,9	0,1	0,023	
	0,672	0,5	0,5	0,005	
	0,643	0,1	0,9	0,000	
0,5	0,704	0,99	0,01	0,605	0,618
	0,672	0,9	0,1	0,491	
	0,530	0,5	0,5	0,132	
	0,389	0,1	0,9	0,003	
0,9	0,701	0,99	0,01	6,456	6,588
	0,643	0,9	0,1	5,322	
	0,389	0,5	0,5	1,609	
	0,134	0,1	0,9	0,055	

Table 2. Results of computational experiments for QS $M/E_2/1$ with a shift.

Input parameter				Average waiting time	
ρ	c_λ	c_μ	t_0	$\bar{W}$	$\bar{W}_{\text{cl}}$
0,1	0,999	0,700	0,01	0,082	0,083
	0,990	0,636	0,1	0,068	
	0,950	0,354	0,5	0,021	
	0,643	0,071	0,9	0,001	
0,5	0,995	0,700	0,01	0,735	0,75
	0,950	0,636	0,1	0,608	
	0,750	0,354	0,5	0,188	
	0,550	0,071	0,9	0,008	
0,9	0,991	0,700	0,01	6,616	6,75
	0,910	0,636	0,1	5,468	
	0,550	0,354	0,5	1,688	
	0,190	0,071	0,9	0,068	

Table 3. Results of computational experiments for QS M/M/1 with a shift.

Input parameter				Average waiting time	
ρ	c_λ	c_μ	t_0	$\bar{W}$	$\bar{W}_{\text{cl}}$
0,1	0,999	0,99	0,01	0,109	0,11
	0,99	0,90	0,1	0,090	
	0,95	0,50	0,5	0,028	
	0,91	0,10	0,9	0,001	
0,5	0,995	0,99	0,01	0,981	1,0
	0,95	0,90	0,1	0,810	
	0,75	0,50	0,5	0,250	
	0,55	0,10	0,9	0,010	
0,9	0,991	0,99	0,01	8,821	9,0
	0,91	0,90	0,1	7,290	
	0,55	0,50	0,5	2,250	
	0,19	0,10	0,9	0,090	

The results of calculations in Tables 1–3 unequivocally confirm all the assumptions that we put forward in paragraph 2 about systems with a time shift. Here, the nature of the change in the coefficients of variations and, consequently, the average waiting time, which is related by a quadratic dependence with the coefficients of variations, is clearly traced. As the shift parameter decreases, the average waiting time in a system with a shift tends to its value in the classical system, and as it grows, it tends to zero.

6 Simulation Results

Tables 4–6 present the results of simulation modeling for the considered QS in the GPSS World system using the corresponding random variable generators, while the Erlang distribution is considered as a special case of the Gamma distribution. These generators provide for a shift of the distribution law to the right by an appropriate amount. In the tables, the average queue length in the system is denoted by $\bar{N}_q$.

The text of the program and the results of the run for the E_2/M/1 system with a shift (case $\rho = 0.9$, $t_0 = 0.5$ — the results are highlighted in gray shown in Fig. 2):

```
10 GENERATE (GAMMA (1,0.5,0.305,2))
20 QUEUE QCHAN
30 SEIZE CHAN
40 DEPART QCHAN
50 ADVANCE (Exponential (11,0.5,0.5))
60 RELEASE CHAN
70 TERMINATE 1
80 START 1000000
```

The results of simulation of the operation of this QS for a wide range of system parameters are shown in Table 4.

```
FACILITY ENTRIES UTIL. AVE. TIME AVAIL. OWNER PEND INTER RETRY DELAY
CHAN      1000000 0.902    1.000      1      0    0     0     0     0

QUEUE    MAX CONT. ENTRY ENTRY(0) AVE.CONT. AVE.TIME AVE.(-0) RETRY
QCHAN    19    0    1000000 234022   1.486     1.649    2.152    0
```

Fig. 2. Results of the run for the $E_2/M/1$ system with a shift.

Table 4. Simulation results for QS $E_2/M/1$ with a shift.

ρ	t_0	$\bar{W}$	$\bar{N}_q$
0,1	0,01	0,030	0,003
	0,1	0,023	0,002
	0,5	0,005	0,000
	0,9	0,000	0,000
0,5	0,01	0,610	0,305
	0,1	0,496	0,248
	0,5	0,134	0,067
	0,9	0,003	0,001
0,9	0,01	6,642	5,991
	0,1	5,476	4,938
	0,5	1.649	1.486
	0,9	0,056	0,050

The text of the program and the results of the run for the $M/E_2/1$ system with a shift (case $\rho = 0.9$, $t_0 = 0.5$ — the results are highlighted in gray shown in Fig. 3):

```
10 GENERATE (Exponential (11,0.5,0.611))
20 QUEUE QCHAN
30 SEIZE CHAN
40 DEPART QCHAN
50 ADVANCE (GAMMA (1,0.5,0.25,2))
60 RELEASE CHAN
70 TERMINATE 1
80 START 1000000
```

```
FACILITY ENTRIES  UTIL. AVE. TIME AVAIL. OWNER PEND INTER RETRY DELAY
CHAN     1000001 0.899    0.999    1   1000001   0     0      0     2

QUEUE     MAX CONT. ENTRY ENTRY(0) AVE.CONT. AVE.TIME AVE.(-0) RETRY
QCHAN     24     3  1000003 183363    1.501     1.669   2.043    0
```

Fig. 3. Results of the run for the M/E2/1 system with a shift.

The results of simulation of the operation of this QS for a wide range of system parameters are shown in Table 5.

Table 5. Simulation results for QS M/E2/1with a shift.

ρ	t_0	$\bar{W}$	$\bar{N}_q$
0,1	0,01	0,082	0,008
	0,1	0,068	0,007
	0,5	0,021	0,002
	0,9	0,001	0,000
0,5	0,01	0,733	0,366
	0,1	0,606	0,303
	0,5	0,187	0,094
	0,9	0,008	0,004
0,9	0,01	6,577	5,921
	0,1	5,389	4,847
	0,5	1.669	1.501
	0,9	0,067	0,061

The text of the program and the results of the run for the M/M/1 system with a shift (case $\rho = 0.9$, $t_0 = 0.5$ — the results are highlighted in gray shown in Fig. 4):

```
10 GENERATE (Exponential (1,0.5,0.611))
20 QUEUE QCHAN
30 SEIZE CHAN
40 DEPART QCHAN
50 ADVANCE (Exponential (1,0.5,0.5))
60 RELEASE CHAN
70 TERMINATE 1
80 START 1000000
```

The results of simulation of the operation of this QS for a wide range of system parameters are shown in Table 6.

Table 6. Simulation results for QS M/M/1 with a shift.

ρ	t_0	$\bar{W}$	$\bar{N}_q$
0,1	0,01	0,110	0,011
	0,1	0,090	0,009
	0,5	0,028	0,003
	0,9	0,001	0,000
0,5	0,01	0,982	0,491
	0,1	0,810	0,405
	0,5	0,250	0,125
	0,9	0,010	0,005
0,9	0,01	8,617	7,749
	0,1	7,320	6,578
	0,5	2.250	2.024
	0,9	0,091	0,081

```
FACILITY  ENTRIES UTIL. AVE. TIME AVAIL. OWNER PEND INTER RETRY DELAY
CHAN      1000001  0.900   1.000     1   1000001  0     0     0     4

QUEUE     MAX CONT. ENTRY ENTRY(0) AVE.CONT. AVE.TIME AVE.(-0) RETRY
QCHAN     31   5   1000005 181827    2.024     2.250     2.751     0
```

Fig. 4. Results of the run for the M/M/1 system with a shift.

7 Conclusion

The results of calculations in Tables 1–3 fully confirm the above assumptions about systems with a time shift. The operation of shifting distribution laws leads to a functional dependence of their parameters and numerical characteristics on the shift parameter. Consequently, the main characteristic of the QS, the average waiting time in the queue, will also depend on the shift parameter.

From Tables 1–3, the relationship between the coefficients of variations in time intervals and the shift parameter is clearly traced. Due to the decrease in the coefficients of variation with an increase in the shift parameter, the average waiting time decreases many times, and vice versa, with a decrease in the shift parameter, the average waiting time approaches its value for a conventional system. Pairwise comparison of the tables of results of analytical and simulation modeling also unambiguously confirms the adequacy of the presented models of systems with a time delay.

QS characteristics that depend on the average waiting time become functionally dependent on the shift parameter. Therefore, the shift parameter becomes a control parameter for adjusting the values of these characteristics. For other systems with a time shift formed by Laplace-transformable distributions, all the above considerations remain valid. In this case, each distribution law used in the QS will have its own implementation features in a system with a shift. For example, for composite distributions, the degrees of polynomials in the expansion numerator will increase.

References

1. Kleinrock, L.: Queueing systems, vol. I: theory. Wiley, New York (1975)
2. Do, T. V., Chakka, R., Sztrik, J.: Spectral Expansion Solution Methodology for QBD-M Processes and Applications in Future Internet Engineering. In: Nguyen, N., van Do, T., le Thi, H. (eds.) Advanced Computational Methods for Knowledge Engineering. Studies in Computational Intelligence, vol. 479, pp. 131–142. Springer, Heidelberg (2016). https://doi.org/10.1007/978-3-319-00293-4-11
3. Ma, X.A., Wang, Y., Zhu, X., Liu, W., Lan, Q., Xiao, W.: Spectral method for two-dimensional ocean acoustic propagation. J. Marine Sci. Eng. **8**(9), 1–19 (2021)
4. Tarasov, V.N., Bakhareva, N.F., Blatov, I.A.: Analysis and calculation of queueing system with delay. Autom. Remote. Control. **11**(76), 1945–1951 (2015)
5. Tarasov, V.N.: Extension of the class of queueing systems with delay. Autom. Remote. Control. **79**(12), 2147–2158 (2018). https://doi.org/10.1134/S0005117918120056
6. Tarasov, V.: Comparison of two queuing systems with ordinary and shifted erlang distributions. In: 2019 IEEE International Scientific-Practical Conference Problems of Infocommunications Science and Technology Proceedings, pp. 899–902. IEEE, Kyiv Ukraine (2019). https://doi.org/10.1109/PICST47496.2019.9061271
7. Tarasov, V.N., Bakhareva, N.F.: Comparative analysis of two queuing systems M/HE2/1 with ordinary and with the shifted input distributions. Radio Electron. Comput. Sci. Control **4**(51), 50–58 (2019)
8. Novitzky, S., Pender, J., Rand, R.H., Wesson, E.: Limiting the oscillations in queues with delayed information through a novel type of delay announcement. Queueing Syst. **3**(95), 281–330 (2020)
9. Novitzky, S., Pender, J., Rand, R.H., Wesson, E.: Nonlinear dynamics in queueing theory: determining the size of oscillations in queues with delay. SIAM J. Appl. Dyn. Syst. **1**(18), 279–311 (2019)
10. Whitt, W.: Approximating a point process by a renewal process: two basic methods. Operation Res. **1**(30), 125–147 (1982)
11. Myskja, A.: An improved heuristic approximation for the GI/GI/1 queue with bursty arrivals. In: 13th International Conference Teletraffic and datatraffic in a Period of Change, pp. 683–688. Elsevier Science Ltd. (1991)
12. Aliev, T.I.: Fundamentals of Modeling Discrete Systems. SPbGU ITMO, SPb (2009)
13. Aliev, T.I.: Approximation of probability distributions in queuing models. Sci. Tech. Bullet. Inf. Technol. Mech. Optics **2**(84), 88–93 (2013)
14. Gromoll, H.C., Terwilliger, B., Zwart, B.: Heavy traffic limit for a tandem queue with identical service times. Queueing Syst. **3**(89), 213–241 (2018)
15. Legros, B.: M/G/1 queue with event-dependent arrival rates. Queueing Syst. **3**(89), 269–301 (2018)
16. Bazhba, M., Blanchet, J., Rhee, C.H.: Queue with heavy-tailed Weibull service times. Queueing Syst. **11**(93), 1–32 (2019)
17. Jacobovic, R., Kella, O.: Asymptotic independence of regenerative processes with a special dependence structure. Queueing Syst. **11**(93), 139–152 (2019)

Analysis of Retrial Queueing System with Two-Way Communication in Different Scenarios Using Simulation

János Sztrik and Ádám Tóth(✉)

University of Debrecen, University Square 1, Debrecen 4032, Hungary
{sztrik.janos,toth.adam}@inf.unideb.hu

Abstract. The purpose of this study is to investigate a finite-source retrial queueing system with two-way communication. Customers, who arrive from a finite source according to an exponential distribution, are referred to as primary customers. If the service unit is available, these customers will receive service immediately, but if not, they are redirected to the orbit and attempt to reach the server again after a random amount of time. The system is unique in that when the server becomes idle, an outgoing call, also known as a secondary customer, is made to the orbit and source with varying parameters. Both primary and secondary customers receive service following an exponential distribution, but with differing rates. This investigation aims to conduct a sensitivity analysis on the performance measures by using different distributions of the customers' retrial time in two separate cases. The results of the comparison will be displayed graphically.

Keywords: Finite-source queuing system · Retrial queues · Two-way communication · Sensitivity analysis · Simulation

1 Introduction

The topic of two-way communication is widely popular due to its ability to be modeled using retrial queueing systems in a variety of real-life scenarios. A prime example is the operation of call centers, where during idle periods, agents engage in activities such as selling, advertising, and promoting products in addition to handling customer calls. One of the most important measures is utilization, and how to optimize the efficiency of the service units or agents which is always a key issue, see for example [1,4,9,13,17]. The characteristic of two-way communication relies on performing calls inside and outside of the system when the server becomes idle. There are two types of outgoing calls that are distinguished:

- One where the server calls a customer from the source for service known as a primary outgoing call,
- And another where the server calls a customer from the orbit referred to as a secondary outgoing call.

A. Dudin et al. (Eds.): ITMM 2022, CCIS 1803, pp. 138–148, 2023.
https://doi.org/10.1007/978-3-031-32990-6_12

In our model, outgoing calls can be made to either the source or the orbit. Exploring the available literature reveals many queuing schemes: in some the incoming customer waits until it is served because the queue size is infinite. In others, the arriving customer at the time of arrival can leave the system observing that the service units are fully occupied. However, in real life, there are various situations where customers do not leave the system, stay close to the service units and try to reach a server again after some random time. In this case, this customer will stay in a so-called virtual waiting room called orbit before launching another attempt to reach a server again. Systems containing an orbit can be modeled easily with retrial queues. Queuing systems with retrial queues are useful tools for modeling various problems that arise in telecommunication systems, such as call centers, telephone switching systems, and computer networks like in [2,8,11]. In the past, researchers investigated infinite source retrial queueing systems with two-way communication, and here are some examples: [3,7,12,16,18,19].

Dragieva and Phung-Duc [6] have investigated the scenario when a secondary outgoing call returns to the source after the service. This paper is the natural continuation of [14] where a more realistic scenario was considered. Instead of sending back the secondary outgoing customers to the source, they will be sent back to the orbit where the call has the opportunity to retry his request for servicing the original incoming call. The motivation for investigating finite source retrial models with two-way communication is based on real-life scenarios in which customers are unable to receive service immediately upon arrival and must go to another location before checking the system again, or the server, once idle, calls for customers.

The uniqueness of this research lies in conducting a sensitivity analysis to assess the impact of various distributions of retrial time on multiple performance measures. The results were generated using a stochastic simulation program based on the SimPack framework ([10]), which is a collection of C/C++ libraries and programs for computer simulation to support various types of simulations, including discrete event simulation, continuous simulation, and multi-model simulation. It provides the flexibility to model any queueing system and perform simulations with custom random number generators to calculate any desired performance measures. The input parameters are presented in a table, and the results of the comparison between different operation modes and distributions are shown through graphical illustrations.

2 System Model

In this section, the considered finite-source retrial queueing model with one server is introduced, which is represented in Fig. 1. Altogether N requests are located in the source, and each of them is capable of generating a primary incoming call toward the server, and the inter-request times are exponentially distributed random variables with parameter λ_1. In the case of an idle server, the service of an incoming customer begins instantaneously that follows an exponential distribution with parameter μ_1. After the successful service, the customers go back to the source. When the incoming customer finds the service unit busy, those customers will not be lost and they are transmitted to the orbit. These will be the secondary incoming jobs from the orbit that may retry to reach the service unit after a random waiting time. The distribution of this period follows different distributions including gamma, hyper-exponential, Pareto, and lognormal, with varying parameters, but all with the same mean value. However, the idle server can also initiate outgoing calls from both the source and the orbit. There are two types of outgoing calls that are distinguished:

- The service unit may call a job from the source to receive service (known as a primary outgoing call) after an exponentially distributed period with rate λ_2,
- Or the service unit may initiate a call from the orbit (referred to as a secondary outgoing call) after an exponentially distributed period with rate ν_2.

The service time of the outgoing customers is exponentially distributed with parameter μ_2. Two scenarios are distinguished when an outgoing call comes from the orbit:

- Case 1: The call has an unserved incoming request so that call is sent back to the orbit after the outgoing service is finished to have its incoming call be served,
- Case 2: Here, the call has also got an unserved incoming request but after the outgoing service is done the service unit serves the incoming request right away. This will result in a two-phase service, first the outgoing call then the incoming one is executed. The call returns to the source after both service phases are finished.

It is assumed that the arrivals of primary incoming calls, the intervals between retries of secondary incoming calls, the service times of both incoming and outgoing calls, and the time it takes to make outgoing calls are all mutually independent.

3 Simulation Results

3.1 First Scenario

SimPack is used to obtain the results as a basic block of our program and it was extended with the desired features. We used a statistical package that can estimate the desired measures. It utilizes the batch means method which is a quite popular method. In brief, the running period is divided into a number of batches (totaling T). In each batch, $s = R-M/T$ observations are conducted. M represents the discarded warm-up period observations that occur at the beginning of the simulation, and R is the length of the simulation. After the initial phase, the average of the entire run is calculated. To obtain meaningful results, the batches should be of sufficient length and the average of each batch should be independent. More detailed information about the used process you can find in these papers: [5,15].

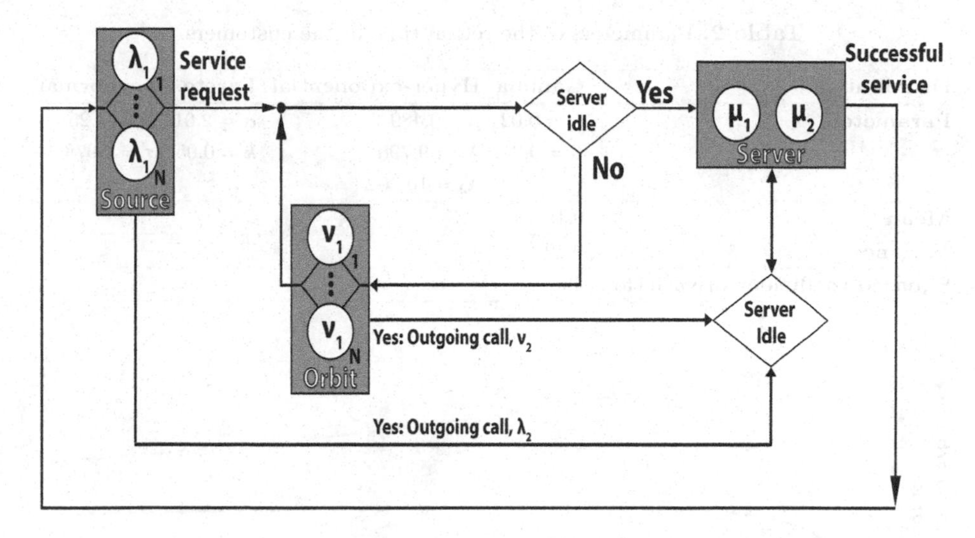

Fig. 1. System model

Throughout the simulations, a confidence level of 99.9% is used, and a relative half-width of the confidence interval of 0.00001 is employed to halt the actual simulation sequence. The size of a batch during the initial transient period cannot be too small, so it is set to 1000. The values of the input parameters used are presented in Table 1.

Table 1. Numerical values of model parameters

N	μ_1	μ_2	λ_2	ν_2
10	1	1	0.2	0.2

The next table (Table 2) contains the parameters of the retrial time of the customers, to achieve a valid comparison parameters are chosen according to having the same mean and variance value. The simulation program was run using various parameter values and the most noteworthy results will be presented in this paper. As shown in the table, the squared coefficient of variation is greater than one in this scenario, allowing for the examination of the impact of specific random variables. We also aim to present results with a different set of parameters when the squared coefficient of variation is less than one.

Table 2. Parameters of the retrial time of the customers

Distribution	Gamma	Hyper-exponential	Pareto	Lognormal
Parameters	$\alpha = 0.02$ $\beta = 0.2$	$p = 0.489$ $\lambda_1 = 9.798$ $\lambda_2 = 10.202$	$\alpha = 2.01$ $k = 0.05$	m =-4.258 $\sigma = 1.978$
Mean	0.1			
Variance	0.49			
Squared coefficient of variation	49			

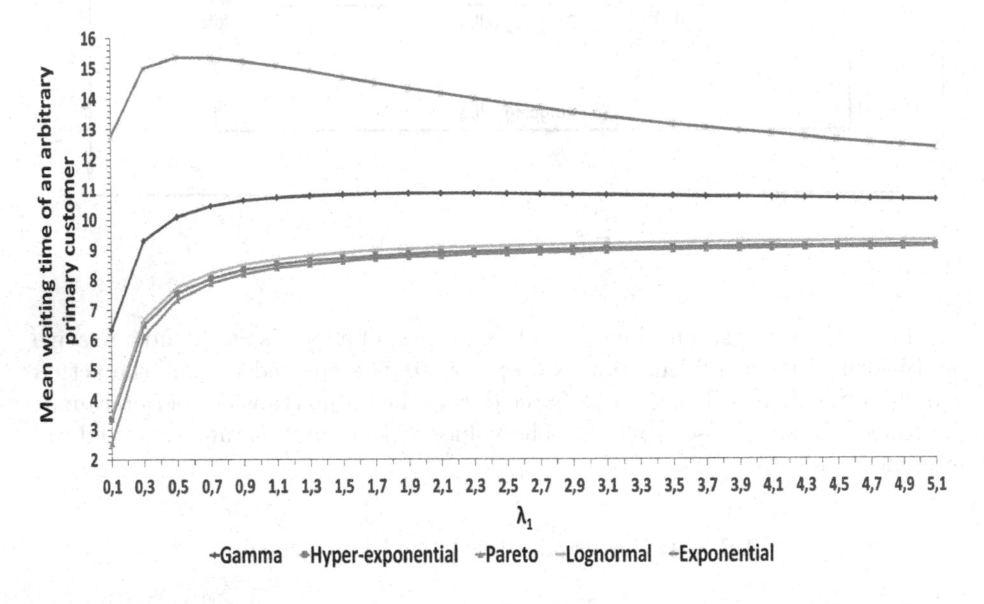

Fig. 2. Mean waiting time of an arbitrary primary customer vs. arrival intensity

The mean waiting time of calls for Case 2 is depicted as a function of the incoming generation rate on Figs. 2 and 3, and comparisons between the different cases are made. Figure 2 demonstrates five cases, the four different distributions, and the exponential case. In the scenario of exponential distribution, the maximum feature is observable, which is a general characteristic of retrial queues when the parameters are set appropriately. Among the distributions applied, the cases of gamma and exponential distribution result in a higher mean waiting time .

On Fig. 3 the comparison of the scenarios is shown using gamma distributed retrial time. The label "No outgoing" indicates that there are only incoming calls in the system, representing a typical finite source retrial system. This figure demonstrates the expected behavior of Cases 1 and 2, showing that Case 2 has a lower mean waiting time, but the lowest values are observed when there are no outgoing calls.

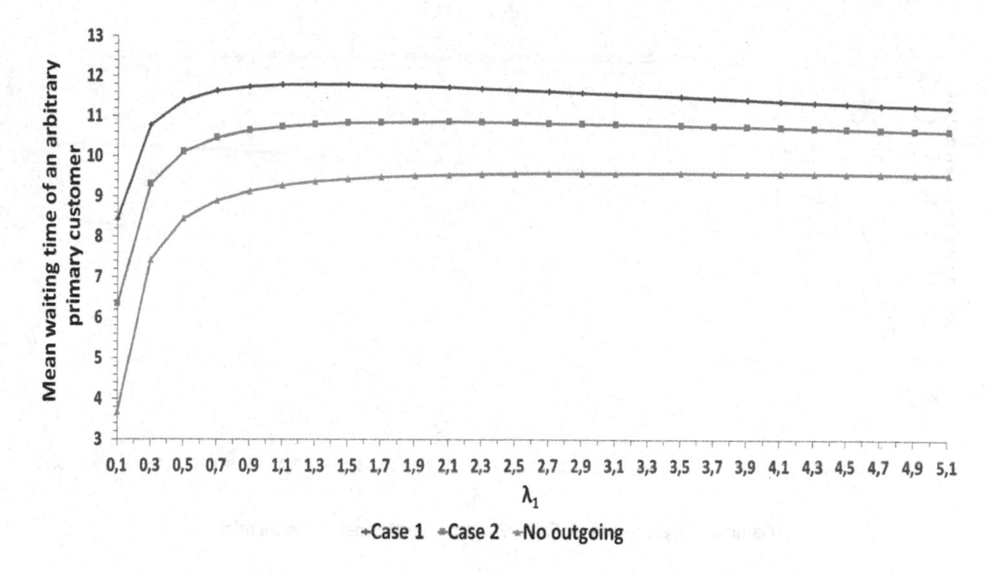

Fig. 3. Comparison of the mean waiting times of the different scenarios

The usage of the service unit is shown in relation to the arrival rate of incoming customers in Fig. 4. Despite having the same mean and variance, there are substantial differences between the different distributions. As the arrival rate increases, the utilization of the service unit also increases. The utilization rate is lower with the gamma and exponential distributions compared to the other distributions, particularly with the Pareto distribution.

Figure 5 demonstrates the comparison of the utilization of the service unit beside various scenarios. Here, it is clear that in cases where there is an outgoing call the server utilization is significantly higher which is true for Case 1 and Case 2 as well. Compared with Fig. 3, it is an optimization problem because there is not a single case where the average waiting time is the lowest and the occupancy rate is the highest. It can be stated that Case 2 can be an optimal solution in terms of mean waiting time and utilization. As λ_1 increases, the value of this performance measure begins to raise and after a certain value, the utilization becomes constant.

3.2 Second Scenario

Having seen the results of the first scenario, we also wondered if different parameter values were used for different distributions. We did not change the mean, but the squared coefficient of variation became less than 1 for the second scenario. Those parameters of each distribution are shown in Table 3, and all the other parameters remain the same (see Table 2). In order to conduct a sensitivity analysis, a hypo-exponential distribution is used instead of a hyper-exponential distribution.

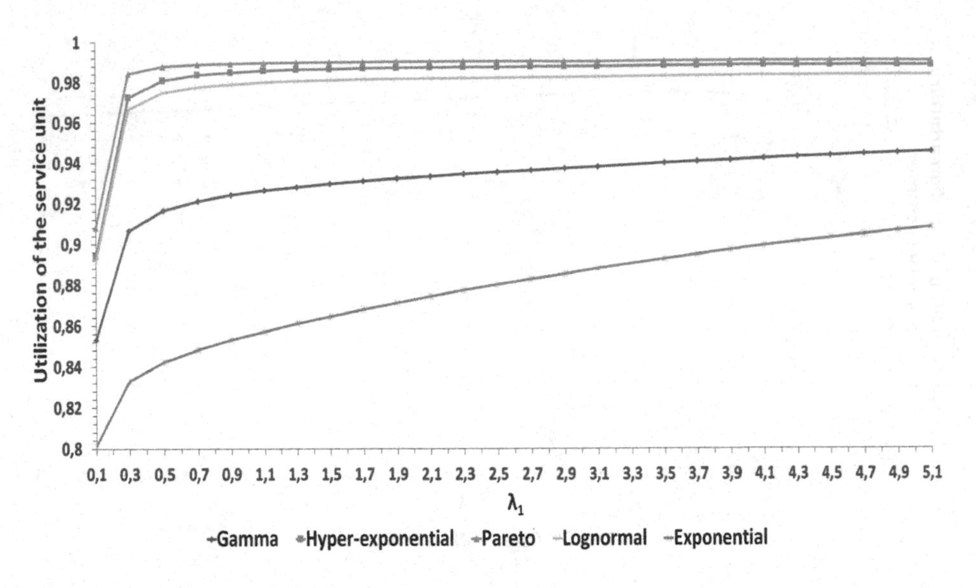

Fig. 4. The utilization of the service unit vs. arrival intensity

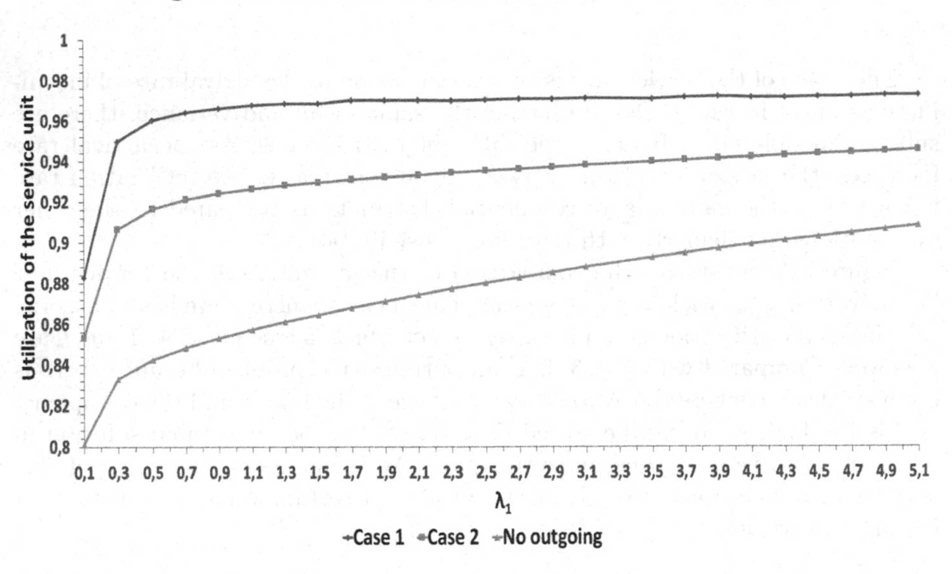

Fig. 5. Comparison of the utilization of the service unit of the different scenarios

In order to observe the similarities and differences between the two scenarios, we go through the same figures but with the modified parameter setting. Initially, we examine the mean waiting time of the calls as a function of the incoming call generation rate for Case 2 in Fig. 6. In this scenario, the plotted curves are perfectly coincident, we got back almost exactly the same results to the hundredth of a percent using all 4 distributions, except the exponential case.

Table 3. Parameters of the retrial time of the customers

Distribution	Gamma	Hypo-exponential	Pareto	Lognormal
Parameters	$\alpha = 1.4706$	$\mu_1 = 50$	$\alpha = 2.572$	$m = -2.562$
	$\beta = 14.706$	$\mu_2 = 12.5$	$k = 0.061$	$\sigma = 0.72$
Mean	0.1			
Variance	0.0068			
Squared coefficient of variation	0.68			

The same trend is observed in this scenario, as the intensity of incoming demand increases, the average waiting time starts to decrease after a certain value that is a characteristic of the retrial queues with appropriate parameters set.

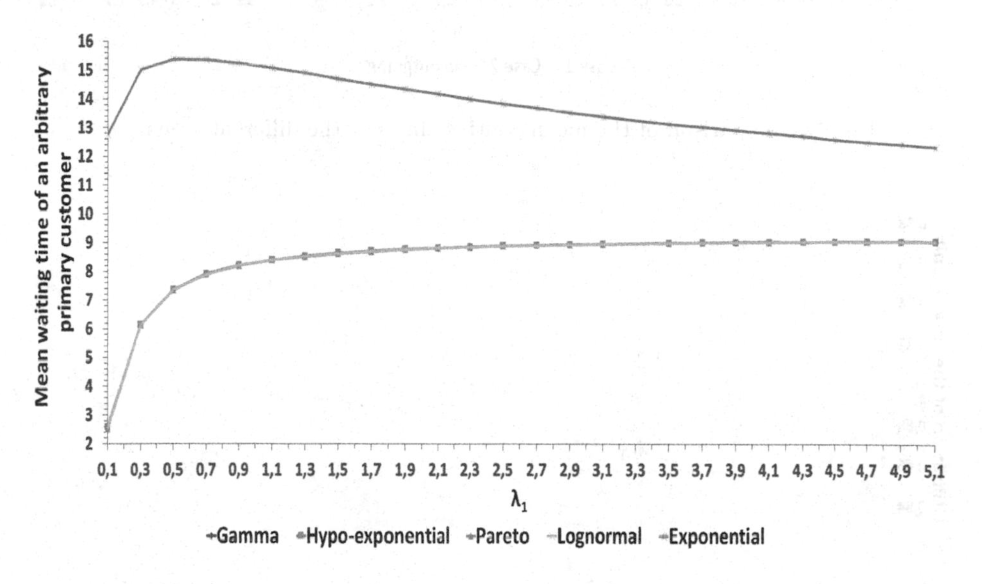

Fig. 6. Mean waiting time of an arbitrary primary customer vs. arrival intensity

Secondly, the comparison of the scenarios is exhibited using gamma distributed retrial time (Fig. 7). With this parameter setting, it is observed that the average waiting times are actually the same not only between the distributions but also between the different operating modes. In fact, external calls are not relevant here as the curves are essentially the same.

The next figure (Fig. 8) shows the utilization of the service unit besides increasing arrival intensity. From the results of the previous two graphs, it is perhaps not surprising that, in addition to the average latency, the server utilization is the same for all the distributions used except the exponential one where the utilization is much less compared with the others.

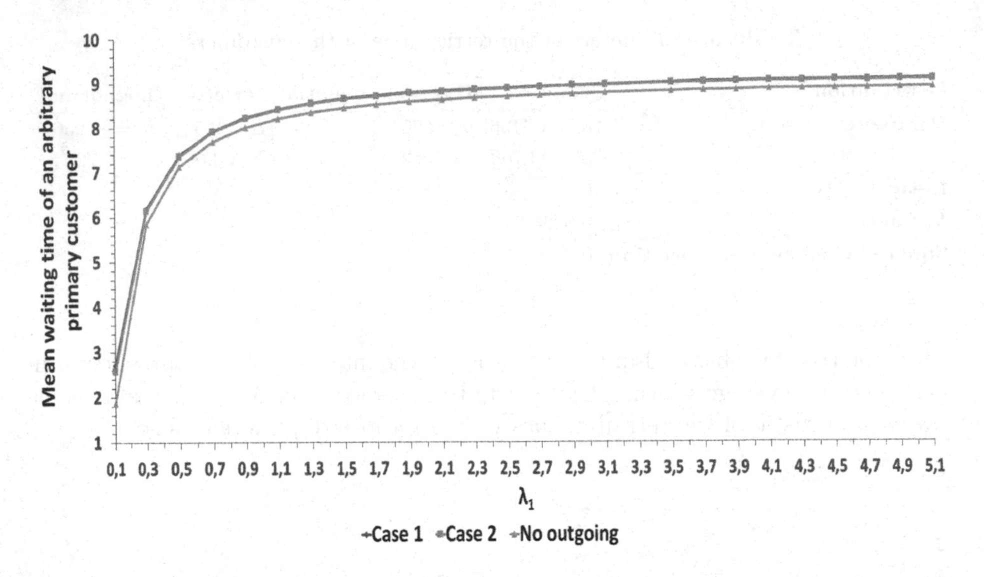

Fig. 7. Comparison of the mean waiting times of the different scenarios

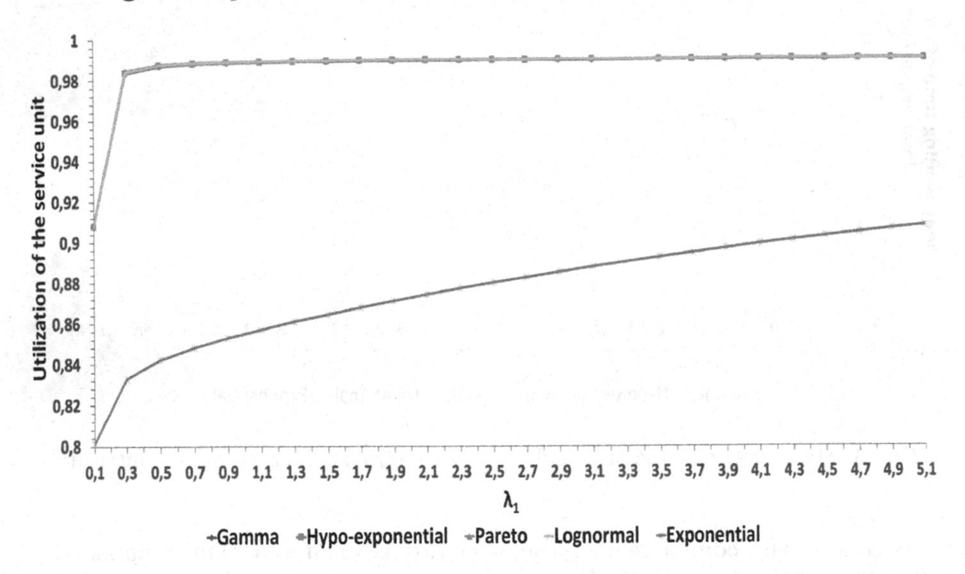

Fig. 8. The utilization of the service unit vs. arrival intensity

To emphasize the effect of this parameter setting, Fig. 9 illustrates the utilization of the service unit using different operation modes. It can be seen from the plotted curves that there is a difference between the cases studied at very low arrival intensities. When there are no external calls, the server utilization is significantly lower, but from lambda=0.3 the results are practically the same. This is not surprising after studying the previous figures.

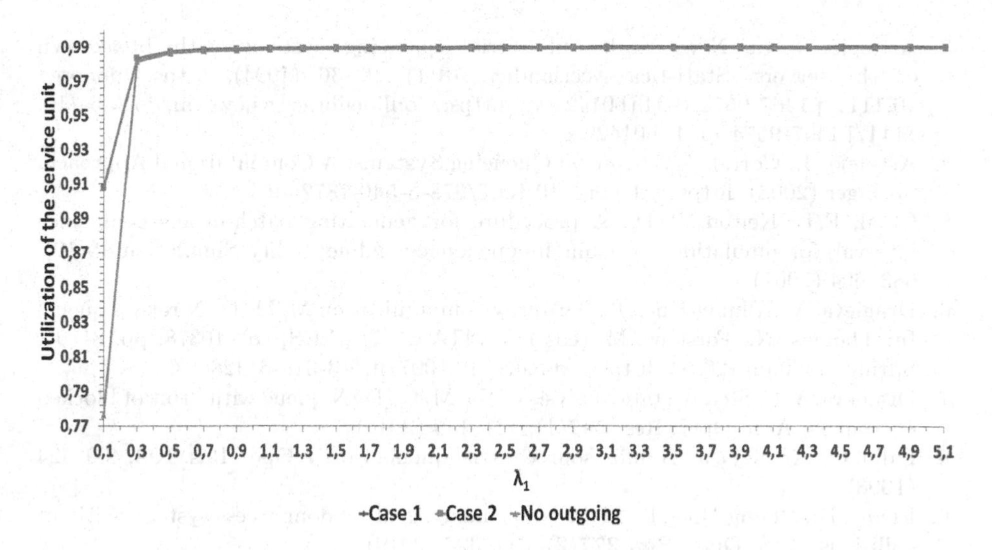

Fig. 9. Comparison of the utilization of the service unit of the different scenarios

4 Conclusion

In this research, we introduce a finite-source retrial queueing system with a two-way communication scheme that uses different distributions for retrial times. We examine multiple scenarios with varying parameters to carry out a sensitivity analysis and focus on the mean waiting time of customers and the utilization of the service unit. The results are obtained through simulations and demonstrated through graphical figures. The figures reveal that slight differences exist among the performance measures when the squared coefficient of variation is greater than one, highlighting the importance of choosing an appropriate distribution. The curves also show the effect of outgoing calls and suggest that in Case 2, the waiting time and utilization are better than in Case 1.

Taking the example of a bank, outgoing calls may be made for signature allocation, both outside and inside the bank as customers wait for transactions. It is more beneficial for the bank to keep a customer waiting inside the bank (Case 2) rather than sending them away or serving their initial request after the signature (Case 1).

Moving forward, we plan to continue our research by exploring other types of finite-source retrial queuing systems with two-way communication or adding a backup service unit.

References

1. Aguir, S., Karaesmen, F., Akşin, O.Z., Chauvet, F.: The impact of retrials on call center performance. OR Spectrum **26**(3), 353–376 (2004)
2. Aksin, Z., Armony, M., Mehrotra, V.: The modern call center: A multi-disciplinary perspective on operations management research. Prod. Oper. Manag. **16**(6), 665–688 (2007)

3. Artalejo, J.R.: New results in retrial queueing systems with breakdown of the servers. Statistica Neerlandica **48**(1), 23–36 (1994). https://doi.org/10.1111/j.1467-9574.1994.tb01429.x, https://onlinelibrary.wiley.com/doi/abs/10.1111/j.1467-9574.1994.tb01429.x
4. Artalejo, J., Corral, A.G.: Retrial Queueing Systems: A Computational Approach. Springer (2008). https://doi.org/10.1007/978-3-540-78725-9
5. Chen, E.J., Kelton, W.D.: A procedure for generating batch-means confidence intervals for simulation: Checking independence and normality. Simulation **83**(10), 683–694 (2007)
6. Dragieva, V., Phung-Duc, T.: Two-way communication M/M/1//N retrial queue. In: Thomas, N., Forshaw, M. (eds.) ASMTA 2017. LNCS, vol. 10378, pp. 81–94. Springer, Cham (2017). https://doi.org/10.1007/978-3-319-61428-1_6
7. Dragieva, V.I.: Steady state analysis of the M/G/1//N queue with orbit of blocked customers. Ann. Oper. Res. **247**(1), 121–140 (2016)
8. Falin, G., Artalejo, J.: A finite source retrial queue. Eur. J. Oper. Res. **108**, 409–424 (1998)
9. Fiems, D., Phung-Duc, T.: Light-traffic analysis of random access systems without collisions. Ann. Oper. Res. **277**(2), 311–327 (2019)
10. Fishwick, P.A.: Simpack: Getting started with simulation programming in c and c++. In: In 1992 Winter Simulation Conference, pp. 154–162 (1992)
11. Gómez-Corral, A., Phung-Duc, T.: Retrial queues and related models. Ann. Oper. Res. **247**(1), 1–2 (2016). https://doi.org/10.1007/s10479-016-2305-2
12. Jinting, W.: Reliability analysis M/G/1 queues with general retrial times and server breakdowns. Progress Natural Sci. **16**(5), 464–473 (2006). https://doi.org/10.1080/10020070612330021, https://www.tandfonline.com/doi/abs/10.1080/10020070612330021
13. Kim, J., Kim, B.: A survey of retrial queueing systems. Ann. Oper. Res. **247**(1), 3–36 (2016)
14. Kuki, A., Sztrik, J., Tóth, Á., Bérczes, T.: A contribution to modeling two-way communication with retrial queueing systems. In: Dudin, A., Nazarov, A., Moiseev, A. (eds.) ITMM/WRQ -2018. CCIS, vol. 912, pp. 236–247. Springer, Cham (2018). https://doi.org/10.1007/978-3-319-97595-5_19
15. Law, A.M., Kelton, W.D.: Simulation Modeling and Analysis. McGraw-Hill Education (1991)
16. Nazarov, A., Phung-Duc, T., Paul, S.: Heavy outgoing call asymptotics for MMP P/M/1/1 retrial queue with two-way communication. In: Dudin, A., Nazarov, A., Kirpichnikov, A. (eds.) Information Technologies and Mathematical Modelling. Queueing Theory and Applications, vol. 800, pp. 28–41. Springer International Publishing, Cham (2017)
17. Pustova, S.: Investigation of call centers as retrial queuing systems. Cybern. Syst. Anal. **46**(3), 494–499 (2010)
18. Sakurai, H., Phung-Duc, T.: Two-way communication retrial queues with multiple types of outgoing calls. TOP **23**(2), 466–492 (2015)
19. Sakurai, H., Phung-Duc, T.: Scaling limits for single server retrial queues with two-way communication. Ann. Oper. Res. **247**(1), 229–256 (2016)

An Explicit Solution for an Inventory Model with Server Interruption and Retrials

E Sandhya[1,2](✉), C. Sreenivasan[3], Smija Skaria[4], and Sajeev S. Nair[4]

[1] Jyothi Engineering College, Cheruthuruthy, Thrissur, Kerala, India
[2] Government College Chittur, University of Calicut, Malappuram, Kerala, India
esandhya1729@gmail.com
[3] Government Victoria College, Palakkad, Kerala, India
[4] Government Engineering College, Thrissur,APJ Abdul Kalam Technological University, Thiruvananthapuram, Kerala, India
sajeev@gectcr.ac.in

Abstract. Customers enter into a single server queuing model in accordance with a Poisson process where inventory is served. The inter service time follows exponential distribution. Upon arrival, finding the server busy the customers enter into an orbit from where they retry for service at a constant retrial rate. While the server serves a customer the service can be interrupted, the inter occurrence time of interruption being exponentially distributed. Following a service interruption the service restarts according to an exponentially distributed time. Inventory is replenished according to (s, S) policy, replenishment being instantaneous. For the model under discussion we assume that no inventory is lost due to server interruption, the customer being served when interruption occurs waits there until his service is completed and no arrivals or retrials are entertained and an order placed if any is cancelled while the server is on interruption. Explicit expression for the steady state distribution is calculated and several performance measures are evaluated explicitly and numerically. Graphs which show the variation of various performance measures with parameter values are also drawn.

Keywords: (s, S) inventory model · Positive lead time · Retrial · Server interruptions · Explicit solution

Introduction

The pioneers in the study of queueing inventory models are Melikov and Molchano [23] and Sigman and Simchi- Levi [25]. In Sigman and Simchi- Levi customers are allowed to join even when there is no inventory in the system. They also discuss the case of non exponential lead time distribution. Later Berman and et al. [3] considered an inventory system where a processing time is required for serving the inventory. Here they considered deterministic service time and

A. Dudin et al. (Eds.): ITMM 2022, CCIS 1803, pp. 149–161, 2023.
https://doi.org/10.1007/978-3-031-32990-6_13

the model was discussed as a dynamic programming model. Berman and Kim [4] and Berman and Sapna [5] later discussed inventory queueing systems with exponential service time distribution and with arbitrary distribution.

There are several papers on inventory queueing models by Krishnamoorthy and his co-authors [6,8,10,13–16,18,24]. They mainly used Matrix Analytic Methods to study these models. In most of the models service time for providing the inventoried item is assumed. Schwarz et al. considered a queuing inventory model with Poisson arrivals and exponentially distributed service and lead times. They could obtain a product form solution for the system steady state. But they assumed that no customers join the system when the inventory level is zero. For a detailed description of papers in inventory queuing models we refer to the papers [19,20]. Melikov et al. [22] studied a Queuing-Inventory System with Two Supply Sources and Destructive Customers. Melikov et al. [21] also carried out a numerical analysis and long run total cost optimization of a perishable queuing inventory systems with delayed feedback. They also did a long run total cost optimization for the problem.

Retrial queuing models are widely used in communication and other fields. Hence they are gaining more and more attention. We refer to the books by Falin and Templeton [7] and Atralejo and Gomez Corral [2] for an extensive analysis of both theory and applications on retrial queues.

The first study of an inventory queing model with positive lead time and retrial of customers was made by Artalejo et al. [1]. Analytical solution to the problem discussed there could be found in Ushakumari [17]. Following these, several papers in this direction emerged. A few among them are the papers by Krishnamoorthy and Islam [9,11], Krishnamoorthy et al. [10,16] and Krishnamoorthy and Jose [12]. These papers are studies on a production inventory model with retrial of customers, analysis of a production inventory model with random shelf times of the items with retrials of the orbiting customers, study of inventory models with positive service time and retrial of customers from an orbit with an intermediate buffer of finite and comparison of different (s, S) inventory models with an orbit of infinite capacity, having/ not having a finite buffer.

1 Mathematical Model

The system under consideration is described as below. We consider a single server queuing model where inventory is served to which customers arrive for service. The number of arrivals of by time t follows a Poisson process with parameter λ. Inventory is replenished according to (s, S) policy, replenishment being instantaneous. Service times follow exponential distribution with parameter μ. Upon arrival, finding the server busy the customers enter into an orbit from where they retry for service at a constant retrial rate. The time between two successive retrials also follow exponential distribution with parameter θ. While the server serves a customer the service can be interrupted, the inter occurrence time of interruption being exponentially distributed with parameter δ_1. Following a service interruption the service restarts after an exponentially distributed time with parameter δ_2. For the model under study the following assumptions are made.

i) No inventory is lost due to server interruption.
ii) The customer being served when interruption occurs waits there until his service is completed.
iii) No arrivals or retrials are entertained when the server is on interruption.
iv) An order placed if any is cancelled while the server is on interruption.

We denote by $N(t)$ the number of the customers in the orbit, $I(t)$ the inventory level and $S(t)$ the server status at time t. Let

$$S(t) = \begin{cases} 0 & \text{if the server is idle} \\ 1 & \text{if the server is busy} \\ 2 & \text{if the server is on interruption} \end{cases}$$

Then $\Omega = X(t) = (N(t), S(t), I(t))$ will be a Markov chain. The state space of this Markov chain can be described as $E = \{(i,j,k) : \ i \geq 0;\ j = 0,1,2;\ s+1 \leq k \leq S\}$. The above state space can be partitioned into levels $L(i)$ where $L(i) = ((i,j,k);\ j = 0,1,2;\ k = s+1, s+2, \ldots, S)\,;\ i \geq 0$ in the lexicographic ordering. The Markov chain Ω described above is a level independent quasi birth death process whose infinitesimal generator matrix is given by $T = \begin{bmatrix} B_0 & A_0 & 0 & 0 & - & - & - \\ A_2 & A_1 & A_0 & 0 & 0 & - & - \\ 0 & A_2 & A_1 & A_0 & 0 & 0 & 0 \\ 0 & 0 & A_2 & A_1 & A_0 & 0 & - \\ - & - & - & - & - & - & - \\ - & - & - & - & - & - & - \end{bmatrix}$. Here all the matrices are of order $3Q \times 3Q$ where $Q = S - s$. The different transitions in the Markov chain $\Omega = X(t) = (N(t), S(t), I(t))$ are given below.

i) Transitions due to arrivals
$(i,0,k) \xrightarrow{\lambda} (i,1,k)$, $(i,1,k) \xrightarrow{\lambda} (i+1,1,k)$; $i \geq 0,\ s+1 \leq k \leq S$
ii) Transitions due to service completion of customers
$(i,1,k) \xrightarrow{\mu} (i-1,0,k-1)$; $i \geq 0,\ s+2 \leq k \leq S$
$(i,1,s+1) \xrightarrow{\mu} (i-1,0,S)$; $i \geq 0$
iii) Transitions due to retrials $(i,0,k) \xrightarrow{\theta} (i-1,1,k)$; $i \geq 1$
iv) Transitions due to interruptions $(i,1,k) \xrightarrow{\delta_1} (i,2,k)$; $i \geq 0$
v) Transitions due to repairs $(i,2,k) \xrightarrow{\delta_1} (i,1,k)$; $i \geq 0$

The diagonal entries of B_0 and A_1 are such that each row sum of T is zero. The matrix B_0 contains the transition rates within level. Similarly the matrices A_0, A_1, A_2 contains the transitions from levels $L(i)$ to $L(i+1)$, $L(i)$ to $L(i)$ and $L(i)$ to $L(i-1)$ respectively.

2 Analysis of the Model

2.1 Stability Condition

Define $A = A_0 + A_1 + A_2$ and $\pi = (\pi(0, s+1), \ldots, \pi(0, S), \pi(1, s+1), \ldots,$ $\pi(1, S), \pi(2, s+1), \ldots, \pi(2, S))$ be the steady state vector of A. We know the QBD process with generator matrix T is stable if and only if $\pi A_0 e < \pi A_2 e$ (see Neuts). Since $A_0 = \begin{bmatrix} 0 & 0 & 0 \\ 0 & \lambda I_Q & 0 \\ 0 & 0 & 0 \end{bmatrix}$ and $A_2 = \begin{bmatrix} 0 & \theta I_Q & 0 \\ 0 & 0 & 0 \\ 0 & 0 & 0 \end{bmatrix}$, the stability condition reduces to
$\lambda\left[\pi(1, s+1) + \ldots + \pi(1, S)\right] < \theta\left[\pi(0, s+1) + \ldots + \pi(0, S)\right]$,
that is $\lambda\left(\frac{\lambda+\theta}{\mu}\right) < \theta$. Thus we have the following theorem for the stability of the system under study.

Theorem 1. *The Markov chain is stable if and only if* $\frac{\lambda}{\theta}\left(\frac{\lambda+\theta}{\mu}\right) < 1$.

2.2 Computation of Steady State Vector

We compute the steady state vector of the model explicitly. Let $\pi = (\pi_0, \pi_1, \pi_2, \ldots)$ be the steady state probability vector of the process Ω, where $\pi_i = (\pi(i, 0, s+1), \ldots, \pi(i, 0, S), \pi(i, 1, s+1), \ldots, \pi(i, 1, S), \pi(i, 2, s+1),$ $\ldots, \pi(i, 2, S))$; $i \geq 0$. Then π satisfies $\pi T = 0$ and $\pi e = 1$. We have the equations $\pi_0 B_0 + \pi_1 A_2 = 0, \quad \pi_i A_0 + \pi_{i+1} A_1 + \pi_{i+2} A_2 = 0;\ i \geq 0$
We first consider a system identical to the above system expect for no inventory is served. This system $\tilde{\Omega} = \tilde{X}(t) = (N(t), S(t))$ will be a Markov chain where $N(t)$ and $S(t)$ is as defined for the original system. The state space of this Markov chain can be described as $\tilde{E} = \{(i, 0), (i, 1), (i, 2)\};\ i \geq 0$. The infinitesimal generator matrix of the process is given by $\tilde{T} = \begin{bmatrix} \tilde{B}_0 & \tilde{A}_0 & 0 & 0 & - & - & - \\ \tilde{A}_2 & \tilde{A}_1 & \tilde{A}_0 & 0 & 0 & - & - \\ 0 & \tilde{A}_2 & \tilde{A}_1 & \tilde{A}_0 & 0 & 0 & 0 \\ 0 & 0 & \tilde{A}_2 & \tilde{A}_1 & \tilde{A}_0 & 0 & - \\ - & - & - & - & - & - & - \\ - & - & - & - & - & - & - \end{bmatrix}$, where $\tilde{B}_0 =$
$\begin{bmatrix} -\lambda & \lambda & 0 \\ \mu & -(\lambda+\mu+\delta_1) & \delta_1 \\ 0 & \delta_2 & -\delta_2 \end{bmatrix}$, $\tilde{A}_1 = \begin{bmatrix} (\lambda+\theta) & \lambda & 0 \\ \mu & -(\lambda+\mu+\delta_1) & \delta_1 \\ 0 & \delta_2 & -\delta_2 \end{bmatrix}$, $\tilde{A}_0 = \begin{bmatrix} 0 & 0 & 0 \\ 0 & \lambda & 0 \\ 0 & 0 & 0 \end{bmatrix}$
and $\tilde{A}_2 = \begin{bmatrix} 0 & \theta & 0 \\ 0 & 0 & 0 \\ 0 & 0 & 0 \end{bmatrix}$. Let $x = (x_0, x_1, \ldots)$, where $x_i = (x(i, 0), x(i, 1), x(i, 2))$ be the steady state probability vector of the process $\tilde{\Omega}$. The steady state equations

are given by $x\tilde{T} = 0$.

$$-\lambda x(0,0) + \mu x(0,1) = 0 \quad (1)$$
$$\lambda x(0,0) - (\lambda + \mu + \delta_1)x(0,1) + \delta_2 x(0,2) + \theta x(1,0) = 0 \quad (2)$$
$$\delta_1 x(0,1) - \delta_2 x(0,2) = 0 \quad (3)$$
$$-(\theta + \lambda)x(i,0) + \mu x(i,1) = 0;\ i \geq 1 \quad (4)$$
$$\lambda x(i-1,1) + \lambda x(i,0) - (\lambda + \mu + \delta_1)x(i,1) + \delta_2 x(i,2) + \theta x(i+1,0) = 0;\ i \geq 1 \quad (5)$$
$$\delta_1 x(i,1) - \delta_2 x(i,2) = 0;\ i \geq 1 \quad (6)$$

We know that $x_i = x_{i-1}R;\ i \geq 1$, where the matrix R satisfies $R^2A_2 + RA_1 + A_0 = 0$. Since the first and third rows of A_0 are zeros, so are that of R. From $R^2A_2 + RA_1 + A_0 = 0$, we obtain

$$\begin{bmatrix} 0 & 0 & 0 \\ 0 & r_1 r_2 \theta & 0 \\ 0 & 0 & 0 \end{bmatrix} + \begin{bmatrix} 0 & 0 & 0 \\ -(\lambda+\theta)r_1 + \mu r_2 & \lambda r_1 - (\lambda + \mu + \delta_1)r_2 + \delta_2 r_3 & r_2\delta_1 - r_3\delta_2 \\ 0 & 0 & 0 \end{bmatrix} + \begin{bmatrix} 0 & 0 & 0 \\ 0 & \lambda & 0 \\ 0 & 0 & 0 \end{bmatrix} = 0$$

We have the following equations. $\mu r_2 = (\lambda + \theta)r_1$, $\delta_1 r_2 = \delta_2 r_3$, $\lambda r_1 - (\lambda + \mu)r_2 + r_1 r_2 \theta + \lambda = 0$. From the above 3 equations we obtain a quadratic in r_2 as $\mu\theta r_2^2 - (\lambda^2 + \mu\theta + \lambda\theta)r_2 + \lambda(\lambda+\theta) = 0$. Clearly the roots of this equation are 1 and $\frac{\lambda(\lambda+\theta)}{\theta\mu}$. It may be noted that the stability condition was $\frac{\lambda(\lambda+\theta)}{\theta\mu} < 1$.

Hence $R = \begin{bmatrix} 0 & 0 & 0 \\ \frac{\lambda}{\theta} & \frac{\lambda(\lambda+\theta)}{\theta\mu} & \frac{\lambda(\lambda+\theta)\delta_1}{\theta\mu\delta_2} \\ 0 & 0 & 0 \end{bmatrix}$. Now from $x_0\tilde{B}_0 + x_1\tilde{A}_2 = 0$, we have the equations $-\lambda x(0,0) + \mu x(0,1) = 0$; $\delta_1 x(0,1) - \delta_2 x(0,2) = 0$. Hence $x_0 = (x(0,0), x(0,1), x(0,2)) = \left(1, \frac{\lambda}{\mu}, \frac{\lambda\delta_1}{\mu\delta_2}\right) x(0,0)$ and from the normalizing condition $x_0(I-R)^{-1}e = 1$, where I is the identity matrix of order 3 and e is column vector of ones we get $x(0,0) = \frac{1 - \frac{\lambda(\lambda+\theta)}{\theta\mu}}{1 + \frac{\delta_1\lambda}{\delta_2\mu}}$. Now the equations $\pi T = 0$ are given by

$$\begin{aligned} &-\lambda\pi(0,0,i) + \mu\pi(0,1,i+1) = 0;\ s+1 \leq i \leq S-1 \\ &-\lambda\pi(0,0,S) + \mu\pi(0,1,s+1) = 0 \end{aligned} \quad (7)$$

$$\lambda\pi(0,0,i) - (\lambda + \mu + \delta_1)\pi(0,1,i) + \delta_2\pi(0,2,i) + \theta\pi(1,0,i) = 0;\ s+1 \leq i \leq S \quad (8)$$
$$\delta_1\pi(0,1,i) - \delta_2\pi(0,2,i) = 0;\ s+1 \leq i \leq S \quad (9)$$

$$\begin{aligned} -(\lambda+\theta)\pi(i,0,j)+\mu\pi(i,1,j+1)=0;\ i\geq 1,\ s+1\leq i\leq S-1 \\ -(\lambda+\theta)\pi(i,0,S)+\mu\pi(i,1,s+1)=0;\ i\geq 1 \end{aligned} \tag{10}$$

$$\begin{aligned} \lambda\pi(i-1,0,i)+\lambda\pi(i,0,i)-(\lambda+\mu+\delta_1)\pi(i,1,i)+ \\ \delta_2\pi(i,2,i)+\theta\pi(i+1,0,i)=0;\ i\geq 1,\ s+1\leq i\leq S \end{aligned} \tag{11}$$

$$\delta_1\pi(i,1,j)-\delta_2\pi(i,2,j)=0;\ i\geq 1,\ s+1\leq j\leq S \tag{12}$$

We assume $\pi(i,j,k)=\frac{1}{Q}x(i,j);\ s+1\leq k\leq S$. Then the first Q equations of $\pi_0B_0+\pi_1A_2=0$ reduces to the first equation of $x_0\tilde{B}_0+x_1\tilde{A}_2=0$; the next Q equations of $\pi_0B_0+\pi_1A_2=0$ reduces to the second equation of $x_0\tilde{B}_0+x_1\tilde{A}_2=0$ and the last Q equations of $\pi_0B_0+\pi_1A_2=0$ reduces to the last equation of $x_0\tilde{B}_0+x_1\tilde{A}_2=0$. Similarly the first Q equations of $\pi_0A_0+\pi_1A_1+\pi_2A_2=0$ reduces to the first equation of $x_0\tilde{A}_0+x_1\tilde{A}_1+x_2\tilde{A}_2=0$; the next Q equations of $\pi_0A_0+\pi_1A_1+\pi_2A_2=0$ reduces to the second equation of $x_0\tilde{A}_0+x_1\tilde{A}_1+x_2\tilde{A}_2=0$ and the last Q equations of $\pi_0A_0+\pi_1A_1+\pi_2A_2=0$ reduces to the last equation of $x_0\tilde{A}_0+x_1\tilde{A}_1+x_2\tilde{A}_2=0$. The intuition behind this is that since the replenishment is instantaneous there is an equal probability for each inventory level to be visited. It is verified that the above values satisfy $\pi T=0$.

3 System Performance Measures

3.1 Expected Number of Interruptions Encountered by a Customer

For computing expected number of interruptions encountered by a customer we consider a Markov process $\{X_1(t),t\geq 0\}=\{(N_1(t),S_1(t)),t\geq 0\}$, where $N_1(t)$ denotes the number of interruptions that has occurred up to time t; $S_1(t)=0$ or 1 according as the service is under interruption or not at time t. The Markov process $\{X_1(t),t\geq 0\}$ has state space $\{0,1,2,\ldots\}\times\{0,1\}\cup\{\Delta\}$, where Δ is an absorbing state which denotes service completion. The infinitesimal generator of the process is given by $\hat{U}=\begin{bmatrix} 0 & 0 & 0 & 0 & 0 & \ldots \\ \hat{B}_{00} & \hat{A}_{00} & \hat{A}_{01} & 0 & 0 & \ldots \\ \hat{A}_2 & 0 & \hat{A}_1 & \hat{A}_0 & 0 & \ldots \\ \hat{A}_2 & 0 & 0 & \hat{A}_1 & \hat{A}_0 & \ldots \\ \hat{A}_2 & 0 & 0 & 0 & \hat{A}_1 & \hat{A}_0 \ldots \\ & & & \cdot & \cdot & \cdot\ \cdot\ \cdot \\ & & & & \cdot & \cdot\ \cdot\ \cdot \\ & & & & \cdot & \cdot\ \cdot\ \cdot \end{bmatrix}$,

$\hat{B}_{00}=[\mu]$, $\hat{A}_{00}=[-(\mu+\delta_1)]$, $\hat{A}_{01}=[\delta_1\ 0]$, $\hat{A}_2=\begin{bmatrix}0\\ \mu\end{bmatrix}$, $\hat{A}_1=\begin{bmatrix}-\delta_2 & \delta_2\\ 0 & -(\mu+\delta_1)\end{bmatrix}$ and $\hat{A}_0=\begin{bmatrix}0 & 0\\ \delta_1 & 0\end{bmatrix}$. If y_k is the probability that absorption occurs with exactly k interruptions, then $y_0=-\hat{A}_{00}^{-1}\hat{B}_{00}=\frac{\mu}{\mu+\delta_1}$

$y_k = (-\hat{A}_{00}^{-1}\hat{A}_{01})(-\hat{A}_1{}^{-1}\hat{A}_0)^{k-1}(-\hat{A}_1{}^{-1}\hat{A}_2) = \frac{\mu}{\mu+\delta_1}\left(\frac{\delta_1}{\mu+\delta_1}\right)^k, k =$

$1, 2, 3, \ldots$
The expected number of interruptions before absorption is given by
$E_1 = \sum_{k=0}^{\infty} k y_k = \left(-\hat{A}_{00}^{-1}\hat{A}_{01}\right)\left[I_2 - \left(-\hat{A}_1^{-1}\hat{A}_0\right)\right]^{-1} e = \frac{\delta_1}{\mu}.$

3.2 Expected Duration of an Interrupted Service

Here we calculate the average duration of an interrupted service. The service process with interruption can be viewed as a Markov process with two transient states 0 and 1, which denote whether the server is interrupted or is busy respectively, and a single absorption state Δ. Let $\hat{X}(t) = \{0, 1, \Delta$ be the corresponding process. The infinitesimal generator matrix of the process is given by $\hat{H} = \begin{bmatrix} \hat{B} & \hat{B}_0 \end{bmatrix}$, where $\hat{B} = \begin{bmatrix} -\delta_2 & \delta_2 \\ \delta_1 & -(\mu + \delta_1) \end{bmatrix}$ and $\hat{B}_0 = \begin{bmatrix} 0 \\ \mu \end{bmatrix}$. The probability distribution of T, the time until absorption is given by $F(x) = 1 - \zeta \exp(\hat{B}x)e$, $x \geq 0$. Its density function $F'(x)$ given by $F'(x) = \zeta \exp(\hat{B}x)\hat{B}_0$. The Laplace-Stieltjes transform $f(s)$ is $f(s) = \zeta(sI - \hat{B})^{-1}\hat{B}_0$. The expected time E_s for service completion is given by $E_s = \zeta(-\hat{B})^{-1}e = \frac{\delta_1 + \delta_2}{\mu\delta_2}$

3.3 Other Performance Measures

1. The probability that the server is busy
   $PSB = \sum_{i=0}^{\infty} \sum_{k=s+1}^{S} \pi(i, 1, k) = \left[\frac{\lambda\theta}{\theta\mu - \lambda^2 - \lambda\theta}\right] x(0, 0)$
2. The probability that the server is on interruption
   $PSI = \sum_{i=1}^{\infty} \sum_{k=s+1}^{S} \pi(i, 2, k) = \left(\frac{\delta_1}{\delta_2}\right) PSB$
3. The probability that the server is idle
   $PSID = \sum_{i=0}^{\infty} \sum_{k=s+1}^{S} \pi(i, 0, k) = \left[1 + \frac{\lambda^2}{\theta\mu - \lambda^2 - \lambda\theta}\right] x(0, 0)$
4. The expected inventory level in the system
   $EIL = \sum_{i=1}^{\infty} \sum_{j=0}^{2} \sum_{k=s+1}^{S} k\pi(i, j, k) = \frac{Q(S + s + 1)}{2}$
5. The expected number of customers in the orbit
   $ENCO = \sum_{i=0}^{\infty} \sum_{j=0}^{2} \sum_{k=s+1}^{S} i\pi(i, j, k)$
6. The expected rate of ordering, $ERO = \sum_{i=1}^{\infty} \mu\pi(i, s + 1)$

4 Numerical Illustration

(See Tables 1, 2, 3, 4, 5 and 6)

Table 1. Effect of regular arrival rate on the various performance measures

$\mu = 9$ $\theta = 5$ $\delta_1 = 2$ $\delta_2 = 3$ $s = 5$ $S = 11$

λ	PSB	PINT	PIDL	EIL	ENCO	ERO
3	0.4545	0.1818	0.3637	8.5	0.8701	0.1928
3.2	0.479	0.1916	0.3294	8.5	1.1112	0.2143
3.4	0.5029	0.2011	0.296	8.5	1.4354	0.2364
3.6	0.5274	0.2109	0.2617	8.5	1.9167	0.2603
3.8	0.5562	0.2201	0.2237	8.5	2.6055	0.2837
4	0.5603	0.2241	0.2156	8.5	3.0369	0.2943
4.2	0.5932	0.2372	0.1696	8.5	5.7195	0.3307
4.4	0.6145	0.2458	0.1397	8.5	10.9939	0.3555
4.6	0.6353	0.2541	0.1106	8.5	52.0843	0.3808

Table 2. Effect of retrial rate on the various performance measures

$\lambda = 3$ $\mu = 9$ $\delta_1 = 2$ $\delta_2 = 3$ $s = 5$ $S = 11$

θ	PSB	PINT	PIDL	EIL	ENCO	ERO
4	0.4545	0.1818	0.3637	8.5	1.1272	0.241
4.2	0.4545	0.1818	0.3637	8.5	1.0575	0.229
4.4	0.4545	0.1818	0.3637	8.5	1.0031	0.2191
4.6	0.4545	0.1818	0.3637	8.5	0.953	0.2096
4.8	0.4545	0.1818	0.3637	8.5	0.909	0.2008
5	0.4545	0.1818	0.3637	8.5	0.8701	0.1928
5.2	0.4545	0.1818	0.3637	8.5	0.8353	0.1854
5.4	0.4545	0.1818	0.3637	8.5	0.8041	0.1785
5.6	0.4545	0.1818	0.3637	8.5	0.7759	0.1721

Table 3. Effect of interruption rate on the various performance measures

$\lambda = 3$ $\mu = 9$ $\theta = 5$ $\delta_2 = 3$ $s = 5$ $S = 11$

δ_1	PSB	PINT	PIDL	EIL	ENCO	ERO
1	0.3999	0.0999	0.5002	8.5	0.8429	0.2076
1.2	0.4117	0.1176	0.4707	8.5	0.8487	0.2045
1.4	0.423	0.1346	0.4424	8.5	0.8543	0.2014
1.6	0.4339	0.1509	0.4152	8.5	0.8598	0.1985
1.8	0.4444	0.1666	0.389	8.5	0.865	0.1956
2	0.4545	0.1818	0.3637	8.5	0.8701	0.1928
2.2	0.4642	0.1964	0.3394	8.5	0.8749	0.1901
4 2.4	0.4736	0.2105	0.3159	8.5	0.8796	0.1875
2.6	0.4827	0.2241	0.2932	8.5	0.8842	0.1849

Table 4. Effect of repair rate on the various performance measures

$\lambda = 3$ $\mu = 9$ $\theta = 5$ $\delta_1 = 2$ $s = 5$ $S = 11$						
δ_2	PSB	PINT	PIDL	EIL	ENCO	ERO
3	0.4545	0.1818	0.3637	8.5	0.8701	0.1928
3.2	0.4482	0.1724	0.3794	8.5	0.8669	0.1945
3.4	0.4426	0.1639	0.3935	8.5	0.8641	0.1961
3.6	0.4374	0.1562	0.4064	8.5	0.8616	0.1975
3.8	0.4328	0.1492	0.418	8.5	0.8592	0.1988
4	0.4285	0.1428	0.4287	8.5	0.8571	0.1999
4.2	0.4246	0.1369	0.4385	8.5	0.8551	0.201
4.4	0.421	0.1315	0.4475	8.5	0.8533	0.202
4.6	0.4177	0.1266	0.4557	8.5	0.8517	0.2029

Table 5. Effect of service rate on the various performance measures

$\lambda = 3$ $\theta = 5$ $\delta_1 = 2$ $\delta_2 = 3$ $s = 5$ $S = 11$						
μ	PSB	PINT	PIDL	EIL	ENCO	ERO
8	0.4999	0.1999	0.3002	8.5	1.1999	0.1846
8.2	0.4901	0.196	0.3139	8.5	1.1176	0.1863
8.4	0.4807	0.1923	0.327	8.5	1.0448	0.188
8.6	0.4716	0.1886	0.3398	8.5	0.9801	0.1897
8.8	0.4629	0.1851	0.352	8.5	0.9222	0.1913
9	0.4545	0.1818	0.3637	8.5	0.8701	0.1928
9.2	0.4464	0.1785	0.3751	8.5	0.823	0.1943
9.4	0.4385	0.1754	0.3861	8.5	0.7803	0.1958
9.6	0.431	0.1724	0.3966	8.5	0.7413	0.1972

Table 6. Effect of reorder level on the various performance measures

$\lambda = 3$ $\mu = 9$ $\theta = 5$ $\delta_1 = 2$ $\delta_2 = 3$ $S = 21$						
s	PSB	PINT	PIDL	EIL	ENCO	ERO
3	0.3333	0.1818	0.4849	12.5	0.8701	0.0642
4	0.3582	0.1818	0.46	13	0.8701	0.068
5	0.3863	0.1818	0.4319	13.5	0.8701	0.0723
6	0.4181	0.1818	0.4001	14	0.8701	0.0771
7	0.4545	0.1818	0.3637	14.5	0.8701	0.0826
8	0.4545	0.1818	0.3637	15	0.8701	0.089
9	0.4545	0.1818	0.3637	15.5	0.8701	0.0964
10	0.4545	0.1818	0.3637	16	0.8701	0.1056

5 Graphical Illustration

(See Figs. 1, 2, 3 and 4)

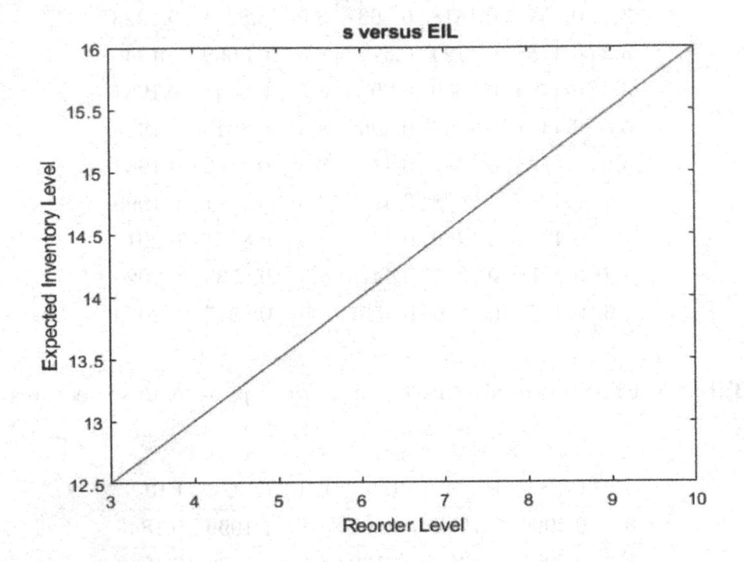

Fig. 1. xx

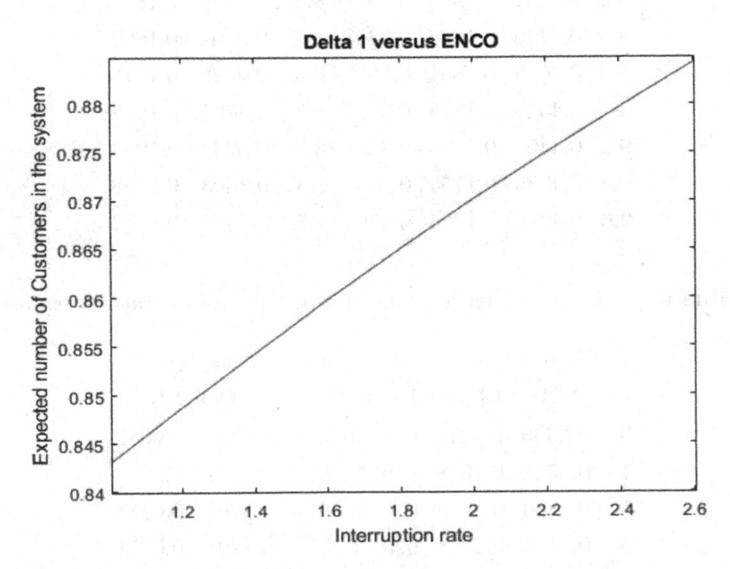

Fig. 2. xxx

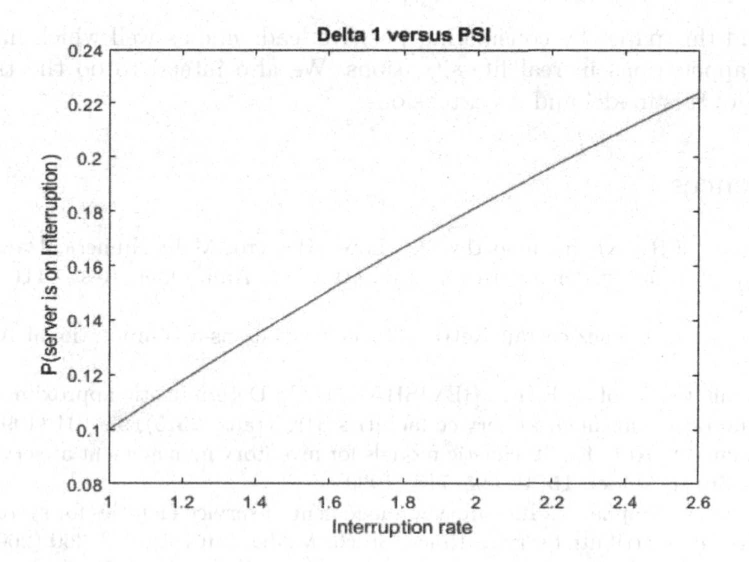

Fig. 3. cccc

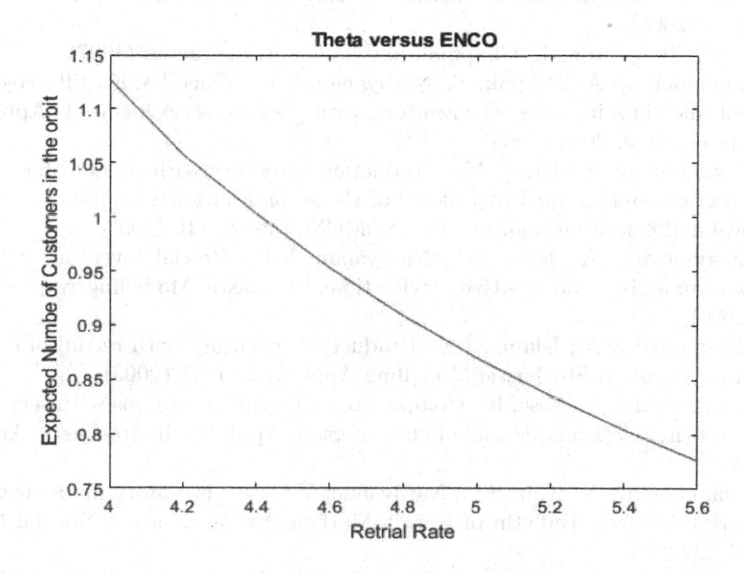

Fig. 4. xxxx

6 Conclusion

In this paper we could derive an explicit expression for the steady state probability vector of an inventory queuing model with retrial and server interruptions. Several other performance measures such as expected waiting time of a customer in the orbit, average duration of an interrupted service and so on can be calculated explicitly. An optimisation of a cost function may be done. We wish to extend this paper by considering positive lead time as well which may have several applications in real life situations. We also intend to do the transient analysis of this model and its extensions.

References

1. Artalejo, J.R., Krishnamoorthy, A., Lopez-Herrero, M.J.: Numerical analysis of (s, s) inventory systems with repeated attempts. Ann. Oper. Res. **141**(1), 67–83 (2006)
2. Artalejo, J.: Gomez-corral. Retrial Queueing systems-a computational Approach (2008)
3. Berman, O., Kaplan, E.H., SHEVISHAK, D.G.: Deterministic approximations for inventory management at service facilities. IIE Trans. **25**(5), 98–104 (1993)
4. Berman, O., Kim, E.: Stochastic models for inventory management at service facilities. Stoch. Model. **15**(4), 695–718 (1999)
5. Berman, O., Sapna, K.: Inventory management at service facilities for systems with arbitrarily distributed service times. Stoch. Model. **16**(3–4), 343–360 (2000)
6. Deepak, T., Krishnamoorthy, A., Narayanan, V.C., Vineetha, K.: Inventory with service time and transfer of customers and/inventory. Ann. Oper. Res. **160**(1), 191–213 (2008)
7. Falin, G., Templeton, J.: Chapman and hall. Retrial Queues (1997)
8. Krishnamoorthy, A., Deepak, T., Narayanan, V.C., Vineetha, K.: Effective utilization of idle time in an (s, s) inventory with positive service tim. J. Appl. Math. Stochastic Anal. 2006 (2006)
9. Krishnamoorthy, A., Islam, M.: Production inventory with random life time and retrial of customers. In: Proceedings of the second national conference on mathematical and computational models, NCMCM. pp. 89–110 (2003)
10. Krishnamoorthy, A., Islam, M., Narayanan, V.C.: Retrial inventory with batch Markovian arrival and positive service time. Stochastic Modelling Appl. **9**(2), 38–53 (2006)
11. Krishnamoorthy, A., Islam, M.E.: Production inventory with retrial of customers in an (s, s) policy. Stochastic Modelling Appl. **6**(2), 1–11 (2003)
12. Krishnamoorthy, A., Jose, K.: Comparison of inventory systems with service, positive lead-time, loss, and retrial of customers. J. Appl. Math. Stochastic Anal. 2007 (2008)
13. Krishnamoorthy, A., Nair, S.S., Narayanan, V.C.: An inventory model with retrial and orbital search, Bulletin of Kerala Mathematics association, Special issue pp. 47–65 (2009)
14. Krishnamoorthy, A., Narayanan, V.C., Deepak, T., Vineetha, P.: Control policies for inventory with service time. Stoch. Anal. Appl. **24**(4), 889–899 (2006)
15. Krishnamoorthy, A., Narayanan, V.C., Islam, M.: Retrial production inventory with map and service times; queues, flows, systems, networks. In: Proceedings of the International Conference "Modern Mathematical Methods of Analysis and Optimization of Telecommunication Networks, pp. 148–156 (2003)
16. Krishnamoorthy, A., Narayanan, V.C., Islam, M.: On production inventory with service time and retrial of customers. In: Proceedings of the 11th International Conference on Analytical and Stochastic Modelling Techniques and Analysis, pp. 238–247. SCS-Publishing House Magdeburg (2004)

17. Krishnamoorthy, A., Ushakumari, P.: On an m/g/1 retrial queue with disaster to the customer in service. In: International Workshop on Retrial Queues, Madrid (1998)
18. Krishnamoorthy, A., Jose, K., Narayanan, V.C.: Numerical investigation of a ph/ph/1 inventory system with positive service time and shortage. Neural Parallel Sci. Comput. **16**(4), 579–591 (2008)
19. Krishnamoorthy, A., Manikandan, R., Lakshmy, B.: A revisit to queueing-inventory system with positive service time. Ann. Oper. Res. **233**(1), 221–236 (2015)
20. Krishnamoorthy, A., Shajin, D., Narayanan, W.: Inventory with positive service time: a survey. Advanced Trends in Queueing Theory. Series of Books "Mathematics and Statistics" Sciences 2, pp. 201–238 (2021)
21. Melikov, A., Krishnamoorthy, A., Shahmaliyevy, M.: Numerical analysis and long run total cost optimization of perishable queuing inventory systems with delayed feedback. Queueing Models Serv. Manag. **2**(1), 83–111 (2019)
22. Melikov, A., Mirzayev, R., Nair, S.: Numerical study of a queuing-inventory system with two supply sources and destructive customers. J. Comput. Syst. Sci. Int. **61**(4), 581–598 (2022)
23. Melikov, A., Molchanov, A.: Stock optimization in transportation/storage systems. Cybern. Syst. Anal. **28**(3), 484–487 (1992)
24. Narayanan, V.C., Deepak, T., Krishnamoorthy, A., Krishnakumar, B.: On an (s, s) inventory policy with service time, vacation to server and correlated lead time. Quality Technol. Quantitative Manage. **5**(2), 129–143 (2008)
25. Sigman, K., Simchi-Levi, D.: Light traffic heuristic for anm/g/1 queue with limited inventory. Ann. Oper. Res. **40**(1), 371–380 (1992)

Asymptotic Analysis of the M/M/1/N−1 Retrial System with Priority and Feedback

S. V. Rozhkova[1,2] and E. Yu. Titarenko[1,2](✉)

[1] National Research Tomsk State University, Lenin Avenue 36, Tomsk 634050, Russia
[2] National Research Tomsk Polytechnic University, Lenin Avenue 30, Tomsk 634050, Russia
{rozhkova,teu}@tpu.ru
https://tsu.ru, https://tpu.ru

Abstract. The retrial queuing system M/M/1/N−1 with instantaneous and delayed feedback is studied. The system has one server and a buffer of capacity N–1. Priority customers come with a certain probability and get into service if the server is free. If the server is busy, they get into the queue. If the buffer is full, then the customer goes into an orbit, where it waits for a random time and tries again to get into service or into the queue. Non-priority customers immediately go into the orbit. After service, the customer either leaves the system, or goes into the orbit or immediately occupies the server again.

An asymptotic analysis method is used to find the stationary distribution of the customers quantity in the orbit. An asymptotic condition is a long delay between customers from the orbit. The paper shows that the asymptotic probability distribution of the customers quantity in the orbit under the condition of an increasing average waiting time in the orbit is Gaussian. Equations for the distribution parameters are obtained.

Keywords: Asymptotic analysis · Priority · Feedback

1 Introduction

Classical queuing theory studies systems with possible loss of customers and systems with waiting space. i.e. systems that have limited or unlimited input buffer. Customers arriving when the server is busy are placed in the buffer and served later when the server becomes free. The study of systems with a finite buffer is of great interest [1]. One of the reasons for this fact is the rapid development of telecommunications networks. In [2], a finite buffer steady state queuing system for three heterogeneous servers is analyzed.

The growth in the amount of information and the number of users leads to high load on network servers. Due to limited buffer capacity, situations when the buffer overflows sometimes happen. Systems with retrial calls (RQ-systems) have a different queue discipline [3,4]. When the server is busy and the buffer is full,

A. Dudin et al. (Eds.): ITMM 2022, CCIS 1803, pp. 162–174, 2023.
https://doi.org/10.1007/978-3-031-32990-6_14

the customers arriving at the system go into the orbit and later repeat attempt to be served. In [5], the system with a finite buffer is studied. The customers visit the buffer when they do not succeed to start the service immediately upon arrival due to the lack of available servers. The customers go into orbit, when the buffer is full.

Since the retries effect is typical for many telecommunication networks, a large number of works are devoted to the study of RQ-systems [6,7]. In addition, retransmission of erroneously transmitted data occurs in real communication networks. In queuing systems, feedback assumes that upon being served, a customer can return to get service. Feedback models have not been studied well. In [8], data packet re-transmission and buffer overflow impact for queuing system performance is studied. Buffer state monitoring is the most commonly used method for improving the performance of a queuing system. Paper [9] considers a model of a system with an infinite buffer, instantaneous feedback by the matrix-geometric method. Paper [10] studies queuing network with feedback, where each node has several queues that share a common waiting space with limited capacity. The feedback customer can put in the tail of the queue.

In networks, heterogeneous traffic (voice, video, computer data, etc.) has different requirements for quality of service indicators. An effective way to meet the conflicting requirements of heterogeneous calls or packets is to use priority schemes. Various types of priority schemes in queuing models have already been explored [11]. Recently, new types of priority schemes [12,13] have been intensively studied. Useful tools for modeling stochastic processes that occur in communication networks, call centers, and queuing systems are queues with repeated calls and feedback [14].

This article considers a retrial system with one server and a buffer containing $N-1$ waiting spaces. A priority incoming customer is immediately submitted for servicing, and when the device is busy, it enters the queue. If the buffer is full, then the customer goes into orbit, where it waits for a random time and tries again to get into service or into the queue. A non-priority incoming customer enters the orbit first. The system takes into account the possibility of re-service in the form of instant and delayed feedback. The system is studied by the method of asymptotic analysis under the condition of a growing average waiting time for customers in orbit.

2 The Mathematical Model

We study a queuing system $M/M/1/N-1$ with retrial calls, priority and feedback (see Fig. 1).

The arrival process of customers is Poisson with parameter λ. A priority customer arrives with probability p. When the server is idle, the customer enters the service immediately. When server is busy, customer waits for service in the buffer if there is a free space. Customers are served in the exact order in which they arrive, that is the type of queuing is FIFO.

If all of $N-1$ places for waiting are occupied, the customer goes into the orbit. A non-priority application immediately goes into the orbit.

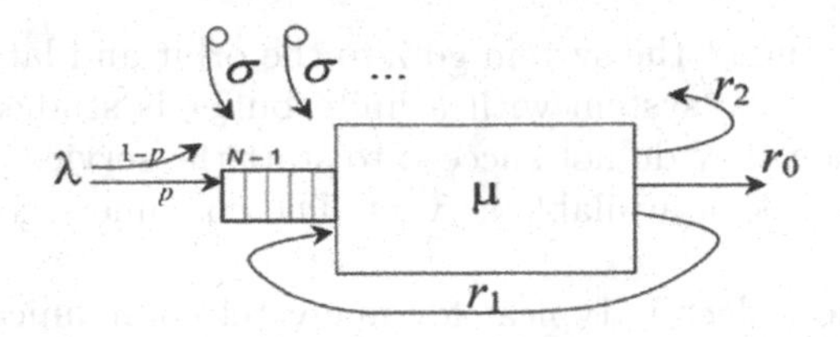

Fig. 1. Queuing system model with retrial calls and feedback

In the orbit, each customer independently of others waits for the time exponentially distributed with parameter σ. Then the customer makes a new effort to get service: it occupies the server if it is idle; it joins the queue if the server is busy and there is a free space in the buffer; it remains in the orbit otherwise.

The service time of customers is exponentially distributed with the parameter μ. A customer, whose service is completed, leaves the system with probability r_0, or immediately occupies the server again with probability r_1, or goes into the orbit with probability r_2, thus $r_0 + r_1 + r_2 = 1$.

Denote by $i(t)$ the number of customers in the orbit at time t. The process $n(t)$ determines the state of the server and the buffer for waiting as follows:

$$n(t) = \begin{cases} 0, & \text{if the server is idle;} \\ n, & \text{if the server is busy and there are } n-1 \text{ customers in the buffer,} \end{cases}$$

where $n = \overline{1, N}$.

The two-dimensional random process $\{i(t), n(t)\}$ is Markovian. We denote the probabilities of the number of customers in the orbit, taking into account the state of the server $P_n(i,t) = P\{i(t) = i,\ n(t) = n\}, n = \overline{0, N};\ i = \overline{0, \infty}$.

It is required to find the stationary probability distribution of the customers quantity in the orbit.

3 Kolmogorov Equations

We write the probability state distribution of the Markov process at time point $t + \Delta t$:

$$\begin{cases} P_0(i, t+\Delta t) = (1 - \lambda\Delta t)(1 - i\sigma\Delta t) P_0(i,t) + r_0\mu\Delta t P_1(i,t) + \\ + \lambda(1-p)P_0(i-1,t) + r_2\mu\Delta t P_1(i-1,t) + o(\Delta t); \\ P_n(i, t+\Delta t) = (1 - \lambda\Delta t)(1 - \mu\Delta t)(1 - i\sigma\Delta t) P_n(i,t) + \\ + r_1\mu\Delta t P_n(i,t) + r_0\mu\Delta t P_{n+1}(i,t) + r_2\mu\Delta t P_{n+1}(i-1,t) + \lambda p\Delta t P_{n-1}(i,t) + \\ + \lambda(1-p)\Delta t P_n(i-1,t) + (i+1)\sigma\Delta t P_0(i+1,t) + o(\Delta t), \quad n = \overline{1, N-1}; \\ P_N(i, t+\Delta t) = (1 - \lambda\Delta t)(1 - \mu\Delta t) P_N(i,t) + r_1\mu\Delta t P_N(i,t) + \\ + \lambda\Delta t P_N(i-1,t) + \lambda p\Delta t P_{N-1}(i,t) + (i+1)\sigma\Delta t P_{N-1}(i+1,t) + o(\Delta t). \end{cases}$$

Then by taking a limit $\Delta t \to 0$, we obtain a system of differential equations

$$\begin{cases} \dfrac{\partial P_0(i,t)}{\partial t} = -(\lambda + i\sigma)\, P_0(i,t) + \lambda(1-p)P_0(i-1,t) + \mu r_0 P_1(i,t) + \\ + \mu r_2 P_1(i-1,t); \\ \dfrac{\partial P_n(i,t)}{\partial t} = -(\lambda + \mu + i\sigma)\, P_n(i,t) + r_1\mu P_n(i,t) + r_0\mu P_{n+1}(i,t) + \\ + r_2\mu P_{n+1}(i-1,t) + \lambda p P_{n-1}(i,t) + \lambda(1-p)P_n(i-1,t) + \\ + (i+1)\sigma P_{n-1}(i+1,t), \quad n = \overline{1, N-1}; \\ \dfrac{\partial P_N(i,t)}{\partial t} = -(\lambda + \mu)\, P_N(i,t) + r_1\mu P_N(i,t) + \lambda P_N(i-1,t) + \\ + \lambda p P_{N-1}(i,t) + (i+1)\sigma P_{N-1}(i+1,t). \end{cases}$$

For a stationary probability distribution $P_n(i) \equiv P_n(i,t)$, we write the system of equations

$$\begin{cases} -(\lambda + i\sigma)P_0(i) + \lambda(1-p)P_0(i-1) + \mu r_0 P_1(i) + \mu r_2 P_1(i-1) = 0; \\ \lambda p P_{n-1}(i) + (i+1)\sigma P_{n-1}(i+1) + \lambda(1-p)P_n(i-1) - (\lambda + \mu r_0 + \mu r_2 + \\ + i\sigma)P_n(i) + \mu r_0 P_{n+1}(i) + \mu r_2 P_{n+1}(i-1) = 0, \quad n = \overline{1, N-1}; \\ (i+1)\sigma P_{N-1}(i+1) - (\lambda + \mu r_0 + \mu r_2)P_N(i) + \lambda p P_{N-1}(i) + \\ + \lambda P_N(i-1) = 0. \end{cases}$$

We consider the partial characteristic functions of the number of customers in the orbit $H_n(u) = \sum\limits_{i=0}^{\infty} e^{jui} P_n(i)$, where $j = \sqrt{-1}$. We rewrite system as

$$\begin{cases} -\lambda H_0(u) + j\sigma \dfrac{\partial H_0(u)}{\partial u} + \lambda(1-p)e^{ju} H_0(u) + \left(\mu r_0 + \mu r_2 e^{ju}\right) H_1(u) = 0; \\ \lambda p H_{n-1}(u) - j\sigma e^{-ju} \dfrac{\partial H_{n-1}(u)}{\partial u} + \lambda(1-p)e^{ju} H_n(u) + j\sigma \dfrac{\partial H_n(u)}{\partial u} - \\ -(\lambda + \mu r_0 + \mu r_2)\, H_n(u) + \left(\mu r_0 + \mu r_2 e^{ju}\right) H_{n+1}(u) = 0, \quad n = \overline{1, N-1}; \\ \lambda p H_{N-1}(u) - j\sigma e^{-ju} \dfrac{\partial H_{N-1}(u)}{\partial u} - \left(\lambda - \lambda e^{ju} + \mu r_0 + \mu r_2\right) H_N(u) = 0. \end{cases} \tag{1}$$

The characteristic function $H(u)$ of the customers quantity in the orbit is expressed as $H(u) = \sum\limits_{n=0}^{N} H_n(u)$.

Summing the equations of the system (1), we write

$$\mu r_2 \sum_{n=1}^{N} H_n(u) + \lambda(1-p) \sum_{n=0}^{N-1} H_n(u) + j\sigma e^{-ju} \sum_{n=0}^{N-1} \frac{\partial H_n(u)}{\partial u} + \lambda H_N(u) = 0. \tag{2}$$

4 First Order Asymptotics

We solve the Eqs. (1), (2) for the characteristic function under the asymptotic condition of the growing average waiting time in the orbit, i.e. we assume that $\sigma \to 0$. The following result can be proved.

Theorem 1. *Let $i(t)$ be the customers quantity in the orbit in the RQ-system $M/M/1/N-1$ with priority and feedback, then the expression holds for the sequence of characteristic functions $\lim_{\sigma\to 0} E\left\{e^{jwi(t)\sigma}\right\} = e^{jw\kappa_1}$, where κ_1 is the solution of the equation*

$$\frac{d^N-1}{d-1}\kappa_1 = (\lambda - \lambda p + \mu r_2)\frac{d^{N+1}-1}{d-1} - \mu r_2 + \lambda p d^N, d(\kappa) = \frac{\lambda p + \kappa_1}{\mu r_0 + \mu r_2}. \quad (3)$$

Proof. We denote $\sigma = \varepsilon$, use the following substitutions in the system (1), (2)

$$u = \varepsilon w, H_n(u) = F_n(w,\varepsilon), n = \overline{0,N}$$

and obtain

$$\begin{cases} -\lambda F_0(w,\varepsilon) + j\dfrac{\partial F_0(w,\varepsilon)}{\partial w} + \lambda(1-p)e^{j\varepsilon w}F_0(w,\varepsilon) + \\ +\left(\mu r_0 + \mu r_2 e^{j\varepsilon w}\right)F_1(w,\varepsilon) = 0; \\ \lambda p F_{n-1}(w,\varepsilon) - je^{-j\varepsilon w}\dfrac{\partial F_{n-1}(w,\varepsilon)}{\partial w} + j\dfrac{\partial F_n(w,\varepsilon)}{\partial w} - \\ -\left(\lambda + \mu r_0 + \mu r_2\right)F_n(w,\varepsilon) + \lambda(1-p)e^{-j\varepsilon w}F_n(w,\varepsilon) + \\ +\left(\mu r_0 + \mu r_2 e^{j\varepsilon w}\right)F_{n+1}(w,\varepsilon) = 0, \quad n = \overline{1,N-1}; \\ \lambda p F_{N-1}(w,\varepsilon) - je^{-j\varepsilon w}\dfrac{\partial F_{N-1}(w,\varepsilon)}{\partial w} - \\ -\left(\lambda - \lambda e^{j\varepsilon w} + \mu r_0 + \mu r_2\right)F_N(w,\varepsilon) = 0; \end{cases} \quad (4)$$

$$\mu r_2 \sum_{n=1}^{N-1} F_n(w,\varepsilon) + \lambda(1-p)\sum_{n=0}^{N-1} F_n(w,\varepsilon) + je^{-j\varepsilon w}\sum_{n=0}^{N-1}\frac{\partial F_n(w,\varepsilon)}{\partial w} + \\ + (\lambda + \mu r_2)F_N(w,\varepsilon) = 0. \quad (5)$$

Let $\varepsilon \to 0, F_n(w) = \lim_{\varepsilon\to 0} F_n(w,\varepsilon)$. The system (4) is transformed into

$$\begin{cases} -\lambda F_0(w) + j\dfrac{\partial F_0(w)}{\partial w} + \lambda(1-p)F_0(w) + (\mu r_0 + \mu r_2)F_1(w) = 0; \\ \lambda p F_{n-1}(w) - j\dfrac{\partial F_{n-1}(w)}{\partial w} + \lambda(1-p)F_n(w) - (\lambda + \mu r_0 + \mu r_2)F_n(w) + \\ + j\dfrac{\partial F_n(w)}{\partial w} + (\mu r_0 + \mu r_2)F_{n+1}(w) = 0, \quad n = \overline{1,N-1}; \\ \lambda p F_{N-1}(w) - j\dfrac{\partial F_{N-1}(w)}{\partial w} - (\mu r_0 + \mu r_2)F_N(w) = 0 \end{cases}$$

and the Eq. (5) into

$$\mu r_2 \sum_{n=1}^{N} F_n(w) + (1-p)\lambda \sum_{n=0}^{N-1} F_n(w) + j \sum_{n=0}^{N-1} \frac{\partial F_n(w)}{\partial w} + \lambda F_N(w) = 0.$$

We find the solution of the system in the form $F_n(w) = R_n \Phi(w) + O(\varepsilon)$ and get the system

$$\begin{cases} -\lambda R_0 + jR_0 \dfrac{\Phi'(w)}{\Phi(w)} + \lambda(1-p)R_0 + (\mu r_0 + \mu r_2) R_1 = 0; \\ \lambda p R_{n-1} - jR_{n-1}\dfrac{\Phi'(w)}{\Phi(w)} + \lambda(1-p)R_n + jR_n \dfrac{\Phi'(w)}{\Phi(w)} - \\ -(\lambda + \mu r_0 + \mu r_2) R_n + (\mu r_0 + \mu r_2) R_{n+1} = 0, \quad n = \overline{1, N-1}; \\ \lambda p R_{N-1} - jR_{N-1}\dfrac{\Phi'(w)}{\Phi(w)} - (\mu r_0 + \mu r_2) R_N = 0; \\ \mu r_2 \sum\limits_{n=1}^{N} R_n + (1-p)\lambda \sum\limits_{n=0}^{N-1} R_n + j\dfrac{\Phi'(w)}{\Phi(w)} \sum\limits_{n=0}^{N-1} R_n + \lambda R_N = 0. \end{cases} \tag{6}$$

It can be seen from the system (6) that $\dfrac{\Phi'(w)}{\Phi(w)}$ does not depend on w, then we can denote $\dfrac{\Phi'(w)}{\Phi(w)} = j\kappa_1$ and write $\Phi(w) = \exp\{jw\kappa_1\}$.

Solving the first $N+1$ equations of the system (6)

$$\begin{cases} -(\lambda p + \kappa_1) R_0 + (\mu r_0 + \mu r_2) R_1 = 0; \\ (\lambda p + \kappa_1) R_{n-1} - (\lambda p + \kappa_1 + \mu r_0 + \mu r_2) R_n + (\mu r_0 + \mu r_2) R_{n+1} = 0, \\ \qquad n = \overline{1, N-1}; \\ (\lambda p + \kappa_1) R_{N-1} - (\mu r_0 + \mu r_2) R_N = 0; \end{cases} \tag{7}$$

taking into account the condition $\sum\limits_{n=0}^{N} R_n = 1$, we derive

$$R_n = \frac{d^n (d-1)}{d^{N+1} - 1}, n = \overline{0, N}.$$

The last equation of the system (6) gives the algebraic equation (3) for κ_1. An algebraic equation of N-th degree has exactly N roots (real or complex). Numerous numerical experiments have shown that among the roots there is only one real positive root, which can be the value of κ_1. *Theorem* 1 *is proved.*

It follows from Theorem 1 that $F_n(w) = R_n e^{j\kappa_1 w}, n = \overline{0, N}$, then the asymptotic characteristic function of the number of customers in the orbit has the following form

$$H(u) = \sum_{n=0}^{N} F_n(w, \varepsilon) \approx \sum_{n=0}^{N} F_n(w) = \exp\{j\kappa_1 w\} = \exp\left\{j\frac{\kappa_1}{\sigma}u\right\}.$$

5 Second Order Asymptotics

The first order asymptotics defines the average value of the customers quantity in the orbit. For further study of the process $i(t)$, we examine the asymptotics of the second order.

Theorem 2. *Let $i(t)$ be the customers quantity in the orbit in the RQ-system $M/M/1/N-1$ with priority and feedback, then the expression holds*

$$\lim_{\sigma\to 0} E\left\{e^{jw\sqrt{\sigma}(i(t)-\kappa_1/\sigma)}\right\} = \exp\left\{\frac{(jw)^2}{2}\kappa_2\right\},$$

where

$$\kappa_2 = \frac{\kappa_1(1-R_N) - \mu r_2 g_0 + (\lambda p + \kappa_1) g_N}{1 - R_N + \mu r_2 \varphi_0 - (\lambda p + \kappa_1)\varphi_N}, \tag{8}$$

and

$$\varphi_n = \frac{(n-N)d^{n+N+1} + (N-n+1)d^{n+N} - (n+1)d^n + nd^{n-1}}{\left(d^{N+1}-1\right)^2(\mu r_0 + \mu r_2)}, n = \overline{0,N}, \tag{9}$$

the functions g_0 and g_N are the solution of the system of equations

$$\begin{cases} -(\lambda p + \kappa_1)g_0 + (\mu r_0 + \mu r_2)g_1 = -\lambda(1-p)R_0 - \mu r_2 R_1; \\ (\lambda p + \kappa_1)g_{n-1} - (\lambda p + \kappa_1 + \mu r_0 + \mu r_2)g_n + (\mu r_0 + \mu r_2)g_{n+1} = \\ = \kappa_1 R_{n-1} - \lambda(1-p)R_n - \mu r_2 R_{n+1}, \quad n = \overline{1, N-1}; \\ (\lambda p + \kappa_1)g_{N-1} - (\mu r_0 + \mu r_2)g_N = \kappa_1 R_{N-1} - \lambda R_N. \end{cases} \tag{10}$$

Proof. In the system of Eqs. (1), (2) we use the substitutions

$$H_n(u) = H_n^{(2)}(u)e^{ju\kappa_1/\sigma}, \; n = \overline{0,N}.$$

Here $H_n^{(2)}(u)$ is the partial characteristic function of the centered random variable $i(t) - \kappa_1/\sigma$. Where $H_n^{(2)}(u)$ is determined by the system

$$\begin{cases} -\left(\lambda + \kappa_1 + \lambda p e^{ju} - \lambda e^{ju}\right)H_0^{(2)}(u) + j\sigma\frac{\partial H_0^{(2)}(u)}{\partial u} + \\ + \left(\mu r_0 + \mu r_2 e^{ju}\right)H_1^{(2)}(u) = 0; \\ \left(\lambda p + \kappa_1 e^{-ju}\right)H_{n-1}^{(2)}(u) - \left(\lambda + \kappa_1 + \lambda p e^{ju} - \lambda e^{ju} + \mu r_0 + \mu r_2\right)H_n^{(2)}(u) - \\ - j\sigma e^{-ju}\frac{\partial H_{n-1}^{(2)}(u)}{\partial u} + j\sigma\frac{\partial H_n^{(2)}(u)}{\partial u} + \left(\mu r_0 + \mu r_2 e^{ju}\right)H_{n+1}^{(2)}(u) = 0, \\ n = \overline{1, N-1}; \\ \left(\lambda p + \kappa_1 e^{-ju}\right)H_{N-1}^{(2)}(u) - j\sigma e^{-ju}\frac{\partial H_{N-1}^{(2)}(u)}{\partial u} - \\ - \left(\lambda - \lambda e^{ju} + \mu r_0 + \mu r_2\right)H_N^{(2)}(u) = 0 \end{cases}$$

and

$$\mu r_2 \sum_{n=1}^{N-1} H_n^{(2)}(u) + j\sigma e^{-ju} \sum_{n=0}^{N-1} \frac{\partial H_n^{(2)}(u)}{\partial u} - \left(\lambda p - \lambda + \kappa_1 e^{-ju}\right) \sum_{n=0}^{N-1} H_n^{(2)}(u) +$$
$$+ (\lambda + \mu r_2) H_N^{(2)}(u) = 0.$$

We denote $\sigma = \varepsilon^2$, use the substitutions

$$u = \varepsilon w, H_n^{(2)}(u) = F_n^{(2)}(w, \varepsilon), n = \overline{0, N}$$

and get the system

$$- \left(\lambda + \kappa_1 + \lambda p e^{ju} - \lambda e^{ju}\right) F_0^{(2)}(w, \varepsilon) + j\varepsilon \frac{\partial F_0^{(2)}(w, \varepsilon)}{\partial w} +$$
$$+ \left(\mu r_0 + \mu r_2 e^{j\varepsilon w}\right) F_1^{(2)}(w, \varepsilon) = 0;$$
$$\left(\lambda p + \kappa_1 e^{-j\varepsilon w}\right) F_{n-1}^{(2)}(w, \varepsilon) - j\varepsilon e^{-j\varepsilon w} \frac{\partial F_{n-1}^{(2)}(w, \varepsilon)}{\partial w} + j\varepsilon \frac{\partial F_n^{(2)}(w, \varepsilon)}{\partial w} -$$
$$- \left(\lambda + \kappa_1 + \mu r_0 + \mu r_2 + \lambda p e^{ju} - \lambda e^{ju}\right) F_n^{(2)}(w, \varepsilon) +$$
$$+ \left(\mu r_0 + \mu r_2 e^{j\varepsilon w}\right) F_{n+1}^{(2)}(w, \varepsilon) = 0, \quad n = \overline{1, N-1};$$
$$\left(\lambda p + \kappa_1 e^{-j\varepsilon w}\right) F_{N-1}^{(2)}(w, \varepsilon) - j\varepsilon e^{-j\varepsilon w} \frac{\partial F_{N-1}^{(2)}(w, \varepsilon)}{\partial w} -$$
$$- \left(\lambda - \lambda e^{j\varepsilon w} + \mu r_0 + \mu r_2\right) F_N^{(2)}(w, \varepsilon) = 0;$$
$$\mu r_2 \sum_{n=1}^{N-1} F_n^{(2)}(w, \varepsilon) + j\varepsilon e^{-j\varepsilon w} \sum_{n=0}^{N-1} \frac{\partial F_n^{(2)}(w, \varepsilon)}{\partial w} -$$
$$- \left(\lambda p - \lambda + \kappa_1 e^{-ju}\right) \sum_{n=0}^{N-1} F_n^{(2)}(w, \varepsilon) + (\lambda + \mu r_2) F_N^{(2)}(w, \varepsilon) = 0.$$

We find the solution of the system as

$$F_n^{(2)}(w, \varepsilon) = \Phi_2(w) \left(R_n + j\varepsilon w f_n\right) + O(\varepsilon^2). \tag{11}$$

Substituting (11) into the system and using the approximation for $\exp\{\pm jw\varepsilon\}$, after some algebraic transformations we obtain

$$
\begin{aligned}
&(-\lambda p+\lambda)R_0-(\lambda p+\kappa_1)\,f_0+\frac{\Phi'{}_2(w)}{w\Phi_2(w)}\cdot R_0+\mu r_2R_1+(\mu r_0+\mu r_2)\,f_1=O(\varepsilon);\\
&-\kappa_1R_{n-1}+(\lambda p+\kappa_1)\,f_{n-1}-\frac{\Phi'{}_2(w)}{w\Phi_2(w)}\cdot R_{n-1}+\frac{\Phi'{}_2(w)}{w\Phi_2(w)}\cdot R_n+(-\lambda p+\lambda)R_0-\\
&-(\lambda+\kappa_1+\mu r_0+\mu r_2)\,f_n+\mu r_2R_{n+1}+(\mu r_0+\mu r_2)\,f_{n+1}=O(\varepsilon),n=\overline{1,N-1};\\
&-\kappa_1R_{N-1}+(\lambda p+\kappa_1)\,f_{N-1}-\frac{\Phi'{}_2(w)}{w\Phi_2(w)}\cdot R_{N-1}+\lambda R_N-(\mu r_0+\mu r_2)\,f_N=\\
&=O(\varepsilon);\\
&(\mu r_2-\kappa_1+\lambda-\lambda p)\sum_{n=1}^{N-1}f_n+\kappa_1\sum_{n=0}^{N-1}R_n+\frac{\Phi'{}_2(w)}{w\Phi_2(w)}\sum_{n=0}^{N-1}R_n-\\
&-(\lambda p-\lambda+\kappa_1)\,f_0+(\lambda+\mu r_2)\,f_N=O(\varepsilon).
\end{aligned}
\tag{12}
$$

It can be seen from the systems (12) that $\frac{\Phi'{}_2(w)}{w\Phi_2(w)}$ does not depend on w, then we can denote $\frac{\Phi'{}_2(w)}{w\Phi_2(w)}=-\kappa_2$ and write $\Phi_2(w)=exp\left\{\frac{(jw)^2}{2}\kappa_2\right\}$. We rewrite system (12) as follows

$$
\begin{aligned}
&-(\lambda p+\kappa_1)\,f_0+(\mu r_0+\mu r_2)\,f_1=\kappa_2R_0-\lambda(1-p)R_0-\mu r_2R_1;\\
&(\lambda p+\kappa_1)\,f_{n-1}-(\lambda p+\kappa_1+\mu r_0+\mu r_2)\,f_n+(\mu r_0+\mu r_2)\,f_{n+1}=\\
&=\kappa_2\,(R_n-R_{n-1})+\kappa_1R_{n-1}-\lambda(1-p)R_n-\mu r_2R_{n+1},\quad n=\overline{1,N-1};\\
&(\lambda p+\kappa_1)\,f_{N-1}-(\mu r_0+\mu r_2)\,f_N=-\kappa_2R_{N-1}+\kappa_1R_{N-1}-\lambda R_N;\\
&-(\lambda p-\lambda+\kappa_1)\,f_0-(\lambda p-\lambda+\kappa_1-\mu r_2)\sum_{n=1}^{N-1}f_n+(\lambda+\mu r_2)\,f_N=\\
&=\kappa_2\,(1-R_N)-\kappa_1\,(1-R_N)
\end{aligned}
\tag{13}
$$

and find the solution of system (13) as

$$f_n=C\cdot R_n+g_n+\kappa_2\varphi_n,\ n=\overline{0,N},$$

where $C\cdot R_n$ is the general solution of the system (7), g_n is the solution of system (10), φ_n is the solution to the following system

$$
\begin{aligned}
&-(\lambda p+\kappa_1)\,\varphi_0+(\mu r_0+\mu r_2)\,\varphi_1=R_0;\\
&(\lambda p+\kappa_1)\,\varphi_{n-1}-(\lambda p+\kappa_1+\mu r_0+\mu r_2)\,\varphi_n+(\mu r_0+\mu r_2)\,\varphi_{n+1}=\\
&=R_n-R_{n-1},\quad n=\overline{1,N-1};\\
&(\lambda p+\kappa_1)\,\varphi_{N-1}-(\mu r_0+\mu r_2)\,\varphi_N=-R_{N-1}.
\end{aligned}
\tag{14}
$$

Differentiating (7) with respect to κ_1 and comparing with (14), we note that

$$\varphi_n=\frac{\partial R_n}{\partial\kappa_1},\ n=\overline{0,N},\ \sum_{n=0}^{N}\varphi_n=0.$$

Hence equality (9) holds. System (10) is a system of $N+1$ linear algebraic equations. It has infinitely many solutions, because the coefficient matrix rank and augmented matrix rank are equal to N. Let $\sum_{n=0}^{N} g_n = 0$, then we can find a unique solution to the system. Last equation (13) leads to (8). *Theorem 2 is proved.*

According to the Theorem 2 the asymptotic probability distribution of the number of customers in the orbit in the RQ-system $M/M/1/N-1$ with priority and feedback is the characteristic function of a Gaussian random variable with expectation κ_1/σ and variance κ_2/σ, which allows us to make the following approximation for the distribution $P(i)$ as

$$P_{apr}(i) = \frac{G(i+0.5) - G(i-0.5)}{1 - G(-0.5)}, \tag{15}$$

where $G(x)$ is the normal distribution function with parameters κ_1/σ and $\sqrt{\kappa_2/\sigma}$.

We note that the results of the theorem 1 and 2 agree with the result obtained in [15] when $p = 1$ and there is no buffer for waiting.

6 Numerical Results

To define the accuracy of the approximation $P_{apr}(i)$, we compute the Kolmogorov distance

$$\Delta = \max_{0 \le k \le N} \left| \sum_{i=0}^{k} (P_{apr}(i) - P(i)) \right|$$

that defines the difference between the asymptotic probability distribution $P_{apr}(i)$ in form (15) and the probability distribution $P(i)$ obtained analytically or by numerical method. However, it is almost impossible to obtain an analytical expression for this characteristic. And the main method for the numerical study of complex queuing systems is the simulation method [16].

We use discrete-event simulation to create a model of RQ-system. There are three types of events for the system under study: the arrival of a customer, the completion of service for a customer, the completion of waiting time in the orbit. State variables are server status and number of customers in the orbit. The result of simulation is the probability distribution of the number of customers in the orbit.

We study the system with parameters $\lambda = 2, r_0 = 0.5, r_1 = 0.3, r_2 = 0.2$. The service intensity μ is obtained taking into account $\mu = \dfrac{\lambda}{\rho r_0}$ while the system load ρ varies from 0.5 to 0.9 and the probability p varies from 0.2 to 0.8.

The probability distributions is obtained by the method of asymptotic analysis and simulation when $\sigma = 0.1$ and $\sigma = 0.01$. It is shown in Fig. 2 and Fig. 3. We can see that the approximation is quite good when parameter σ is lower.

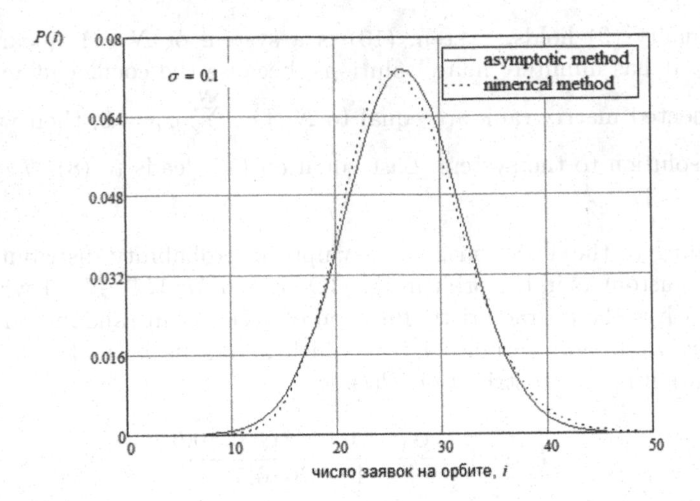

Fig. 2. The probability distributions obtained with the different methods ($\sigma = 0.1$).

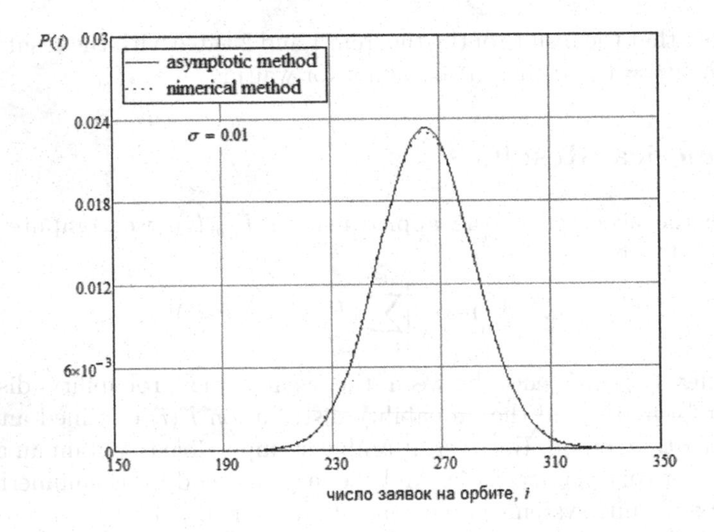

Fig. 3. The probability distributions obtained with the different methods ($\sigma = 0.01$).

Table 1. Kolmogorov distances

p	rho	$\sigma = 1$	$\sigma = 0.1$	$\sigma = 0.01$
1	0.5	0.144	0.030	0.019
	0.9	0.192	0.080	0.037
0.8	0.5	0.122	0.019	0.017
	0.9	0.167	0.075	0.029
0.2	0.5	0.065	0.012	0.007
	0.9	0.163	0.093	0.044

Table 1 shows the Kolmogorov distances between the asymptotic distribution and simulated distribution for different values of the parameters p, σ and ρ.

We suppose that an approximation is applicable if its Kolmogorov distance is less than 0.05. So we can conclude that the accuracy of the approximation has a wide application. And the accuracy increases with a decrease in the parameter σ and decreases with an increase in the system load ρ.

7 Conclusion

In this paper, we study the $M/M/1/N-1$ queuing system with priority, feedback and retries. It is shown that the asymptotic probability distribution of the number of customers in the orbit under the condition of increasing average waiting time in orbit is Gaussian with parameters κ_1/σ and $\sqrt{\kappa_2/\sigma}$. Equations are obtained for finding the distribution parameters. Numerous numerical experiments have shown that the equations make it possible to find the only positive value of the distribution parameters. The accuracy of the obtained approximations is determined in comparison with the results of simulation modeling. Numerical examples are given for various sets of parameters of the system under study.

References

1. Balsamo, S., De Nilto Persone, V., Inverardi, P.: A review on queueing network models with finite capacity queues for software architectures performance prediction. Perform. Eval. **51**, 269–288 (2003)
2. Misra, C., Swain, P.K.: Performance analysis of finite buffer queueing system with multiple heterogeneous servers, **5966**, 180–183 (2010). https://doi.org/10.1007/978-3-642-11659-9_19
3. Kim, J., Kim, B.: A survey of retrial queueing systems. Ann. Oper. Res. **247**(1), 3–36 (2015). https://doi.org/10.1007/s10479-015-2038-7
4. Phung-Duc, T.: Retrial queueing models. A survey on theory and applications. arXiv preprint arXiv:1906.09560 (2019)
5. Kim, C.S., Klimenok, V., Dudin, A.: Retrial queueing system with correlated input, finite buffer, and impatient customers. In: Dudin, A., De Turck, K. (eds.) ASMTA 2013. LNCS, vol. 7984, pp. 262–276. Springer, Heidelberg (2013). https://doi.org/10.1007/978-3-642-39408-9_19
6. Vishnevskii, V.M., Dudin, A.N.: Queueing systems with correlated arrival flows and their applications to modeling telecommunication networks. Autom. Remote. Control. **78**(8), 1361–1403 (2017). https://doi.org/10.1134/S000511791708001X
7. Stepanov, S.N.: The Theory of Teletraffic. Concepts, Models, Applications. NTI Hot Line-Telecom (2015)
8. Rindzevicius, R., Tervydis, P., Zvironiene, A., Navickas, Z.: Feedback impact analysis of queueing system with stable states. In: Fourth International Multi-Conference on Computing in the Global Information Technology, Cannes/La Bocca, France, 2009, pp. 63–68 (2009). https://doi.org/10.1109/ICCGI.2009.17

9. Melikov, A., Divya, V., Aliyeva, S.: Analyses of feedback queue with positive server setup time and impatient calls. In: Information Technologies and Mathematical Modeling: Proceedings of the XIX International Conference named after A. F. Terpugov, Tomsk. NTL Publishing House, pp. 77–81 (2021)
10. Shanmugasundaram, S., Vanitha, S.: Single server Markovian feedback queueing network with shared buffer and multi-queue nodes. AIP Conf. Proc. **2261**, 030062 (2020). https://doi.org/10.1063/5.0024489
11. Lee, Y., Choi, B.D.: Queuing system with multiple delay and loss priorities for ATM networks. Inf. Sci. **138**(1–4), 7–29 (2001)
12. Demoor, T., Fiems, D., Walraevens, J., Brunnel, H.: Partially shared buffers with full or mixed priority. J. Ind. Manag. Optim. **7**(3), 735–751 (2011)
13. Youngjin, O., Kim, C.S., Melikov, A.Z.: A space merging approach to the analysis of the performance of queueing models with finite buffers and priority jumps. Ind. Eng. Manag. Syst. **12**(3), 274–280 (2013)
14. Melikov, A.Z., Sztrik, J., Aliyeva, S.H.: Analysis of retrial queues with delayed feedback. Miskolc Math. Notes **22**(2), 769–782 (2021)
15. Nazarov, A.A., Rozhkova, S.V., Titarenko, E.Y.: Asymptotic analysis of RQ-system with feedback and batch poisson arrival under the condition of increasing average waiting time in orbit. In: Vishnevskiy, V.M., Samouylov, K.E., Kozyrev, D.V. (eds.) DCCN 2020. CCIS, vol. 1337, pp. 327–339. Springer, Cham (2020). https://doi.org/10.1007/978-3-030-66242-4_26
16. Averill, M., Law, W., Kelton, D.: Simulation Modeling and Analysis. McGraw-Hill, New York (2000)

Modelling of a MAP/PH(1),PH(2)/2 Production Inventory System with Multiple Servers and Production Vacations

Beena P.[1] and K. P. Jose[2(✉)]

[1] Sree Neelakanta Govt.Sanskrit College, Pattambi, Palakkad 679306, Kerala, India
[2] Department of Mathematics, St.Peter's College, Kolenchery 682311, Kerala, India
kpjspc@gmail.com

Abstract. A $MAP/PH(1), PH(2)/2$ multiple vacations production inventory model is considered in which the MAP is followed between consecutive arrivals, with the production facility and two heterogeneous servers taking advantage of multiple vacations. Servers have a phase type distributed service time, and the time to produce an item is exponentially distributed. The reloading of the inventory is done as per the (s, S) policy. Once the stock level of goods reaches S, the production unit goes on vacation and if the stock level of goods drops to s when returning after the vacation, the production unit will start production until the inventory level reaches S. The production unit will go back to another vacation if the stock level of goods is greater than s, even after returning from vacation. The algorithmic solution to the model is obtained by Matrix Geometric Method. This article analyzes the impact of the production vacation parameter on different combinations of arrival and service on measures of different activities of the system and cost function. Some numerical examples are presented at different parameter rates.

Keywords: Markovian arrival process · Multiple servers · Multiple vacations · Cost analysis · Production vacation

1 Introduction

Production machinery is always subject to timely monitoring and maintenance. Production vacation is the cessation of production for a short period of random length. To reduce holding costs, it is more desirable to produce items at a certain level and stop production when that level is reached. The main strategy of most manufacturing sectors is to reduce production costs and achieve customer satisfaction. In the production process, the strategy of most manufacturing companies is to produce products based on customer demand, stock items at that level, and then start production based on demand. But if the goods are not available, customers will not come back later and orders will be lost. Hence an

A. Dudin et al. (Eds.): ITMM 2022, CCIS 1803, pp. 175–188, 2023.
https://doi.org/10.1007/978-3-031-32990-6_15

optimum production level is required. In all these circumstances it is better to give a short break for the production unit. In some cases, the servers may also be unable to provide service. Servers may take vacation due to system outages or other tasks assigned to them.

Levy and Yechiali [10] were the first to study the queueing systems with one or more vacations. Amalesh Kumar Manna [11] discussed a single item imperfect production inventory system in which the demands depend upon the rate of advertisement and the rate of producing defective items depends on the production rate. The important calculations are the optimum production rate and production running time of the system. Goyal and Giri [4] analyzed a production-inventory problem in which the demand, production, and deterioration rates vary with time and shortages are allowed. A production control inventory model for deteriorating items in which two different rates of production are discussed by Ajantha Roy and Samantha [15]. They reached the conclusion that the two different production rates will yield more profit and customer satisfaction than the single-production rate. Dequan Yue and Yaling Qin [21] analyzed a production inventory system with production time and production vacation times exponentially distributed and service time positive. Krishnamoorthy et al. [7] considered a production inventory system in which the production time of each item and service time follow Erlang distributions. More on vacation models can be learned from the works of Teghem [20], Doshi [3], as well as Takagi's monograph [19]. Jose and Salini [12] discussed a production inventory model with varying execution rates. A Markovian service queue attached to (s, S) production inventory system was analyzed by Baek and Moon [1]. A $MAP/PH(1), PH(2)/2$ system under (s, S) policy with multiple server vacation and phase type distributed service times for the two heterogeneous servers were investigated by Suganya et al. [17]. Sugapriya et al. [18] discussed a retrial inventory system with multiple server vacation and stock dependent arrival processes.

Sivasankari and Panayappan [16] analyzed two production inventory models with and without shortages of defective items. Jose and Salini [6] compared the efficiency of two production inventory systems with different rates of production. Krishnakumar et al. [8] discussed a multiple server queueing system with heterogeneous vacation rates and vacation policy as Bernoulli's scheduled vacation. Latouche et al. [9] have detailed PH distributions and QBD in their studies. Jose and Beena [5] analyzed a multi-server production inventory system with the retrial of customers. Beena and Jose [2] analyzed a production inventory system with multiple servers in which different vacation strategies are considered for the heterogeneous servers. In a multi-server queue Neuts and Takahashi [14] discussed the asymptotic behaviour of the stationary distribution. To the best of my knowledge, there was no study reported on the production inventory system with the vacation for the heterogeneous servers and manufacturing unit.

This article is organized as follows. Model description and analysis is given in Sects. 2 and 3. Sections 4 and 5 describe the measures of effectiveness and cost analysis. Numerical experiments and concluding remarks are presented in Sect. 6 and 7.

2 Model Description

An (s, S) production inventory model that produces a single type of product is considered, in which customers enter into the system one at a time, according to the Markovian arrival process (MAP) with D_0 and D_1 as its parametric matrices having dimension l. The servers recommended a heterogeneous phase type distributed service time with representation $(\alpha, S)_m$ and $(\beta, T)_n$ respectively. The machine can make only one item at a time and the time required for the machine to build an item follows an exponential distribution with the rate η. The server will go on multiple vacations when there are no customers in the system or the stock runs out or there are no stock and customers in the system. Servers returning after vacation will restart service only when the stock level is positive and customers are waiting on the system. Otherwise, the servers will go back to vacation and this pattern will continue until the customer level and stock level are positive. The duration of the vacation time of both servers is exponentially distributed with parameters θ_1 and θ_2. Once the stock level of goods reaches S, the production unit goes on vacation with an exponentially distributed rate θ. On return from the vacation, if the stock level of goods drops to s, the production unit will start production until the inventory level reaches S. The production unit will go back to another vacation of random duration having the same distribution rate, if the stock level of goods is greater than s, even after returning from vacation.

3 Analysis

- $N(t), I(t), J_0(t), J_1(t), J_2(t)$ are respectively indicate the number of buyers in the system, level of stock, phase of the arrival process, and phases of the service process of servers 1 and 2 at time $t \geq 0$.
- C(t) indicates the status of servers 1 and 2 where

$$C(t): \begin{cases} 0, & \text{if both the servers are on vacation} \\ 1, & \text{if server 1 is busy and server 2 is on vacation} \\ 2, & \text{if server 1 is on vacation and server 2 is busy} \\ 3, & \text{if both servers are busy} \end{cases}$$

- J(t) denotes production status where

$$J(t): \begin{cases} 0, \text{ if the production unit is taking vacation} \\ 1, \text{ if the production process is ON mode} \end{cases}$$

- I is the identity matrix of appropriate dimension.
- e is a column vector of suitable dimension with each of its entries are one.
- $e_1 = e_{l(2S+1)+lm(2S-1)+ln(2S-1)+lmn(2S-2)}$.
- $e_2 = e_{l(2S+1)+lm(2S-1)+ln(2S-1)}$.
- $e_3 = e_{l(2S-s)}$.

The process $\{X(t) = (N(t), C(t), J(t), I(t), J_0(t), J_1(t), J_2(t)), t \geq 0\}$ is a continuous time Markov chain with state space

$$\Omega = l(0) \cup l(1) \cup l(2) \cup l(3),$$
$$l(0) = (i, 0, 0, k, j_0) | 0 \leq k \leq S \;\cup\; (i, 0, 1, k, j_0) | 0 \leq k \leq S-1; i \geq 0$$
$$l(1) = (i, 1, 0, k, j_0, j_1) | 1 \leq k \leq S \;\cup\; (i, 1, 1, k, j_0, j_1) | 1 \leq k \leq S-1; i \geq 1$$
$$l(2) = (i, 2, 0, k, j_0, j_2) | 1 \leq k \leq S \;\cup\; (i, 2, 1, k, j_0, j_2) | 1 \leq k \leq S-1; i \geq 1$$
$$l(3) = (i, 3, 0, k, j_0, j_1, j_2) | 2 \leq k \leq S \;\cup\; (i, 3, 1, k, j_0, j_1, j_2) | 2 \leq k \leq S-1; i \geq 2$$

where $1 \leq j_0 \leq l, 1 \leq j_1 \leq m, 1 \leq j_2 \leq n$.

The conversions in the Markov chain are described as follows.

a) conversions due to the arrival of customers :

$$(i, 0, 0, k, j_0) \xrightarrow{D_1} (i+1, 0, 0, k, j_0); i \geq 0, 0 \leq k \leq S$$
$$(i, 0, 1, k, j_0) \xrightarrow{D_1} (i+1, 0, 1, k, j_0); i \geq 0, 0 \leq k \leq S-1$$
$$(i, 1, 0, k, j_0, j_1) \xrightarrow{D_1 \otimes I_m} (i+1, 1, 0, k, j_0, j_1); i \geq 1, 1 \leq k \leq S$$
$$(i, 1, 1, k, j_0, j_1) \xrightarrow{D_1 \otimes I_m} (i+1, 1, 1, 1, k, j_0, j_1); i \geq 1, 1 \leq k \leq S-1$$
$$(i, 2, 0, k, j_0, j_2) \xrightarrow{D_1 \otimes I_n} (i+1, 2, 0, k, j_0, j_2); i \geq 1, 1 \leq k \leq S$$
$$(i, 2, 1, k, j_0, j_2) \xrightarrow{D_1 \otimes I_n} (i+1, 2, 1, k, j_0, j_2); i \geq 1, 1 \leq k \leq S-1$$
$$(i, 3, 0, k, j_0, j_1, j_2) \xrightarrow{D_1 \otimes I_m \otimes I_n} (i+1, 3, 0, k, j_0, j_1, j_2); i \geq 2, 2 \leq k \leq S$$
$$(i, 3, 1, k, j_0, j_1, j_2) \xrightarrow{D_1 \otimes I_m \otimes I_n} (i+1, 3, 1, k, j_0, j_1, j_2); i \geq 2, 2 \leq k \leq S-1$$

b) conversions due to the completion of service:

$$(1, 1, 0, k, j_0, j_1) \xrightarrow{I_l \otimes S^0} (0, 0, 0, k-1, j_0); 1 \leq k \leq S$$
$$(1, 1, 1, k, j_0, j_1) \xrightarrow{I_l \otimes S^0} (0, 0, 1, k-1, j_0); 1 \leq k \leq S-1$$
$$(1, 2, 0, 1, j_0, j_2) \xrightarrow{I_l \otimes T^0} (0, 0, 0, 0, j_0)$$
$$(1, 2, 0, k, j_0, j_2) \xrightarrow{I_l \otimes T^0} (0, 0, 0, k-1, j_0); 2 \leq k \leq S$$
$$(1, 2, 1, k, j_0, j_2) \xrightarrow{I_l \otimes T^0} (0, 0, 1, k-1, j_0); 1 \leq k \leq S-1$$

$For \;\; i \geq 2$

$$(i,1,0,1,j_0,j_1) \xrightarrow{I_l \otimes S^0} (i-1,0,0,0,j_0)$$
$$(i,1,0,k,j_0,j_1) \xrightarrow{I_l \otimes S^0\alpha} (i-1,1,0,k-1,j_0,j_1); 2 \le k \le S$$
$$(i,1,1,1,j_0,j_1) \xrightarrow{I_l \otimes S^0} (i-1,0,1,0,j_0)$$
$$(i,1,1,k,j_0,j_1) \xrightarrow{I_l \otimes S^0\alpha} (i-1,1,1,k-1,j_0,j_1); 2 \le k \le S-1$$
$$(i,2,0,1,j_0,j_2) \xrightarrow{I_l \otimes T^0} (i-1,0,0,0,j_0)$$
$$(i,2,0,k,j_0,j_2) \xrightarrow{I_l \otimes T^0\beta} (i-1,2,0,k-1,j_0,j_2); 2 \le k \le S$$
$$(i,2,1,1,j_0,j_2) \xrightarrow{I_l \otimes T^0} (i-1,0,1,0,j_0)$$
$$(i,2,1,k,j_0,j_2) \xrightarrow{I_l \otimes T^0\beta} (i-1,2,1,k-1,j_0,j_2); 2 \le k \le S-1$$
$$(i,3,0,2,j_0,j_1,j_2) \xrightarrow{I_l \otimes S^0 \otimes I_n} (i-1,2,0,1,j_0,j_2)$$
$$(i,3,0,2,j_0,j_1,j_2) \xrightarrow{I_l \otimes I_m \otimes T^0} (i-1,1,0,1,j_0,j_1)$$
$$(i,3,0,j,j_0,j_1,j_2) \xrightarrow{I_l \otimes [S^0\alpha \oplus T^0\beta]} (i-1,1,0,1,j_0,j_1); 2 \le k \le S$$
$$(i,3,1,2,j_0,j_1,j_2) \xrightarrow{I_l \otimes S^0 \otimes I_n} (i-1,2,1,1,j_0,j_2)$$
$$(i,3,1,2,j_0,j_1,j_2) \xrightarrow{I_l \otimes I_m \otimes T^0} (i-1,2,1,1,j_0,j_2)$$
$$(i,3,1,k,j_0,j_1,j_2) \xrightarrow{I_l \otimes [S^0\alpha \oplus T^0\beta]} (i-1,3,1,k-1,j_0,j_1,j_2); 3 \le k \le S-1$$

c) conversions due to the completion of production of an item:

$$(i,0,1,k,j_0) \xrightarrow{\eta I_l} (i,0,1,k+1,j_0); i \ge 0, 0 \le k \le S-2$$
$$(i,0,1,S-1,j_0) \xrightarrow{\eta I_l} (i,0,0,S,j_0); i \ge 0$$
$$(i,1,1,k,j_0,j_1) \xrightarrow{\eta I_{lm}} (i,1,1,k+1,j_0,j_1); i \ge 1, 1 \le k \le S-2$$
$$(i,1,1,S-1,j_0,j_1) \xrightarrow{\eta I_{lm}} (i,1,0,S,j_0,j_1); i \ge 1$$
$$(i,2,1,k,j_0,j_2) \xrightarrow{\eta I_{ln}} (i,2,1,k+1,j_0,j_2); i \ge 1, 1 \le k \le S-2$$
$$(i,2,1,S-1,j_0,j_2) \xrightarrow{\eta I_{ln}} (i,2,0,S,j_0,j_2); i \ge 1$$
$$(i,3,1,k,j_0,j_1,j_2) \xrightarrow{\eta I_{lmn}} (i,3,1,k+1,j_0,j_1,j_2); i \ge 2, 2 \le k \le S-2$$
$$(i,3,1,S-1,j_0,j_1,j_2) \xrightarrow{\eta I_{lmn}} (i,3,0,S,j_0,j_1,j_2); i \ge 2$$

d) conversions due to the completion of vacation of servers:

$$(i,0,0,k,j_0) \xrightarrow{I_l \otimes \theta_1 \alpha} (i,1,0,k,j_0,j_1); i \geq 1, 1 \leq k \leq S$$

$$(i,0,1,k,j_0) \xrightarrow{I_l \otimes \theta_1 \alpha} (i,1,1,k,j_0,j_1); i \geq 1, 1 \leq k \leq S-1$$

$$(i,0,1,k,j_0) \xrightarrow{I_l \otimes \theta_2 \beta} (i,2,1,k,j_0,j_2); i \geq 1, 1 \leq k \leq S-1$$

$$(i,1,0,k,j_0,j_1) \xrightarrow{I_l \otimes I_m \otimes \theta_2 \beta} (i,3,0,k,j_0,j_1,j_2); i \geq 2, 2 \leq k \leq S$$

$$(i,1,1,k,j_0,j_1) \xrightarrow{I_l \otimes I_m \otimes \theta_2 \beta} (i,3,1,k,j_0,j_1,j_2); i \geq 2, 1 \leq k \leq S-1$$

$$(i,2,0,k,j_0,j_2) \xrightarrow{I_p \otimes \theta_1 \alpha \otimes I_n} (i,3,0,k,j_0,j_1,j_2); i \geq 2, 1 \leq k \leq S$$

$$(i,2,1,k,j_0,j_2) \xrightarrow{I_l \otimes \theta_1 \alpha \otimes I_n} (i,3,1,k,j_0,j_1,j_2); i \geq 2, 1 \leq k \leq S-1$$

e) conversions due to the completion of vacation of manufacturing unit:

$$(i,0,0,k,j_0) \xrightarrow{\theta I_l} (i,0,1,k,j_0); i \geq 0, 0 \leq k \leq s$$

$$(i,1,0,k,j_0,j_1) \xrightarrow{\theta I_{lm}} (i,1,1,k,j_0,j_1); i \geq 1, 1 \leq k \leq s$$

$$(i,2,0,k,j_0,j_2) \xrightarrow{\theta I_{ln}} (i,2,1,k,j_0,j_2); i \geq 1, 1 \leq k \leq s$$

$$(i,3,0,k,j_0,j_1,j_2) \xrightarrow{\theta I_{lmn}} (i,3,1,k,j_0,j_1,j_2); i \geq 2, 2 \leq k \leq s$$

f) conversions that make no change in the first four coordinates of Ω:

$$(i,0,0,0,j_0) \xrightarrow{D_0 - \theta I_l} (i,0,0,0,j_0); i \geq 0$$

$$(i,0,0,k,j_0) \xrightarrow{D_0 - (\theta + \theta_1 + \theta_2) I_l} (i,0,0,k,j_0); i \geq 0, 1 \leq k \leq s$$

$$(i,0,0,k,j_0) \xrightarrow{D_0 - (\theta_1 + \theta_2) I_l} (i,0,0,k,j_0); i \geq 0, s+1 \leq k \leq S$$

$$(i,0,1,0,j_0) \xrightarrow{D_0 - \eta I_l} (i,0,1,0,j_0); i \geq 0$$

$$(i,0,1,k,j_0) \xrightarrow{D_0 - (\theta 1 + \theta_2 + \eta) I_l} (i,0,1,k,j_0); i \geq 0, 1 \leq k \leq S-1$$

$$(i,1,0,1,j_0,j_1) \xrightarrow{D_0 \oplus S - \theta I_{lm}} (i,1,0,1,j_0,j_1); i \geq 1, 1 \leq k \leq S-1$$

$$(i,1,0,k,j_0,j_1) \xrightarrow{D_0 \oplus S - (\theta + \theta_2) I_{lm}} (i,1,0,k,j_0,j_1); i \geq 1, 2 \leq k \leq s$$

$$(i,1,0,k,j_0,j_1) \xrightarrow{D_0 \oplus S - \theta_2 I_{lm}} (i,1,0,k,j_0,j_1); i \geq 2, s+1 \leq k \leq S$$

$$(1,1,0,k,j_0,j_1) \xrightarrow{D_0 \oplus S - \theta I_{lm}} (1,1,0,k,j_0,j_1); 2 \leq k \leq s$$

$$(1,1,0,k,j_0,j_1) \xrightarrow{D_0 \oplus S} (1,1,0,k,j_0,j_1); s+1 \le k \le S$$

$$(i,1,1,1,j_0,j_1) \xrightarrow{D_0 \oplus S - \eta I_{lm}} (i,1,1,1,j_0,j_1); i \ge 1$$

$$(i,1,1,k,j_0,j_1) \xrightarrow{D_0 \oplus S - (\eta+\theta_2) I_{lm}} (i,1,1,k,j_0,j_1); i \ge 2, 1 \le k \le S-1$$

$$(1,1,1,k,j_0,j_1) \xrightarrow{D_0 \oplus S - \eta I_{lm}} (1,1,1,k,j_0,j_1); i \ge 1, 1 \le k \le S-1$$

$$(i,2,0,1,j_0,j_2) \xrightarrow{D_0 \oplus T - \theta I_{ln}} (i,2,0,1,j_0,j_2)$$

$$(i,2,0,k,j_0,j_2) \xrightarrow{D_0 \oplus T - (\theta+\theta_1) I_{ln}} (i,2,0,k,j_0,j_2); i \ge 1, 2 \le k \le s$$

$$(i,2,0,k,j_0,j_2) \xrightarrow{D_0 \oplus T - \theta_1 I_{ln}} (i,2,0,k,j_0,j_2); i \ge 1, s+1 \le k \le S$$

$$(i,2,1,1,j_0,j_2) \xrightarrow{D_0 \oplus T - \eta I_{ln}} (i,2,1,1,j_0,j_2); i \ge 1$$

$$(i,2,1,k,j_0,j_2) \xrightarrow{D_0 \oplus T - (\eta+\theta_1) I_{ln}} (i,2,1,k,j_0,j_2); 2 \le k \le S-1$$

$$(i,3,0,k,j_0,j_1,j_2) \xrightarrow{D_0 \otimes S \otimes T - \theta I_{lmn}} (i,3,0,k,j_0,j_1,j_2); i \ge 2, 2 \le k \le s$$

$$(i,3,0,k,j_0,j_1,j_2) \xrightarrow{D_0 \otimes S \otimes T} (i,3,0,k,j_0,j_1,j_2); i \ge 2, s+1 \le k \le S$$

$$(i,3,1,k,j_0,j_1,j_2) \xrightarrow{D_0 \otimes S \otimes T - \eta I_{lmn}} (i,3,1,k,j_0,j_1,j_2); i \ge 2, 2 \le k \le S-1$$

The infinitesimal generator of the process Q can be conveniently written in a block partitioned matrix as

$$Q = \begin{bmatrix} B_{00} & B_{01} & 0 & 0 & 0 & \dots \\ B_{10} & B_{11} & B_{12} & 0 & 0 & \dots \\ 0 & B_{21} & A_1 & A_0 & 0 & \dots \\ 0 & 0 & A_2 & A_1 & A_0 & \dots \\ \vdots & \vdots & \vdots & \vdots & \vdots & \ddots \end{bmatrix}$$

where A_0, A_1, A_2 are square matrices of order $8S-4$.

3.1 Stability Analysis

To prove the stability of the system, define a transition rate matrix $A = A_0 + A_1 + A_2$. A is irreducible so there exists a steady state probability vector $\boldsymbol{\pi} = (\pi^{[c]}, c = 0,1,2,3)$ which confirms $\boldsymbol{\pi A} = 0$ and $\boldsymbol{\pi e} = 1$, where $\pi^{[c]} = \{(\pi^{[c,0]}, \pi^{[c,1]}), c = 0,1,2,3\}$. The necessary and sufficient condition required by the system to reach the stability condition is $\boldsymbol{\pi A_0 e} < \boldsymbol{\pi A_2 e}$ (refer Nuets [13]) where,

$$\boldsymbol{\pi A_0 e} = \begin{cases} \pi^{[0,0]}[D_1 I_S]e_{lS} + \pi^{[0,1]}[D_1 I_{S-1}]e_{l(S-1)} \\ +\pi^{[1,0]}[(D_1 I_S)e_{lS}] \otimes e_m + \pi^{[1,1]}[(D_1 I_{S-1})e_{l(S-1)}] \otimes e_m \\ +\pi^{[2,0]}[(D_1 I_S)e_{lS}] \otimes e_n + \pi^{[2,1]}[(D_1 I_{S-1})e_{l(S-1)}] \otimes e_n \\ +\pi^{[3,0]}[(D_1 I_{S-1})e_{l(S-1)}] \otimes e_{mn} + \pi^{[3,1]}[(D_1 I_{S-2})e_{l(S-2)}] \otimes e_{mn} \end{cases}$$

$$\boldsymbol{\pi A_2 e} = \begin{cases} \pi^{[1,0]}[(e_l \otimes S^0) \otimes e_S] + \pi^{[1,1]}[e_l \otimes S^0 \otimes e_{(S-1)}] \\ +\pi^{[2,0]}[(e_l \otimes T^0) \otimes e_S] + \pi^{[2,1]}[(e_l \otimes T^0) \otimes e_{S-1}] \\ +\pi^{[3,0]}[e_l \otimes (S^0 \oplus T^0) \otimes e_{S-1}] + \pi^{[3,1]}[e_l \otimes (S^0 \oplus T^0) \otimes e_{S-2}] \end{cases}$$

3.2 Steady State Probability Vector

The transition probability vector of the infinitesimal generator matrix is $\mathbf{x} = (\boldsymbol{x_0}, \boldsymbol{x_1}, \dots)$ where, $\boldsymbol{x_0} = (y_{0,0,0,0}, \dots y_{0,0,0,S}, y_{0,0,1,0}, \dots y_{0,0,1,S-1})$

$$\boldsymbol{x_1} = \begin{cases} (y_{1,0,0,0}, \dots, y_{1,0,0,S}, y_{1,0,1,0}, \dots, y_{1,0,1,S-1}, y_{1,1,0,1}, \dots, y_{1,1,0,S}, \\ y_{1,1,1,1}, \dots, y_{1,1,1,S-1}, y_{1,2,0,1}, \dots, y_{1,2,0,S}, y_{1,2,1,1}, \dots, y_{1,2,1,S-1}) \end{cases}$$

$$\text{for } i \geq 2, \boldsymbol{x_i} = \begin{cases} (y_{i,0,0,0}, \dots, y_{i,0,0,S}, y_{i,0,1,0}, \dots, y_{i,0,1,S-1}, y_{i,1,0,1}, \dots, y_{i,1,0,S}, \\ y_{i,1,1,1}, \dots, y_{i,1,1,S-1}, y_{i,2,0,1}, \dots, y_{i,2,0,S}, y_{i,2,1,1}, \dots, y_{i,2,1,S-1} \\ y_{i,3,0,2}, \dots, y_{i,3,0,S}, y_{i,3,1,2}, \dots, y_{i,3,1,S-1}) \end{cases}$$

The sub vectors of $\mathbf{x}$ can be derived by using

$$\boldsymbol{x_i} = \boldsymbol{x_{i-1}} * \boldsymbol{R}, \qquad i = 3, 4, 5 \dots \tag{1}$$

where R is the minimal non-negative solution to the matrix quadratic equation $R^2 A_2 + RA_1 + A_0 = 0$. The rate matrix R can be computed from

$R = -A_0(A_1)^{-1} - R^2 A_2 (A_1)^{-1}$. R is approximated by the successive substitution method developed by Neuts [13] namely

$R_0 = 0, R_{n+1} = -A_0(A_1)^{-1} - R_n^2 A_2 (A_1)^{-1}$, $n = 0, 1, 2, \dots$. The process is continued until the successive difference in the value of R is less than a specified tolerance criterion. The vectors $\boldsymbol{x_0}$, $\boldsymbol{x_1}$, $\boldsymbol{x_2}$ and $\boldsymbol{x_i}, i \geq 3$ can be calculated using Eqs. 1,2,3,4 and 5.

$$\boldsymbol{x_0} B_{00} + \boldsymbol{x_1} B_{10} = 0 \tag{2}$$

$$\boldsymbol{x_0} B_{01} + \boldsymbol{x_1} B_{11} + \boldsymbol{x_2} B_{21} = 0 \tag{3}$$

$$\boldsymbol{x_1} B_{12} + \boldsymbol{x_2}[A_1 + RA_2] = 0 \tag{4}$$

and the normalizing equation is

$$\boldsymbol{x_0} e_3 + \boldsymbol{x_1} e_1 + \boldsymbol{x_2}(I - R)^{-1} e_2 = 1 \tag{5}$$

4 Measures of Effectiveness

(i) Expected number of buyers in the system,

$$E_{EC} = \boldsymbol{x_1 e_2} + \boldsymbol{x_2}[2(I - R)^{-1} + R(I - R)^{-2}]e_1$$

(ii) Expected number of departures after completing service,

$$E_{EDS}=\sum_{i=1}^{\infty}\sum_{k=1}^{S}\Big[y_{i,1,0,k}(I_l\otimes S^0)e+y_{i,2,0,k}(I_l\otimes T^0)e\Big]$$

$$+\sum_{i=1}^{\infty}\sum_{k=1}^{S-1}\Big[y_{i,1,1,k}(I_l\otimes S^0)e+y_{i,2,1,k}(I_l\otimes T^0)e\Big]$$

$$+\sum_{i=2}^{\infty}\sum_{k=1}^{S}y_{i,3,0,k}(I_l\otimes(S^0\oplus T^0))e+\sum_{i=2}^{\infty}\sum_{k=2}^{S-1}y_{i,3,1,k}(I_l\otimes(S^0\oplus T^0))e$$

(iii) Expected inventory level,

$$E_{EI}=\sum_{i=0}^{\infty}\sum_{k=1}^{S}\sum_{j_0=1}^{l}ky_{i,0,0,k,j_0}+\sum_{i=0}^{\infty}\sum_{k=1}^{S-1}\sum_{j_0=1}^{l}ky_{i,0,1,k,j_0}$$

$$+\sum_{i=1}^{\infty}\sum_{k=1}^{S}\sum_{j_0=1}^{l}\sum_{j_1=1}^{m}ky_{i,1,0,k,j_0,j_1}+\sum_{i=1}^{\infty}\sum_{k=1}^{S-1}\sum_{j_0=1}^{l}\sum_{j_1=1}^{m}ky_{i,1,1,k,j_0,j_1}$$

$$+\sum_{i=1}^{\infty}\sum_{k=1}^{S}\sum_{j_0=1}^{l}\sum_{j_2=1}^{n}ky_{i,2,0,k,j_0,j_2}+\sum_{i=1}^{\infty}\sum_{k=1}^{S-1}\sum_{j_0=1}^{l}\sum_{j_2=1}^{n}ky_{i,2,1,k,j_0,j_2}$$

$$+\sum_{i=2}^{\infty}\sum_{k=1}^{S}\sum_{j_0=1}^{l}\sum_{j_1=1}^{m}\sum_{j_2=1}^{n}ky_{i,3,0,k,j_0,j_1,j_2}+\sum_{i=2}^{\infty}\sum_{k=2}^{S-1}\sum_{j_0=1}^{l}\sum_{j_1=1}^{m}\sum_{j_2=1}^{n}ky_{i,3,1,k,j_0,j_1,j_2}$$

(iv) Mean Production rate,

$$E_{EPR}=\eta\left[\sum_{i=0}^{\infty}\sum_{k=0}^{S-1}y_{i,0,1,k}+\sum_{i=1}^{\infty}\sum_{k=1}^{S-1}y_{i,1,1,k}+\sum_{i=1}^{\infty}\sum_{k=1}^{S-1}y_{i,2,1,k}+\sum_{i=2}^{\infty}\sum_{k=2}^{S-1}y_{i,3,1,k}\right]$$

5 Cost Analysis

The expected total cost (T_{cost}) of the system per unit per unit time is given by

$$T_{cost}=c_1E_{EC}+c_2E_{EDS}+c_3E_{EI}+c_4E_{EPR}$$

where, c_1= holding cost of customers per unit per unit time, c_2 cost due to service per unit per unit time, c_3= holding cost of inventory per unit per unit time, c_4= cost of manufacturing per unit per unit time.

6 Numerical Experiments

Variations on the model for different combinations of arrival and service processes are discussed. These processes are normalized to have a specific arrival rate. The first three arrival processes are particular cases of renewal processes and the correlation between the time of arrival of customers is 0. The arrival processes denoted as MAP(C-) have correlated arrivals with a correlation coefficient of −0.3255 and the arrivals corresponding to MAP(C+) have a positive correlation coefficient of 0.3255.

1. Hyper-exponential of order 2 (HEX):

$$D_0 = \begin{bmatrix} -5.5 & 0 \\ 0 & -0.55 \end{bmatrix}, D_1 = \begin{bmatrix} 4.67855 & 0.82145 \\ 0.467855 & 0.082145 \end{bmatrix}$$

2. Erlang of order 2 (ER):

$$D_0 = \begin{bmatrix} -4.6924 & 4.6924 \\ 0 & -4.6924 \end{bmatrix}, D_1 = \begin{bmatrix} 0 & 0 \\ 4.6924 & 0 \end{bmatrix}$$

3. Exponential (EX):

$$D_0 = [-2.3462], D_1 = [2.3462]$$

4. Map with negative correlation ($MAP(C-)$):

$$D_0 = \begin{bmatrix} -1.195 & 1.195 & 0 \\ 0 & -128 & 0 \\ 0 & 0 & -128 \end{bmatrix}, D_1 = \begin{bmatrix} 0 & 0 & 0 \\ 1.5 & 0 & 126.5 \\ 126.5 & 0 & 1.5 \end{bmatrix}$$

5. Map with positive correlation ($MAP(C+)$):

$$D_0 = \begin{bmatrix} -1.195 & 1.195 & 0 \\ 0 & -128 & 0 \\ 0 & 0 & -128 \end{bmatrix}, D_1 = \begin{bmatrix} 0 & 0 & 0 \\ 126.5 & 0 & 1.5 \\ 1.5 & 0 & 126.5 \end{bmatrix}$$

Three different PH- distributions are considered for the service time distribution of each server and these processes are normalized so that servers 1 and 2 have a specific service rates.

1. Erlang distribution (ER):

$$\alpha = \begin{bmatrix} 1 & 0 \end{bmatrix}, \beta = \begin{bmatrix} 1 & 0 \end{bmatrix}, S = \begin{bmatrix} -5.1613 & 5.1613 \\ 0 & -5.1613 \end{bmatrix}, T = \begin{bmatrix} -5.0746 & 5.0746 \\ 0 & -5.0746 \end{bmatrix}$$

2. Hyper exponential (HEX):

$$\alpha = \begin{bmatrix} 0.8 & 0.2 \end{bmatrix}, \beta = \begin{bmatrix} 0.7 & 0.3 \end{bmatrix}, S = \begin{bmatrix} -4.2665 & 0 \\ 0 & -1 \end{bmatrix}, T = \begin{bmatrix} -7.437 & 0 \\ 0 & -1 \end{bmatrix}$$

3. Exponential distribution (EX):

$$\alpha = [1], \beta = [1], S = [-2.5806], T = [-2.5373]$$

The optimal value of the expected total cost obtained by varying the production vacation parameter in different combinations of arrival and service distributions to both the servers are recorded in Tables 1, 2, 3, 4 and 5.

In all different combinations of service, the lowest expected costs are obtained when we give EX service distribution to the first server and ER distribution to the second server in $MAP(C-)$ arrival (see Table 1). The lowest expected cost is 232.72 at $\theta = 7$. The optimum expected cost in $MAP(C+)$ is 221.54 at $\theta = 8$ and it is obtained when ER service distributions are assumed to both servers (Table 2). Out of these expected costs, the lowest expected cost in EX arrival is 219.02 at $\theta = 6$ and this value is obtained when HEX service is given to the first server and ER service distribution to the second server (from Table 3). After analyzing Table 4, in ER arrival the optimum value of the expected cost

Table 1. T_{cost} versus θ in MAP(C-) $(S, s, \lambda, \mu_1, \mu_2, \theta_1, \theta_2, \eta) = (20, 4, 2.3462, 2.5806, 2.5373, 6, 10, 5.3)(c_1, c_2, c_3, c_4) = (10, 4, 10, 40)$

	EX			HEX			ER		
θ	EX	ER	HEX	HEX	ER	EX	EX	ER	HEX
6	233.53	232.79	235.41	236.96	233.83	234.24	233.73	232.97	235.64
7	233.38	**232.72**	235.16	236.69	**233.65**	234.12	**233.65**	**232.96**	235.41
8	**233.36**	232.75	235.06	236.59	233.67	**234.12**	233.67	233.04	235.34
9	233.40	232.81	**235.03**	**236.56**	233.69	234.17	233.73	233.13	**235.34**
10	233.45	232.89	235.05	236.57	233.71	234.23	233.80	233.23	235.37
11	233.51	232.96	235.08	236.59	233.73	234.29	233.87	233.32	235.41
12	233.56	233.03	235.11	236.63	233.81	234.35	233.94	233.40	235.46
13	223.62	233.10	235.15	236.66	233.91	234.41	234.01	233.47	235.50

Table 2. T_{cost} versus θ in MAP(C+) $(S, s, \lambda, \mu_1, \mu_2, \theta_1, \theta_2, \eta) = (20, 4, 2.3462, 2.5806, 2.5373, 6, 10, 5.3)(c_1, c_2, c_3, c_4) = (10, 4, 10, 35)$

	EX			HEX			ER		
θ	EX	ER	HEX	HEX	ER	EX	EX	ER	HEX
7	223.60	222.80	225.62	225.83	222.97	223.76	222.41	221.60	224.45
8	223.41	222.68	225.33	225.51	222.79	223.52	222.27	**221.54**	224.17
9	223.35	**222.67**	225.18	225.35	**222.73**	223.42	**222.24**	221.57	224.04
10	**223.34**	222.70	225.12	225.28	222.74	**223.39**	222.26	221.62	223.99
11	223.36	222.74	**225.10**	225.25	222.76	223.39	222.30	221.69	**223.99**
12	223.40	222.80	225.11	**225.25**	222.80	223.42	222.35	221.76	224.00
13	223.44	222.85	225.13	225.26	222.85	223.45	222.41	221.82	224.03

Table 3. T_{cost} versus θ in EX arrival $(S, s, \lambda, \mu_1, \mu_2, \theta_1, \theta_2, \eta) = (20, 4, 2.3162, 2.5806, 2.5373, 6, 10, 5.3)(c_1, c_2, c_3, c_4) = (10, 4, 10, 40)$

	EX			HEX			ER		
θ	EX	ER	HEX	HEX	ER	EX	EX	ER	HEX
5	223.87	221.64	220.63	219.70	219.18	222.94	222.39	220.21	220.73
6	223.25	221.17	220.29	219.43	**219.02**	222.33	221.84	219.88	220.30
7	222.99	220.99	**220.20**	**219.39**	219.05	222.07	221.63	**219.81**	220.17
8	222.89	220.96	220.22	219.43	219.13	221.97	**221.56**	219.84	**220.15**
9	**222.86**	**220.68**	220.27	219.49	219.23	**221.95**	221.56	219.89	220.18
10	222.87	221.01	220.33	219.57	219.32	221.96	221.59	219.97	220.23
11	222.89	221.06	220.39	219.64	219.39	221.99	221.63	220.03	220.29
12	222.92	221.11	220.45	219.70	219.47	222.02	221.67	220.09	220.34

Table 4. T_{cost} versus θ in ER arrival $(S, s, \lambda, \mu_1, \mu_2, \theta_1, \theta_2, \eta) = (20, 4, 2.3462, 2.5806, 2.5373, 6, 10, 5.3)(c_1, c_2, c_3, c_4) = (10, 4, 10, 35)$

	EX			HEX			ER		
θ	EX	ER	HEX	HEX	ER	EX	EX	ER	HEX
5	235.99	237.31	235.29	234.51	236.48	233.72	233.16	234.52	233.99
6	235.55	236.83	234.92	234.18	236.01	233.48	**232.99**	234.21	233.72
7	235.39	236.64	234.79	**234.09**	236.83	**233.45**	233.02	**234.14**	**233.68**
8	**235.36**	**236.58**	**234.79**	234.11	**236.78**	233.51	233.11	234.17	233.72
9	235.39	236.59	234.83	234.16	235.79	233.59	233.22	234.23	233.79
10	235.43	236.61	234.88	234.23	235.82	233.68	233.32	234.30	233.88
11	235.48	236.65	234.94	234.29	235.86	233.76	233.41	234.38	233.96
12	235.54	236.69	235.00	234.36	235.91	233.83	233.49	234.44	234.03

Table 5. T_{cost} versus θ in HEX arrival $(S, s, \lambda, \mu_1, \mu_2, \theta_1, \theta_2, \eta) = (20, 4, 2.3462, 2.5806, 2.5373, 6, 10, 5.3)(c_1, c_2, c_3, c_4) = (10, 4, 10, 40)$

	EX			HEX			ER		
θ	EX	ER	HEX	HEX	ER	EX	EX	ER	HEX
5	253.61	252.19	256.76	257.80	253.92	255.10	252.71	251.12	256.15
6	253.16	251.84	256.19	257.24	253.48	254.60	252.34	**250.90**	255.59
7	**253.07**	**251.84**	255.97	257.01	**253.39**	**254.45**	**252.33**	250.99	255.39
8	253.14	251.97	**255.93**	**256.95**	253.46	254.47	252.45	251.20	**255.38**
9	253.26	252.14	255.98	256.98	253.59	254.56	252.62	251.43	255.45
10	253.41	252.31	256.06	257.05	253.73	254.68	252.79	251.64	255.55
11	253.55	252.48	256.16	257.13	253.88	254.79	252.95	251.83	255.66
12	253.68	252.63	256.26	257.22	254.01	254.91	253.10	252.00	255.77

is reached when ER service distribution is assumed to the first server and EX service distribution to the second server and the value is 232.99 at $\theta = 6$. Analyzing Table 5, the optimum value of the expected cost is reached when ER service distribution is assumed to both the servers and the value is 250.90 at $\theta = 6$.

Concluding Remarks

A two-server production inventory system with vacations to both production and servers is considered. In the present scenario, it is highly recommended to study the impact of providing vacation to both production and servers. Further studies can be made by introducing the Markovian Production Process and Phase type distributed lead time.

References

1. Baek, J.W., Moon, S.K.: A production-inventory system with a Markovian service queue and lost sales. J. Korean Stat. Soc. **45**(1), 14–24 (2016)
2. Beena, P., Jose, K.P.: A map/ph(1), ph(2)/2 production inventory model with inventory dependent production rate and multiple servers. In: AIP Conference Proceedings, vol. 2261, no. 1, p. 030052 (2020). https://doi.org/10.1063/5.0017008, https://aip.scitation.org/doi/abs/10.1063/5.0017008
3. Doshi, B.T.: Queueing systems with vacations-a survey. Queueing Syst. **1**(1), 29–66 (1986)
4. Goyal, S., Giri, B.C.: The production-inventory problem of a product with time varying demand, production and deterioration rates. Eur. J. Oper. Res. **147**(3), 549–557 (2003)
5. Jose, K.P., Beena, P.: On a retrial production inventory system with vacation and multiple servers. Int. J. Appl. Comput. Math. **6**(4), 1–17 (2020)
6. Jose, K.P., Nair, S.S.: Analysis of two production inventory systems with buffer, retrials and different production rates. J. Ind. Eng. Int. **13**(3), 369–380 (2017). https://doi.org/10.1007/s40092-017-0191-0
7. Krishnamoorthy, A., Nair, S.S., Narayanan, V.C.: Production inventory with service time and interruptions. Int. J. Syst. Sci. **46**(10), 1800–1816 (2015)
8. Kumar, B.K., Rukmani, R., Thangaraj, V.: Analysis of map/ph (1), ph (2)/2 queue with Bernoulli vacations. Int. J. Stochastic Anal. **2008** (2008)
9. Latouche, G., Ramaswami, V.: A logarithmic reduction algorithm for quasi-birth-death processes. J. Appl. Probab. **30**(3), 650–674 (1993)
10. Levy, Y., Yechiali, U.: An m/m/s queue with servers' vacations. INFOR: Inf. Syst. Oper. Res. **14**(2), 153–163 (1976)
11. Manna, A.K., Dey, J.K., Mondal, S.K.: Imperfect production inventory model with production rate dependent defective rate and advertisement dependent demand. Comput. Ind. Eng. **104**, 9–22 (2017)
12. Nair, S.S., Jose, K.P.: A MAP/PH/1 production inventory model with varying service rates. Int. J. Pure Appl. Math. **117**(12), 373–381 (2017)
13. Neuts, M.F.: Matrix-geometric solutions to stochastic models. In: Steckhan, H., Bühler, W., Jäger, K.E., Schneeweiß, C., Schwarze, J. (eds.) DGOR, vol. 1983, pp. 425–425. Springer, Heidelberg (1984). https://doi.org/10.1007/978-3-642-69546-9_91

14. Neuts, M.F., Takahashi, Y.: Asymptotic behavior of the stationary distributions in the gi/ph/c queue with heterogeneous servers. Zeitschrift für Wahrscheinlichkeitstheorie und verwandte Gebiete **57**(4), 441–452 (1981)
15. Roy, A., Samanta, G.: Inventory model with two rates of production for deteriorating items with permissible delay in payments. Int. J. Syst. Sci. **42**(8), 1375–1386 (2011)
16. Sivashankari, C., Panayappan, S.: Production inventory model for two-level production with deteriorative items and shortages. Int. J. Adv. Manuf. Technol. **76**(9–12), 2003–2014 (2015)
17. Suganya, C., Sivakumar, B.: Map/ph (1), ph (2)/2 finite retrial inventory system with service facility, multiple vacations for servers. Int. J. Math. Oper. Res. **15**(3), 265–295 (2019)
18. Sugapriya, C., et al.: Analysis of stock-dependent arrival process in a retrial stochastic inventory system with server vacation. Processes **10**(1), 176 (2022)
19. Takagi, H.: Time-dependent process of m/g/1 vacation models with exhaustive service. J. Appl. Probab. **29**(2), 418–429 (1992)
20. Teghem, J., Jr.: Control of the service process in a queueing system. Eur. J. Oper. Res. **23**(2), 141–158 (1986)
21. Yue, D., Qin, Y.: A production inventory system with service time and production vacations. J. Syst. Sci. Syst. Eng. **28**(2), 168–180 (2019)

A Discrete Time Perishable Inventory System Under (s, S) policy and Age-Dependent-Requests

Jijo Joy[1] and K. P. Jose[2](✉)

[1] St. Aloysius College Edathua-689573, Alappuzha, Kerala, India
[2] St. Peter's College, Kolenchery-682371, Ernakulam, Kerala, India
kpjspc@gmail.com

Abstract. A discrete time perishable inventory model under (s, S) policy with demand is age-dependent is considered in this paper. By assuming demand process to be a Bernoulli and lead time and service time geometric we define a suitable cost function for the model. An arriving customer will accept or reject an item depending on its manufacturing date and expiry date. We analyse the model using the Matrix Analytic Method. Numerical experiments are also performed to find the optimal (s, S) pair for the model for a set of pre-assigned parameter values.

Keywords: Bernoulli Process · Perishable Inventory · Age-Dependent Demand · Matrix Analytic Method

Introduction

A healthy society needs daily consumable items to be fresh and free from preservatives. Over a period of time, such items lose their nutrition and flavour. Customers choice is for newly baked or freshly packed products. Our model has applications in firms dealing with drugs, baked products, processed fish and meat, canned food, etc.

The study of perishable items was first started by Whitin(1957) [17] for fashion goods. The concept of exponential decay in inventory management was introduced by Ghare(1963) [6]. Deteriorating inventory model with positive lead time was first studied by Nahmias(1979) [12], who considered a Heuristic (Q, r) model with exponential decay and random demand. Dave and Patel(1981) [3] considered a perishable inventory model by assuming demand as a linear function of time. Pal M.(1989) [13] studied an inventory model of decaying items with positive lead time and without fixed ordering cost.

Varghese(2004) [16] developed an (s, S) continuous review perishable inventory model with items subject to external disaster. Assuming inter-arrival times to be an iid random variable they derived the steady state and transient probabilities of inventory level. Manuel et al.(2007) [11] developed an (s, S) continuous review perishable inventory model with MAP arrivals and service time following

A. Dudin et al. (Eds.): ITMM 2022, CCIS 1803, pp. 189–199, 2023.
https://doi.org/10.1007/978-3-031-32990-6_16

PH distributions. They derived the joint probability distribution of inventory level and the number of customers in the system. Gurler and Ozkaya(2008) [7] analyzed a continuous time (s, S) perishable inventory with random shelf life, zero lead time and renewal arrivals. They derived the expected cost rate function expression for the batch and unit demands. Hamadi et al.(2015) [8] considered a finite capacity perishable inventory system with impatient customers. By considering service times to be exponential and demand Poisson, they determined the rates of service which minimizes the long-run expected total cost. Sangeetha and Sivakumar (2020) [15] studied a continuous time perishable inventory system with Markovian arrivals and lead times following PH distribution. They used a linear programming algorithm to determine the stationary optimal policy.

Recently, Adrian(2021) [10] studied a model in which items have both freshness and physical decay. The other parameters of consideration of the paper include the date of expiration, salvage cost, and deterioration cost. The proposed model optimises the total profit by considering the optimal price, replenishment cycle time, and order quantity. An EOQ model whose demand factors are displayed volume and expiry date were proposed by Chen J. and et al. (2016) [2]. This model was an attempt to optimize annual total profit and to determine optimal shelf size and replenishment time. The theoretical aspects are highlighted using numerical examples. A single item inventory model with decaying items considering lifetime of products was developed by Dobson et al.(2007) [4]. Assuming demand rate as a linearly decreasing function of the age of the products, they obtained the optimal cycle length of the retailer analytically. Considering demand as a function of three factors, price, displayed stock and freshness Feng et al.(2017) [5] developed an inventory model. They succeeded in displaying total profit as strictly concave with respect to the three decision variables. Also, they derived the optimal cycle time, selling price, and ending inventory level which is non-zero. In a recent work on continuous time deteriorating inventory system, Reshmi and Jose(2020) [14] considered perishability as a function of linear rate. Assuming lead time to be exponential and allowing retrial of customers, they derived the stationary probability vector using Matrix Analytic Method. Jose and Reshmi(2021) [9] analysed a continuous time perishable inventory model with production and retrial of customers. They succeeded in finding the algorithmic solution of the problem and numerical calculations to study the effect of various parameters on performance measures.

Discrete time queueing models receive significant attention of researchers due to its wide range of applications in data transfer and telecommunication systems. In the discrete time inventory model the concept of priority queue was introduced by Anilkumar and Jose(2020) [1]. By assuming geometric service time and lead time, they derived the stability condition for the system.

This paper is an extension of the concept of age dependent demand in perishable inventory model. We assume that the customer will choose an item or not depending on the remaining lifetime of items in the inventory. It is assumed that the probability of buying an item is high in the first few days of the date of manufacturing. The rest of the paper is arranged as follows. In Sect. 1 Mathematical

Modelling and analysis are discussed. Sections 2 and 3 deal with Stability Condition and Steady State Analysis of the model. System performance measures and Cost analysis were discussed in Sects. 4 and 5. Numerical illustration and (s, S) pair are presented in Sects. 6 and 7. Finally, graphical representation is presented in Sect. 8.

1 Mathematical Modelling and Analysis

In this model, we study a discrete-time (s, S) perishable inventory model with positive service time and positive lead time. The arrival is assumed to be a Bernoulli process with parameter a. Both service time and lead time are assumed to be geometric with parameters b and c respectively. Lead time is assumed to be positive. When the lead time is realized we will discard all the items already in the inventory so that the remaining new items have a common lifetime m. A customer's demand is for exactly one item and the customer is not allowed to join the system when the inventory level is zero.

We hypothesize that the demand depends on age. The age of the fresh items in the inventory is assumed to be m units and after which the item becomes spoiled. We assume that the customer's decision to buy a product depends on the freshness of the product. When the age of the product increases the customer has the tendency to not buy. We assume that when the product is fresh the probability that the customer chooses them is high and there is a high probability that a customer not buy a product whose expiry date is near. For an item with remaining lifetime i, where $i = 1, 2, 3, \cdots, m$, we assume that the customer will choose the item with probability β_i and not choose the item with probability $1 - \beta_i$.

Notations

X_m : Number of customers in the queue at an epoch m .

Y_m : Inventory level at the epoch m .

Z_m : The remaining life of the product at an epoch m .

$\mathbf{e}$: $(1, 1, 1, ..., 1)'$, column vector of 1's of size $Sm + 1$.

Then $\Psi = \{(X_m, Y_m, Z_m); m = 0, 1, 2, 3, ..\}$ is a Quasi-Birth-Death Process on the state space

$$\mathcal{E} = \{(i', j', k'), i' \geq 0, 0 \leq j' \leq S, 1 \leq k' \leq m\}\}$$

. Now the transition probability matrix of the process is

$$\mathcal{P} = \begin{array}{c} \\ 0 \\ 1 \\ 2 \\ 3 \\ \vdots \end{array} \begin{array}{c} \begin{array}{ccccc} 0 & 1 & 2 & 3 & \ldots \end{array} \\ \begin{pmatrix} C_1 & C_0 & & & \\ A_2 & A_1 & A_0 & & \\ & A_2 & A_1 & A_0 & \\ & & A_2 & A_1 & A_0 \\ & & & \ddots & \ddots & \ddots \\ & & & & \end{pmatrix} \end{array}$$

where the blocks C_0, C_1, A_0, A_1, A_2 are square matrices whose $(j,k)^{th}$ element with phase i is given below. For convenience we use the following notations $j^* = j-(i-1)m-1, i^* = (i-1)m, \Delta_1 = \beta_{j^*}b+\bar{\beta}_{j^*}, \Delta_2 = (ab+\bar{a}\bar{b})\beta_{j^*}+a\bar{\beta}_{j^*}, \Delta_3 = a\bar{\beta}_{j^*} + \bar{a}\bar{b}\beta_{j^*}$ and $\bar{x} = 1-x$, for $x = a, b, c, \beta_j$.

$$C_0(j,k) = \begin{cases} \bar{c}, & \text{if } j=1; k=1 \\ c, & \text{if } j=1; k=Sm+1 \\ \bar{a}c, & \text{if } j=2,\cdots,sm+1; k=Sm+1 \\ \bar{a}\bar{c}, & \text{if } i=1,2,\cdots,s; j=(i-1)m+2; k=1 \\ \bar{a}\bar{c}, & \text{if } i=1,\cdots,s; j=(i-1)m+2,\cdots,im+1; k=j-1 \\ \bar{a}, & \text{if } i=s+1,\cdots,S; j=(i-1)m+2; k=1 \\ \bar{a}, & \text{if } i=s+1,\cdots,S; j=(i-1)m+2,\cdots,im+1; k=j-1 \\ 0, & \text{otherwise} \end{cases}$$

$$C_1(j,k) = \begin{cases} ac, & \text{if } j=2,\cdots,sm+1; k=Sm+1 \\ a\bar{c}, & \text{if } i=1,\cdots,s; j=(i-1)m+2; k=1 \\ a\bar{c}, & \text{if } i=1,\cdots,s; j=(i-1)m+2,\cdots,im+1; k=j-1 \\ a, & \text{if } i=s+1,\cdots,S; j=(i-1)m+2; k=1 \\ a, & \text{if } i=s+1,\cdots,S; j=(i-1)m+2,\cdots,im+1; k=j-1 \\ 0, & \text{otherwise} \end{cases}$$

$$A_2(j,k) = \begin{cases} \bar{c}, & \text{if } j=1; k=1 \\ \bar{a}c, & \text{if } i=1,\cdots,s; j=(i-1)m+2; k=Sm+1 \\ \bar{a}c\Delta_1, & \text{if } i=1,\cdots,s; j=i^*+3,\cdots,im+1; k=Sm+1 \\ \bar{a}\bar{c}, & \text{if } i=1,\cdots,s; j=(i-1)m+2; k=1 \\ \bar{a}\beta_{j-1}b\bar{c}, & \text{if } i=2,\cdots,m; j=i+1; k=1 \\ \bar{a}\bar{\beta}_{j^*}\bar{c}, & \text{if } i=1,\cdots,s; j=i^*+3,\cdots,im+1; k=j-1 \\ \bar{a}\bar{\beta}_{j^*}, & \text{if } i=s+1,\cdots,S; j=i^*+3,\cdots,im+1; k=j-1 \\ \bar{a}, & \text{if } i=s+1,\cdots,S; j=i^*+2; k=1 \\ \bar{a}\beta_{j^*}b\bar{c}, & \text{if } i=2,\cdots,s; j=i^*+3,\cdots,im+1; k=j-m-1 \\ \bar{a}\beta_{j^*}b, & \text{if } i=s+1,\cdots,S; j=i^*+3,\cdots,im+1; k=j-m-1 \\ 0, & \text{otherwise} \end{cases}$$

$$A_1(j,k) = \begin{cases} c, & \text{if } j = 1; k = Sm+1 \\ ac, & \text{if } i = 1,\cdots,s; j = i^*+2; k = Sm+1 \\ c\Delta_2, & \text{if } i = 1,\cdots,s; j = i^*+3,\cdots,im+1; k = Sm+1 \\ a\bar{c}, & \text{if } i = 1,\cdots,s; j = i^*+2; k = 1 \\ a\beta_{j-1}b\bar{c}, & \text{if } i = 2,\cdots,m; j = i+1; k = 1 \\ \Delta_3\bar{c}, & \text{if } i = 1,\cdots,s; j = i^*+3,\cdots,im+1; k = j-1 \\ \Delta_3, & \text{if } i = s+1,\cdots,S; j = i^*+3,\cdots,im+1; k = j-1 \\ a, & \text{if } i = s+1,\cdots,S; j = i^*+2; k = 1 \\ a\beta_j^* b\bar{c}, & \text{if } i = 2,\cdots,s; j = i^*+3,\cdots,im+1; k = j-m-1 \\ a\beta_j^* b, & \text{if } i = s+1,\cdots,S; j = i^*+3,\cdots,im+1; k = j-m-1 \\ 0, & \text{otherwise} \end{cases}$$

$$A_0(j,k) = \begin{cases} a\beta_j^*\bar{b}c, & \text{if } i = 1,\cdots,s; j = i^*+3,\cdots,im+1; k = Sm+1 \\ a\beta_j^*\bar{b}\bar{c}, & \text{if } i = 1,\cdots,s; j = i^*+3,\cdots,im+1; k = j-1 \\ a\beta_j^*\bar{b}, & \text{if } i = s+1,\cdots,S; j = i^*+3,\cdots,im+1; k = j-1 \\ 0, & \text{otherwise} \end{cases}$$

2 Stability Condition

In this model when the inventory is zero, no customer is allowed to join the system. Hence the inventory level is independent of queue length. The stability of the considered queueing inventory model is the same as that of ordinary discrete-time Geo/Geo/1 queue with inter-arrival time and service time being geometrically distributed with parameters say a and b respectively. Then, the system is stable if $a < b$.

3 Steady-State Analysis

Assume that the QBD is aperiodic and positive recurrent. Denote by π its stationary probability vector. It is the unique solution of the system $\pi P = \pi$ and $\pi \mathbf{e} = 1$, where $\mathbf{e}$ is a column vector of ones of appropriate order.
Let π be partitioned by levels as $\pi = (\pi_\mathbf{0}, \pi_\mathbf{1}, \pi_\mathbf{2}, \pi_\mathbf{3}\ldots,)$. Then π_i has the matrix geometric form $\pi_i = \pi_1 R^{i-1}, i \geq 2$. Where R is the minimal non negative solution of the matrix quadratic equation

$$R^2 A_2 + RA_1 + A_0 = R$$

where π_0 and π_1 are obtained by solving the equations

$$\pi_0(C_0 - I) + \pi_1 A_2 = 0$$

and

$$\pi_0 C_1 + \pi_1(A_1 + RA_2 - I) = 0$$

with the normalizing condition

$$\pi_\mathbf{0}\mathbf{e} + \pi_1(I-R)^{-1}\mathbf{e} = 1$$

where $\mathbf{e}$ is a column vector of 1's of size $(Sm+1)$.

4 System Performance Measures

We partition the components of π as $\pi = (\pi_0, \pi_1, \pi_2 \ldots)$ and $\pi_{\rm i} = (\pi_{i0}, \pi_{i11} \ldots, \pi_{i1m}, \pi_{i21} \ldots, \pi_{i2m}, \ldots, \pi_{iS1} \ldots, \pi_{iSm})(i \geq 0)$. The performance measures of the system under steady state are

1. Expected number of customers in the system, EC, is given by

$$EC = \sum_{i=1}^{\infty} i\pi_i e$$

2. Expected inventory level, EI, is given by

$$EI = \sum_{i=1}^{\infty}\sum_{j=1}^{S}\sum_{k=1}^{m} j\pi_{ijk}$$

3. Expected reorder rate, ER, is given by

$$ER = \sum_{i=0}^{\infty}\sum_{j=s+1}^{S} \pi_{ij1} + \sum_{i=0}^{\infty}\sum_{k=1}^{m} \beta_k b\pi_{i(s+1)k}$$

4. Expected replinishment rate, ERR, is given by

$$ERR = c\sum_{i=0}^{\infty}\sum_{j=0}^{s}\sum_{k=1}^{m} \pi_{ijk}$$

5. Probability that the inventory level zero, $(PIL)_0$, is given by

$$(PIL)_0 = \sum_{i=0}^{\infty} \pi_{i0}$$

6. Expected loss rate of customers, EL, is given by

$$EL = a\sum_{i=0}^{\infty} \pi_{i0}$$

7. Expected number of customers waiting in the system when the invetory level is zero, $EW0$, is given by

$$EW0 = \sum_{i=1}^{\infty} iy_{i0}$$

8. Expected rate of departure after completing the service, ED, is given by

$$ED = \sum_{i=1}^{\infty}\sum_{j=1}^{S}\sum_{k=2}^{m} b\pi_{ijk} + \sum_{i=1}^{\infty}\sum_{j=1}^{S} \beta_1\pi_{ij1}$$

9. Expected perishable Quantity, EPQ, is given by

$$EPQ = \sum_{i=1}^{\infty}\sum_{j=1}^{S} j\bar{\beta}_1\pi_{ij1} + \sum_{i=1}^{\infty}\sum_{j=1}^{S}(j-1)\beta_1\pi_{ij1} + \sum_{j=1}^{S} j\pi_{0j1}$$

5 Cost Analysis

Define the expected total cost of the system per unit time as

$$ETC = [c_o + Sc_1](ER) + c_2(EI) + c_3(EW) + c_4(EL) + c_5(EPQ)$$

where,

c_0 : The fixed ordering cost
c_1 : Procurement cost/unit
c_2 : Holding cost of inventory/unit/unit time
c_3 : Holding cost of customers/unit/unit time
c_4 : Cost due to loss of customers/unit/unit time
c_5 : Disposal cost/unit Perished

6 Numerical Illusration

The following tables illustrate the numerical results obtained for various performance measures with respect to the different parameters are considered. The relationship between various performance measures with parameters is indicated below.

6.1 Effect of a on Various Performane Measures

Table 1 shows that as the expected arrival rate increases, the expected reorder rate, expected replenishment rate, and expected rate of departure increases, but expected inventory level and expected perishale quantity decrease.

Table 1. Effect of a on the model. $b = 0.55, c = 0.4 s = 5, S = 10, m = 15$

r=.3

a	ER	EI	EW	EL	EPQ	EC	ERR	ED	ETC
0.54	0.06591	6.9988	0.07930	0.068315	0.36405	1.084	0.06052	0.26416	36.47263
0.55	0.06598	6.9974	0.08090	0.069316	0.36244	1.1143	0.06064	0.26452	36.47245
0.56	0.06604	6.996	0.08270	0.070295	0.36076	1.147	0.06077	0.26491	36.47231
0.57	0.06611	6.9945	0.08473	0.071251	0.35901	1.1825	0.06090	0.26531	36.47220
0.58	0.06618	6.993	0.08703	0.072181	0.35718	1.2212	0.06104	0.26573	36.47214
0.59	0.06624	6.9914	0.08963	0.073085	0.35527	1.2635	0.06119	0.26618	**36.47213**
0.60	0.06631	6.9898	0.09259	0.073961	0.35327	1.3099	0.061351	0.26664	36.47220
0.61	0.06637	6.988	0.09596	0.074805	0.35117	1.3611	0.06151	0.26713	36.47238
0.62	0.06643	6.9862	0.09981	0.075617	0.34897	1.4179	0.06168	0.26765	36.47271

6.2 Effect of b on Various Performane Measures

Table 2 shows that as the expected service rate increases, the expected reorder rate, expected replenishment rate and expected rate of departure increases but expected inventory level, expected perishability quantity, and the expected loss of customers decrease.

Table 2. Effect of b on the model. $a = 0.4, c = 0.4 s = 5, S = 10, m = 15$

b	ER	EI	EW	EL	EPQ	EC	ERR	ED	ETC
0.54	0.06446	7.0306	0.06967	0.05283	0.38419	0.82875	0.05917	0.25533	19.012
0.55	0.06490	7.0176	0.06917	0.05261	0.38146	0.81809	0.05929	0.26037	19.005
0.56	0.06536	7.0051	0.06867	0.05238	0.37874	0.80769	0.05941	0.26543	19.000
0.57	0.06583	6.9931	0.06818	0.05215	0.37605	0.79754	0.05953	0.27051	18.997
0.58	0.06631	6.9816	0.06770	0.05192	0.37338	0.78762	0.05965	0.27562	**18.996**
0.59	0.06681	6.9705	0.06723	0.05168	0.37073	0.77794	0.05978	0.28075	18.997
0.60	0.06732	6.9599	0.06677	0.05144	0.36811	0.76848	0.05990	0.2859	18.999
0.61	0.06784	6.9497	0.06631	0.05120	0.36551	0.75925	0.06003	0.29108	19.003
0.62	0.06838	6.94	0.06585	0.05095	0.36293	0.75022	0.06017	0.29628	19.008

6.3 Effect of c on Various Performane Measures

Table 3 shows that as the replenishment rate increases, the expected inventory level, expected reorder rate, expected replenishment rate, and expected rate of departure increases, but the expected number of customers waiting and expected loss of customers decrease.

Table 3. Effect of c on the model. $a = 0.4, b = 0.5 s = 5, S = 10, m = 15$

c	ER	EI	EW	EL	EPQ	EC	ERR	ED	ETC
0.43	0.06348	7.1582	0.06760	0.05034	0.3988	0.8783	0.05936	0.2377	22.23355
0.44	0.06368	7.1798	0.06632	0.04932	0.3998	0.8797	0.05954	0.2385	22.23349
0.45	0.06387	7.2006	0.06509	0.04833	0.4008	0.881	0.05972	0.2392	22.23346
0.46	0.06406	7.2206	0.06391	0.04739	0.4018	0.8822	0.05990	0.2398	**22.23345**
0.47	0.06424	7.2398	0.06277	0.04648	0.4027	0.8834	0.06006	0.2405	22.23346
0.48	0.06441	7.2583	0.06168	0.04561	0.4037	0.8846	0.06022	0.2411	22.23348
0.49	0.06458	7.2761	0.06063	0.04477	0.4045	0.8857	0.06038	0.2417	22.23352
0.50	0.06474	7.2933	0.05961	0.04396	0.4054	0.8868	0.06053	0.2423	22.23356

7 (s, S) Pair

For fixed parameter values, the optimal (s, S) pair is given in the following talble.

Table 4. (s, S) pair of the model for fixed parameter values $a = 0.3, b = 0.8, c = 0.6, m = 15, c_0 = 50, c_1 = 15, c_2 = 4, c_3 = 3, c_4 = 15, c_5 = 1$

s S	14	15	16	17	18
6	52.5199	**52.2080**	53.0113	54.9641	57.9154
7	58.8076	57.2699	56.7648	57.4316	59.2979
8	66.5386	63.8129	62.0200	61.3216	61.8519
9	75.8634	71.8685	68.8181	66.7700	65.8785
10	87.4268	81.6039	77.1983	73.8233	71.5200

8 Graphical Illustration

We draw the graphs of ETC by varying the parameters a, b and c by keeping other parameters constant. These graphs indicate that in the long run the company will have the expected total cost as shown in the graphs. The minimum value of the expected total cost and corresponding values of a, b and c can be calculated. The minimum values for Fig. 1, Fig. 2 and Fig. 3 are $(0.59, 36.47213)$, $(0.58, 18.996)$ and $(0.46, 22.23345)$. From Fig. 4, the optimum (s, S) pair is $(6, 15)$ with expected total cost 52.2080.

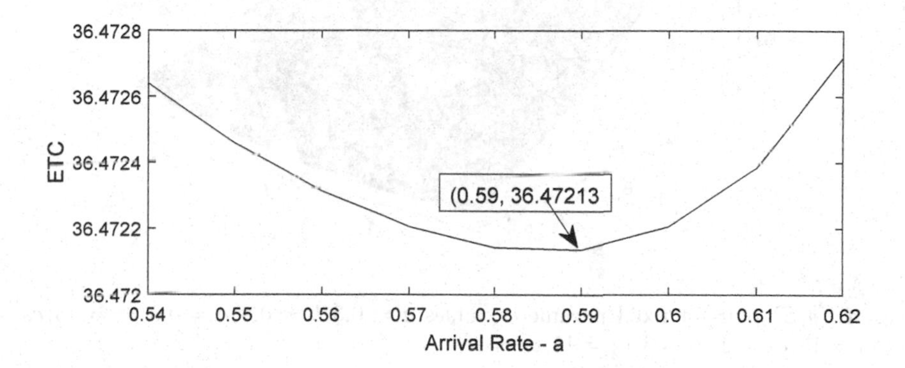

Fig. 1. a vs ETC, $b = 0.55, c = 0.4 s = 5, S = 10, m = 15$

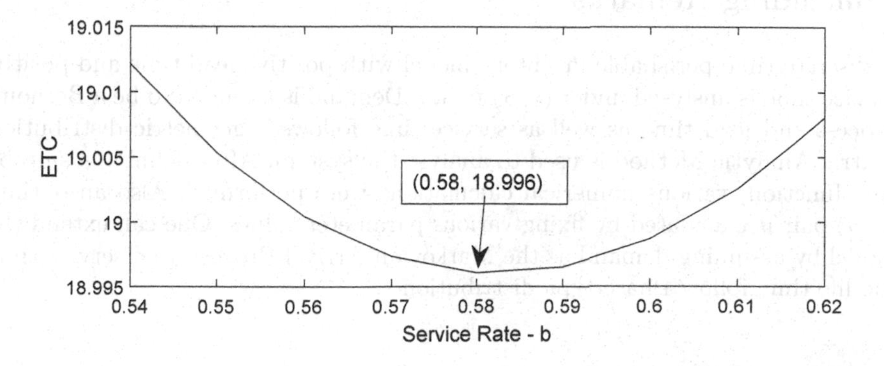

Fig. 2. b vs ETC, $a = 0.4, c = 0.4 s = 5, S = 10, m = 15$

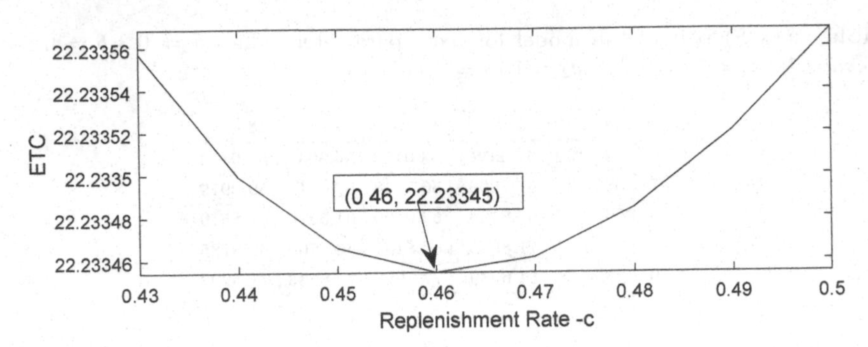

Fig. 3. c vs ETC, $a = 0.4, b = 0.5s = 5, S = 10, m = 15$

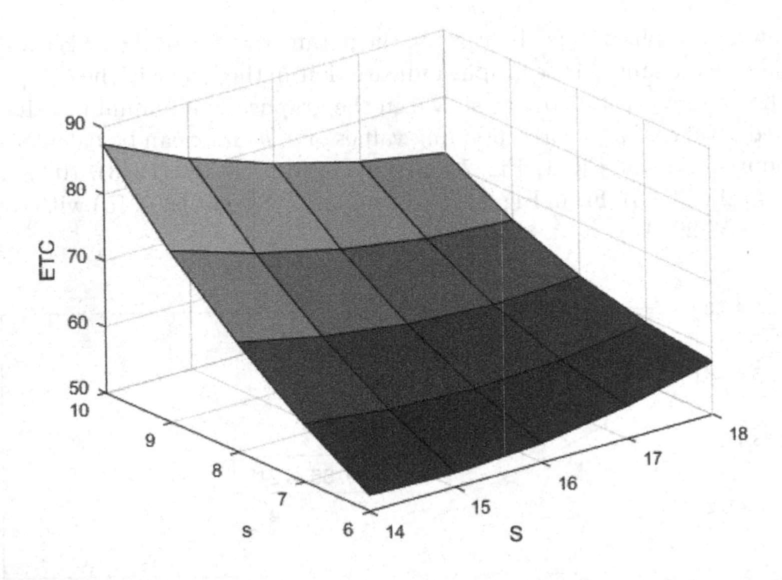

Fig. 4. (s, S) pair for fixed parameter values $a = 0.3, b = 0.8, c = 0.6, m = 15, c_0 = 50, c_1 = 15, c_2 = 4, c_3 = 3, c_4 = 15, c_5 = 1$

Concluding Remarks

A discrete time perishable inventory model with positive lead time and positive service time is anslysed under (s, S) policy. Demand is assumed to be a Bernoulli process and lead time as well as service time follows a geometric distribution. Matrix Analytic Method is used to analyse the system. After defining a suitable cost function, various numerical calculations were performed. Also an optimal (s, S) pair is calculated by fixing various parameter values. One can extend this model by assuming demand as the Markovian Arrival Process, and service time, the life time follow Phase-type distribution.

References

1. Anilkumar, M.P., Jose, K.P.: A Geo/Geo/1 inventory priority queue with self induced interruption. Int. J. Appl. Comput. Math. **6**(4), 1–14 (2020)
2. Chen, J., Min, J.T., Tenf, J.T., Li, F.: Inventory and shelf space optimization for fresh produce with expiration date under freshness-and-stock-dependent demand rate. J. Oper. Res. Soc. **67**(6), 884–896 (2016)
3. Dave, U., Patel, L.K.: (T, S_i) policy inventory model for deteriorating items with time proportional demand. J. Oper. Res. Soc. **2**, 137–142 (1981)
4. Dobson, D., Pinker, I.J., Yildiz, O.: An EOQ model for perishable goods with age-dependent demand rate. Eur. J. Oper. Res. **257**(1), 84–88 (2017)
5. Feng, L., Chan, Y.L., Cardenas-Barron, L.D.: Pricing and lot-sizing policies for perishable goods when the demand depends on selling price, displayed stocks, and expiration date. Int. J. Prod. Econ. **185**, 11–20 (2017)
6. Ghare, P.M., Scharder, G.F.: A model for exponentially decaying inventory. J. Ind. Eng. **14**, 238–243 (1963)
7. Gurler, U., Ozkaya, B.Y.: Analysis of (s, S) policy for perishables with a random shelf life. IEEE Trans. **410**, 759–781 (2008)
8. Hamadi, H.M., Sangeetha, N., Sivakumar, B.: Optimal control of service parameter for a perishable inventory system maintained at service facility with impatient customers. Ann. Oper. Res. **23**, 3–23 (2015)
9. Jose, K.P., Reshmi, P.S.: A production inventory model with deteriorating items and retrial demands. Opsearch **58**(1), 71–82 (2021)
10. Adrian, M.-L., et al.: An inventory model for perishable items with price-, stock-, and Time-dependent demand rate considering shelf-life and non linear holding costs. Math. Prob. Eng. (2021). https://doi.org/10.1155/2021/6630938
11. Manuel, P., Sivakumar, B.: A perishable inventory system with service facilities. J. Syst. Sci. Syst. Eng. **16**, 62–73 (2007)
12. Nahmias, S., Wang, S.S.: A Heuristic lot size reorder point model for decaying inventories. Mgmt. Sci. **25**(1), 90–97 (1979)
13. Pal, M.: The (S-1, S) inventory model for deteriorating items with exponential lead time. Calcutta Statist. Assoc. Bull. **8**, 149–150 (1989)
14. Reshmi, P.S., Jose, K.P.: MAP/PH/1 perishable inventory system with dependent retrial loss. Int. J. Appl. Comput. Math. **6**(6), 1–11 (2020)
15. Sangeetha, N., Sivakumar, B.: Optimal service rates of a perishable inventory system with service facility. Int. J. Math. Oper. Res. **16**(4), 515–550 (2020)
16. Varghese, T.V., Krishnamoorthy, A.: Inventory system subject to disaster with general inter arrival times. Stoc. Anal. Appl. **22**(5), 1315–1326 (2004)
17. Whitin, T.M.: Theory of inventory management. Princeton University Press, Princeton (1957)

N-Policy on a Retrial Inventory System

K. P. Jose[1(✉)] and N. J. Thresiamma[2]

[1] PG & Research Department of Mathematics, St. Peter's College, Kolenchery 682311, Kerala, India
kpjspc@gmail.com

[2] Government Polytechnic College, Muttom 685527, Kerala, India

Abstract. The concept of $N-$ policy in a retrial inventory system with positive lead time is described in this paper. The customer's arrival follows the Poisson distribution. Service time is distributed exponentially. When the customer arrives, he joins a buffer of finite size S. The server begins to provide service only when there are N customers in the buffer, and it shuts down when the buffer is empty or the inventory reaches zero. The client exits the system or enters an orbit of unlimited capacity when the buffer is full. The customer may attempt again from orbit. Replenishment orders are placed when the inventory level reaches s. The lead time and inter-retrial time follow exponential distribution. The Lyapunov test function is used to determine stability. The Matrix- Analytic Method is employed to compute steady state probability vectors. Various steady state system performance measures are derived, and the long-run expected cost is calculated. Several numerical examples are presented to provide insight into the behavior of the cost function.

Keywords: Retrial inventory · N- Policy · Neuts-Rao truncation method

1 Introduction

In many real-world systems, deciding when the server should start the service is critical because frequent setups inevitably raise operating costs. Over the recent decade, there has been a lot of interest in queuing scenarios that apply the N-policy concept to control queues. The so-called N-policy is characterized by the fact that customers are accumulated and an idle server is turned ON whenever there are N ($N \geq 1$) or more customers in the system, and that the server is turned OFF when the system becomes empty. It is widely accepted because of its applicability for modeling purposes in production and manufacturing systems, as well as in computing and telecommunication systems.

The concept of N-policy was first introduced by Yadin and Naor [11] in 1963 in queuing literature to minimize the total operational cost in a cycle. During the past two decades, queuing systems with N-policy have been studied extensively by several authors. A considerable number of works in this area were

A. Dudin et al. (Eds.): ITMM 2022, CCIS 1803, pp. 200–211, 2023.
https://doi.org/10.1007/978-3-031-32990-6_17

completed in the early 1990s. Motivated by these applications more studies on queuing models under N-Policy have been done during the late 1990s and 2000s. A comprehensive review of N-policy till 2014 can be found in Jayachitra and Albert [1]. Tao et.al [8] studied $M/M/1$ Retrial Queue with Working Vacation Interruption and Feedback under N-Policy. Yen et al. [12]studied optimization analysis of the N-policy M/G/1 queue with working breakdowns. Even though a lot of works on N-policy is there in queuing theory, only a few works are carried out on inventory. Krishnamoorthy et al.. [5] were the first to introduce N-policy in (s, S) inventory system with positive service time. They assumed that the lead time is zero and showed that the cost function is separately convex in the variables S and N. Suganya et al. [7]analyzed the impatient demands caused by the N-policy server to a perishable inventory system. In the steady state, they obtain the joint probability distribution of the level of inventory and the number of customers in the system. Thresiamma and Jose [9] introduced the N policy on the production inventory system with positive service time. They observed that introducing the N policy to production inventory will reduce the total cost considerably.

Artalejo et al. [1] were the first to study inventory policies with positive lead-time and retrial of customers who could not get service during their earlier attempts to access the service station. They compared and analysed the efficiency of a generalized truncated model with a finite truncated model. A retrial inventory system with two modes of service were examined by Rejitha and Jose [6]. They made the assumption that customers were arrive according to the Markovian Arrival Process and that service time is phase type distributed. Salini and Jose [3] compared two production inventory systems with buffer, retrials and different production rates. They studied two models one with finite buffer capacity and the other with varying buffer capacity and found out that the model with buffer of varying capacity is efficient for practical purposes in the given range of parameter values. Salini and Jose [4] analyzed a MAP/PH/1 production inventory model with the retrial of customers and varying service rates. A single server perishable inventory system with Poisson arrivals and retrial needs were examined by Jose and Reshmi [2]. In this paper, we introduce N-policy to a retrial inventory system with a finite buffer.

The remaining content of this paper is divided into four parts. In Sect. 2, the model under study is described. The system's stability and steady state proabilty is discussed in Sect. 3. Numerical and graphical illustrations are provided in Sect. 4. Conclusions and suggested further research are provided in Sect. 5.

2 Mathematical Modelling and Analysis of the Problem

Consider an (s, S) retrial inventory system with a finite buffer of size S and orbit of infinite capacity. An arriving customer joins the buffer. The server starts service only when N or more customers are in the buffer, and it shuts down when the buffer is empty or the inventory drops to zero. When the buffer overflows, the customer either leaves the system or enters the orbit. From the orbit, the customers reattempt the service. When the inventory level depletes to s, an order for replenishment is placed. The following assumptions are made for modeling this problem.

- The Arrival of customers forms a Poisson distribution with rate λ.
- The lead time follows an exponential distribution with rate β
- The inter -retrial time is an exponential distribution with linear rate $i\theta$, when there are i customers in the orbit.
- An arriving customer who finds the buffer full is directed to an orbit with probability γ and is lost forever with probability $(1-\gamma)$.
- A retrial customer in the orbit finds the buffer full and returns to the orbit with probability δ and is lost forever with probability $(1-\delta)$.

The following notations are used in this model.

$$N(t) : \text{Number of customers in the orbit at time } t.$$
$$M(t) : \text{Number of customers in the buffer at time t.}$$
$$I(t) : \text{Inventory level at time} t.$$
$$C(t) : \begin{cases} 0 \text{ if server is idle at time t,} \\ 1 \text{ if server 1 is busy at time t.} \end{cases}$$

$$\Delta = -(\lambda\gamma + \beta + i\theta(1-\delta)).$$
$$\Omega = -(\lambda\gamma + \mu + \beta + i\theta(1-\delta)).$$
$$\Psi = -(\lambda + i\theta).$$
$$\Phi = -(\lambda + \beta + i\theta).$$
$$\kappa = -(\lambda + \mu + \beta + i\theta).$$
$$\omega = -(\lambda\gamma + \mu + i\theta(1-\delta)).$$
$$\Sigma = -(\lambda + \mu + i\theta).$$
$$X(t) = \{(N(t), C(t), I(t), M(t)) : t \geq 0.\}$$

Then $X(t)$ is a Continuous Time Markov Chain on the state space

$$\{(i,0,0,m) : i \geq 0; m = 0, 1, ...S.\}$$
$$\bigcup\{(i,0,k,m) : i \geq 0; k = 1, ..., S : m = 0, 1, ...N-1.\}$$
$$\bigcup\{(i,1,k,m) : i \geq 0; 1 \leq k \leq S; m = 1, 2, ...S.\}$$

The infinitesimal generator G of the process is a block tri-diagonal matrix and has the form:

$$G = \begin{bmatrix} A_{1,0} & A_0 & & & \\ A_{2,1} & A_{1,1} & A_0 & & \\ & A_{2,2} & A_{1,2} & A_0 & \\ & & A_{2,3} & A_{1,3} & A_0 \\ & & & \ddots & \ddots & \ddots \end{bmatrix}$$

where the block matrices $A_0, A_{1,i}$ and $A_{2.i}$; $(i \geq 0.)$ are square matrices of order $S(N+S+1)+1$ and has the following form.

$$A_{1,i} = \begin{array}{c} \underline{0,0} \\ \underline{0,1} \\ \\ \underline{0,s} \\ \underline{0,s+1} \\ \underline{0,S} \\ \underline{1,1} \\ \underline{1,2} \\ \underline{1,s} \\ \underline{1,s+1} \\ \underline{1,S} \end{array} \begin{bmatrix} G_0 & & & & & C_o & & & & & D_0 \\ & G1 & & & & C_1 & K_1 & & & & \\ & & \ddots & & & \vdots & & \ddots & & & \\ & & & G_1 & & C1 & & & K_1 & & \\ & & & & G_2 & & & & & K_1 & \\ & & & & & \ddots & & & & & \ddots \\ & & & & & G_2 & & & & & K_1 \\ L_1 & & & & & & G_3 & & & & C_2 \\ & L_0 & & & & & L_2 & G_3 & & & C_2 \\ & & \ddots & & & & & \ddots & \ddots & & \vdots \\ & & & L_0 & & & & & L_2 & G_3 & C_2 \\ & & & & L_0 & & & & & L_2 & G_4 \\ & & & & & L_0 & & & & & L_2 \; G_4 \end{bmatrix}$$

$$A_{2,i} = \begin{array}{c} \underline{0,0} \\ \underline{0,1} \\ \underline{0,S} \\ \underline{1,1} \\ \underline{1,S} \end{array} \begin{bmatrix} V_0 & & & & \\ & V1 & & D & \\ & & \ddots & & \ddots \\ & & & V_1 & & D \\ & & & & V_2 & \\ & & & & & \ddots \\ & & & & & & V_2 \end{bmatrix}, \quad A_0 = \begin{array}{c} \underline{0,0} \\ \underline{0,S} \\ \underline{1,1} \\ \underline{1,S} \end{array} \begin{bmatrix} B_0 & & & \\ & B_1 & & \\ & & \ddots & \\ & & & B_1 \end{bmatrix},$$

$$B_0 = \begin{bmatrix} 0 & & \\ & \ddots & \\ & & 0 \\ & & & \lambda\gamma \end{bmatrix}_{S+1}, \quad B_1 = \begin{bmatrix} 0 & & \\ & \ddots & \\ & & 0 \\ & & & \lambda\gamma \end{bmatrix}_{S}, \quad V_0 = \begin{bmatrix} 0 & i\theta & & \\ & \ddots & \ddots & \\ & & 0 & i\theta \\ & & & i\theta(1-\delta) \end{bmatrix}_{S+1},$$

$$V_1 = \begin{bmatrix} 0 & i\theta & & \\ & \ddots & \ddots & \\ & & 0 & i\theta \\ & & & 0 \end{bmatrix}_{N}, \quad V_2 = \begin{bmatrix} 0 & i\theta & & \\ & \ddots & \ddots & \\ & & 0 & i\theta \\ & & & i\theta(1-\delta) \end{bmatrix}_{S}, G_0 = \begin{bmatrix} \Phi & \lambda & & \\ & \ddots & \ddots & \\ & & \Phi & \lambda \\ & & & \Delta \end{bmatrix}_{S+1},$$

$$G_1 = \begin{bmatrix} \Phi & \lambda & & \\ & \ddots & \ddots & \\ & & \Phi & \lambda \\ & & & \Phi \end{bmatrix}_{N}, \quad G_2 = \begin{bmatrix} \Psi & \lambda & & \\ & \ddots & \ddots & \\ & & \Psi & \lambda \\ & & & \Psi \end{bmatrix}_{N}, \quad G_3 = \begin{bmatrix} \kappa & \lambda & & \\ & \ddots & \ddots & \\ & & \kappa & \lambda \\ & & & \Omega \end{bmatrix}_{S},$$

$$G_4 = \begin{bmatrix} \Sigma\ \lambda & & \\ & \ddots\ \ddots & \\ & & \Sigma\ \lambda \\ & & \omega \end{bmatrix}_S, \qquad L_2 = \begin{bmatrix} 0 & & & \\ \mu & & & \\ & \mu & & \\ & & \mu & 0 \end{bmatrix}_S, \qquad L_0 = \begin{bmatrix} \mu & O \\ O & O \end{bmatrix}_{S_X N},$$

$$C_0 = \begin{bmatrix} H \\ O \end{bmatrix}_{S+1_X N}, \qquad D_0 = \begin{bmatrix} O & O \\ O & J \end{bmatrix}_{S+1};$$

$$H = \beta I_N,$$

$$J = \beta I_{S+1-N}, \qquad C_1 = \beta I_N, \qquad C_2 = \beta I_S, \qquad L_1 = [\mu I_S\ O_{S_X 1}]_{S_X S+1},$$

$$K_1(N,N)_{N_X S} = \begin{cases} \lambda, \\ 0; otherwise, \end{cases} \qquad D(p,q)_{N_X S} = \begin{cases} i\theta, if p = q = N \\ 0; otherwise \end{cases}$$

3 Steady State Analysis

3.1 System Stability

Define the test function $\phi(s)$ as, $\phi(s) = i$, ifsis a state in level$i$.The mean drift y_s for any s belonging to the level $i \geq 1$ is given by

$$y_s = \sum_{p \neq s} q_{sp}(\phi(p) - \phi(s)) = \sum_u q_{su}(\phi(u) - \phi(s))$$
$$+ \sum_v q_{sv}(\phi(v) - \phi(s)) + \sum_w q_{sw}(\phi(w) - \phi(s))$$

where u, v, w vary over the states belonging to the levels $(i-1)$, i & $(i+1)$ respectively.

$$\phi(u) = i - 1, \phi(v) = i, \&\phi(w) = i + 1$$

so that

$$y_s = -\sum_u q_{su} + \sum_w q_{sw}$$

Here mean drift y_s is given by

$$y_s = \begin{cases} -i\theta(1-\delta) + \lambda\gamma; \text{if the buffer is full,} \\ -i\theta; \text{otherwise.} \end{cases}$$

Since $(1-\delta) > 0$, for any $\epsilon > 0$, we can find n' large enough that $y_s < -\epsilon$ for any s belonging to the level $i \geq n$. Hence by Tweedi's [10] result, the system under consideration is stable.

3.2 Steady State Probability Vector

As $X(t)$ is a Level Dependant Quasi Birth Death Process, we obtain a truncation level n using Neuts-Rao truncation. So we choose n in such a way that

$$\|\eta(n) - \eta(n+l)\| < \epsilon$$

where ϵ is an arbitrarily small value and $\eta(n)$ is the spectral radius of $R(n)$. Let

$$\mathbf{x} = (x(0), x(1), ..., x(n-1), x(n), ...)$$

be the steady state probability vector. Under the stability condition, $x(i)$'s are given by

$$x(n+r-1) = x(n-1)R^r, (r \geq 1)$$

where R is the unique non negative solution of the equation

$$R^2A_2 + RA_1 + A_0 = 0;$$

for which the spectral radius is less than 1. The vectors $x(0), x(1), ..., x(n-1)$ are obtained by solving the equations

$$\begin{aligned} x(0)A_{1,0} + x(1)A_{2,1} &= 0, \\ x(i-1)A_0 + x(i)A_{1,i} + x(i+1)A_{2,i+1} &= 0; (1 \leq i \leq n-2), \\ x(n-2)A_0 + x(n-1)(A_{1,n-1} + RA_2) &= 0. \end{aligned}$$

subject to the normalizing condition

$$\left(\sum_{i=0}^{n-2} x(i) + x(n-1)(I-R)^{-1}\right) e = 1.$$

Partition $\mathbf{x}$ according to the levels as
$x(i) = (x_{i,0,0,0}, x_{i,0,0,1}, ..., x_{i,0,0,S}, x_{i,0,1,0}, x_{i,0,1,1}, ..., x_{i,0,1,N-1}, ..., x_{i,0,S,0}, x_{i,0,S,1}, ..., x_{i,0,S,N-1}, x_{i,1,1,1}, x_{i,1,1,2}, ..., x_{i,1,1,S}, x_{i,1,2,1}, x_{i,1,2,2}, ..., x_{i,1,2,S}, ..., x_{i,1,S,1}, x_{i,1,S,2}, ..., x_{i,1,S,S})$; $x(i)$ contain $S(N+S+1)+1$ elements.

3.3 System Performance Measures

(i) Expected inventory Level,

$$EI = \sum_{i=0}^{\infty}\sum_{j=1}^{S}\sum_{k=0}^{N-1} jx_{i,0,j,k} + \sum_{i=0}^{\infty}\sum_{j=1}^{S}\sum_{k=1}^{S} jx_{i,1,j,k}.$$

(ii) Expected Number of customers in the orbit,

$$EC = \left(\sum_{i=1}^{\infty} ix(i)\right) e = \left(\sum_{i=1}^{n-1} ix(i) + x(n)(n(I-R)^{-1} + R(I-R)^{-2})\right) e.$$

(iii) Expected Number of customers in the buffer,

$$EB = \sum_{i=0}^{\infty}\sum_{k=0}^{S} kx_{i,0,0,k} + \sum_{i=0}^{\infty}\sum_{j=1}^{S}\sum_{k=1}^{N-1} kx_{i,0,j,k} + \sum_{i=0}^{\infty}\sum_{j=1}^{S}\sum_{k=1}^{S} kx_{i,1,j,k}.$$

(iv) Expected number of departures after completing service,

$$EDS = \mu \sum_{i=0}^{\infty}\sum_{j=1}^{S}\sum_{k=1}^{S} x_{i,1,j,k}.$$

(v) Expected Number of external customers lost before entering orbit,

$$EL1 = (1-\gamma)\lambda \left(\sum_{i=0}^{\infty} x_{i,0,0,S} + \sum_{i=0}^{\infty}\sum_{j=1}^{S} x_{i,1,j,S} \right).$$

(vi) Expected Number of customers lost due to retrials,

$$EL2 = \theta(1-\delta) \left(\sum_{i=1}^{\infty} ix_{i,0,0,S} + \sum_{i=1}^{\infty}\sum_{j=1}^{S} ix_{i,1,j,S} \right).$$

(vii) Overall rate of retrials,

$$ORR = \theta \left(\sum_{i=1}^{\infty} ix(i) \right) e.$$

(viii) Successful rate of retrials,

$$SRR = \theta \left(\sum_{i=1}^{\infty}\sum_{k=0}^{S-1} ix_{i,0,0,k} + \sum_{i=1}^{\infty}\sum_{j=0}^{S}\sum_{k=0}^{N-1} ix_{i,0,j,k} + \sum_{i=1}^{\infty}\sum_{j=1}^{S}\sum_{k=1}^{S-1} ix_{i,1,j,k} \right).$$

(ix) Expected Re order rate,

$$ERO = \mu \sum_{i=0}^{\infty}\sum_{k=1}^{S} x_{i,1,s+1,k}.$$

3.4 Cost Function

The aim is to obtain an adaptive (s, S) policy according to specified cost requirements. Since the objective cost function is not specified explicitly, we describe it as a combination of pertinent system attributes. By minimizing the total cost, one can identify the ideal values of s, N, and S. The long-run cost function for this model is defined as

$$ETC = (C + (S-s)C_1)ERO + C_2EI + C_3EC + C_4EB + C_5EL_1 + C_6EL_2 + (C_7 - C_8)EDS.$$

where,

C = Fixed cost.
C_1 =procurement cost/unit.
C_2 = holding cost of inventory /unit/unit time.
C_3 =holding cost of the customers in the orbit/unit/unit time.
C_4 =holding cost of the customer in the buffer/unit/unit time.
C_5 =cost due to loss of primary customers/unit/unit time.
C_6 =cost due to loss of retrial customers/unit/unit time.
C_7 =cost due to service/unit/unit time.
C_8 = Revenue from service/unit/unit time.

4 Numerical and Graphical Illustrations

This section provides a description of the numerical experiments that were conducted to investigate the effects of changes in various parameters on the performance measures and expected total cost.

Table 1 displays how N has an impact on performance measures and expected total cost. The expected number of consumers in the buffer increases noticeably with an increase in N, as seen in Table 1. Consumers are more likely to be lost as N increases since there are more customers waiting in the orbit and buffer. Because of this, there is a modest increase in the expected number of clients lost due to retry and prior to entering orbit. The overall retrial rate and successful retrial rate both rise. The plot of the expected total cost variation with respect to N is shown in Fig. 1, and it appears to be convex in nature.

Table 1. Performance measures for different values of N.

N	EI	EC	EB	EL1	EL2	EDS	ORR	SRR	ERO	ETC
1	6.3108	0.0447	2.7922	0.0062	0.0015	0.7896	0.0268	0.0121	0.0106	4.0825
2	6.2058	0.0486	2.9992	0.0066	0.0016	0.7754	0.0291	0.0133	0.0103	4.0360
3	6.1308	0.053	3.2121	0.0072	0.0017	0.7647	0.0318	0.0147	0.0101	4.0183
4	6.0818	0.058	3.4339	0.0078	0.0019	0.7571	0.0348	0.0164	0.0099	4.0093
5	6.0572	0.0639	3.6674	0.0085	0.0021	0.7523	0.0383	0.0184	0.0097	4.0140
6	6.0567	0.0706	3.9145	0.0094	0.0023	0.7502	0.0424	0.0208	0.0095	4.0369
7	6.0807	0.0785	4.1779	0.0105	0.0026	0.7508	0.0471	0.0235	0.0095	4.1521
8	6.1321	0.0876	4.4603	0.0118	0.0029	0.7542	0.0526	0.0268	0.0095	4.0839
9	6.2123	0.0983	4.765	0.0134	0.0032	0.7604	0.059	0.0307	0.0096	4.2428

$S = 20 : s = 3 : \lambda = 1.5 : \mu = 1.8 : \beta = 1.2 : \delta = 0.9 : \gamma = 0.9 : \theta = 0.6 : C = 10 :$
$C_1 = 3 : C_2 = 1 : C_3 = 0.1 : C_4 = 0.1 : C_5 = 0.1 : C_6 = 0.1 : C_7 = 2 : C_8 = 6.$

Table 2 displays the effect of s-the reorder level -on performance measures and expected total cost. The expected inventory rises as s. Due to the low likelihood of a server being idle from a lack of inventory, the expected number of customers waiting will therefore decrease. Hence there is little possibility of losing customers. The expected total cost's fluctuation with respect to s is plotted in Fig. 2, and it appears to be convex in shape. It is clear from the graph that when $N = 4$, the cost is at its lowest.

Table 2. Performance measures for different values of s.

s	EI	EC	EB	EL1	EL2	EDS	ORR	SRR	ERO	ETC
0	4.9762	0.1466	4.0326	0.0160	0.0052	0.7160	0.0879	0.0494	0.0376	5.1634
1	5.3724	0.0917	3.6985	0.0112	0.0031	0.7363	0.0550	0.0282	0.0233	4.3677
2	5.7444	0.0689	3.5278	0.0089	0.0023	0.7498	0.0414	0.0200	0.0149	4.0595
3	6.0818	0.0580	3.4339	0.0078	0.0019	0.7571	0.0348	0.0164	0.0099	4.0092
4	6.3903	0.0523	3.3790	0.0072	0.0017	0.7597	0.0314	0.0146	0.0064	4.0685
5	6.6772	0.0492	3.3442	0.0068	0.0016	0.7592	0.0295	0.0136	0.0041	4.2035
6	6.9484	0.0472	3.3198	0.0066	0.0015	0.7564	0.0283	0.0130	0.0025	4.3913
7	7.2076	0.0460	3.3006	0.0064	0.0015	0.7519	0.0276	0.0127	0.0016	4.6117
8	7.4569	0.0451	3.2836	0.0063	0.0015	0.7460	0.0270	0.0125	0.0010	4.8507

$S = 20 : N = 4 : \lambda = 1.5 : \mu = 1.8 : \beta = 1.2 : \delta = 0.9 : \gamma = 0.9 : \theta = 0.6 : C = 10 :$
$C_1 = 3 : C_2 = 1 : C_3 = 0.1 : C_4 = 0.1 : C_5 = 0.1 : C_6 = 0.1 : C_7 = 2 : C_8 = 6,$

Table 3. Performance Measures for different values of S.

S	EI	EC	EB	EL1	EL2	EDS	ORR	SRR	ERO	ETC
19	4.8288	0.6414	5.2455	0.0618	0.0249	0.5994	0.3848	0.1411	0.0063	10.4291
20	4.7907	0.6115	5.2005	0.056	0.0237	0.5716	0.3669	0.1343	0.0056	10.4265
21	4.7561	0.5841	5.1587	0.051	0.0226	0.5462	0.3505	0.1281	0.0051	10.4249
22	4.7247	0.559	5.1202	0.0466	0.0217	0.5229	0.3354	0.1225	0.0046	10.4242
23	4.6959	0.536	5.0846	0.0427	0.0208	0.5015	0.3216	0.1173	0.0042	10.4244
24	4.6697	0.5147	5.0517	0.0393	0.0199	0.4818	0.3088	0.1125	0.0039	10.4253
25	4.6456	0.4949	5.0211	0.0363	0.0192	0.4636	0.297	0.1081	0.0035	10.4267
26	4.6233	0.4767	4.9926	0.0336	0.0184	0.4467	0.286	0.1041	0.0033	10.4285

$S = 20 : N = 4 : \lambda = 1.5 : \mu = 1.55 : \beta = 1.2 : \delta = 0.9 : \gamma = 0.9 : \theta = 0.6 : C = 10 :$
$C_1 = 3 : C_2 = 10 : C_3 = 0.1 : C_4 = 0.1 : C_5 = 0.1 : C_6 = 0.1 : C_7 = 2 : C_8 = 6$

The impact of S—the maximum inventory level—on performance indicators and the expected total cost is shown in Table 3. The expected total cost variation with respect to S is represented in Fig. 3, and it is also a convex graph. Figure 4 depicts the variation of ETC in relation to (s,S). The changes of expected total cost with respect to (s,N) and (N,S), respectively, are depicted in Figs. 5 and 6.

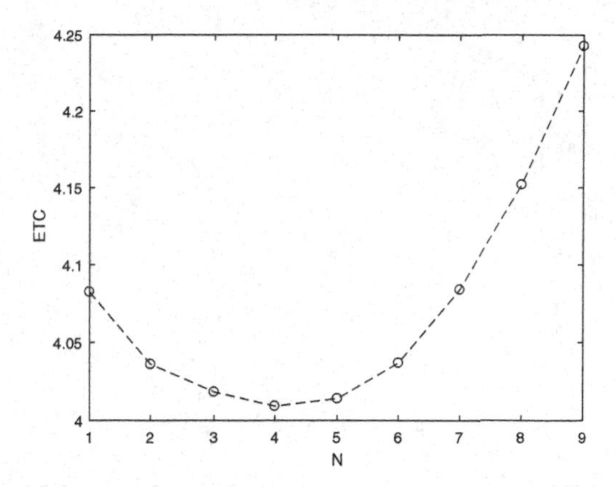

Fig. 1. variation of ETC with respect to N $S = 20 : s = 3 : \lambda = 1.5 : \mu = 1.8 : \beta = 1.2 : \delta = 0.9 : \gamma = 0.9 : \theta = 0.6$.

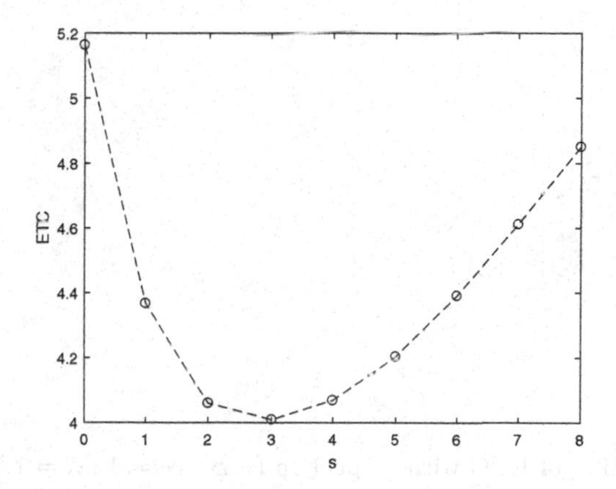

Fig. 2. Variation of ETC with respect to s $S = 20 : N = 4 : \lambda = 1.5 : \mu = 1.8 : \beta = 1.2 : \delta = 0.9 : \gamma = 0.9 : \theta = 0.6$.

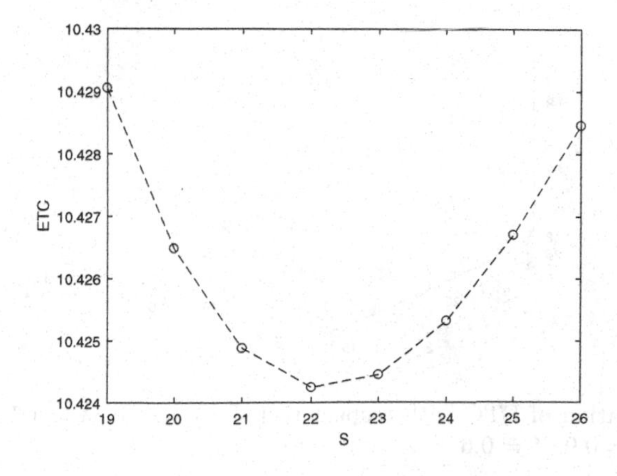

Fig. 3. $N = 4 : s = 3 : \lambda = 1.5 : \mu = 1.55 : \beta = 1.2 : \delta = 0.9 : \gamma = 0.9 : \theta = 0.6$

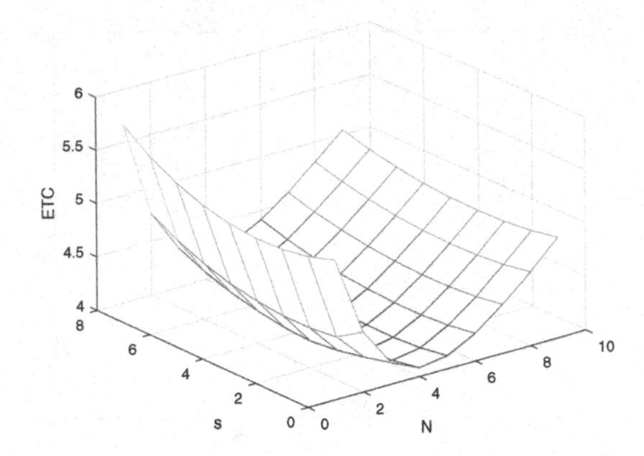

Fig. 4. Variation of ETC with respect to (s, N) $S = 20 : \lambda = 1.5 : \mu = 1.8 : \beta = 1.2 : \delta = 0.9 : \gamma = 0.9 : \theta = 0.6$

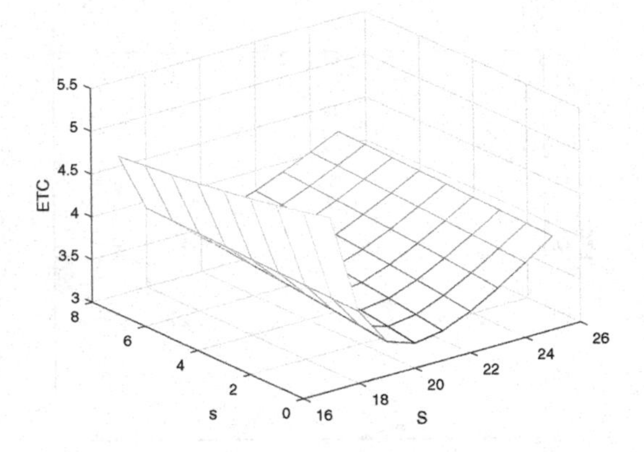

Fig. 5. The variation of ETC with respect to (s, S) $N = 4 : \lambda = 1.5 : \mu = 1.55 : \beta = 1.2 : \delta = 0.9 : \gamma = 0.9 : \theta = 0.6$.

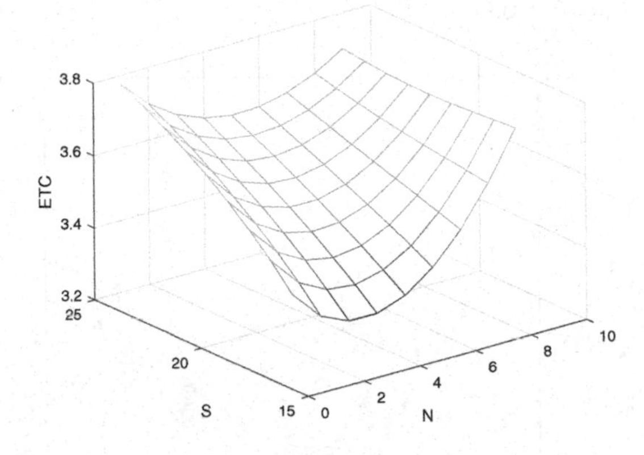

Fig. 6. The variation of ETC with respect to (N, S) $s = 3 : \lambda = 1.5 : \mu = 1.55 : \beta = 1.2 : \delta = 0.9 : \gamma = 0.9 : \theta = 0.6$

5 Concluding Remarks

The concept of the N policy into a retrial inventory model with a finite buffer is covered in the paper. The exponential distribution is considered for the inter arrival time as well as service time. Matrix- Analytic Method is used to find the stationary probability vector, which makes it easier to obtain some key performance measures. A suitable cost function is constructed and the optimal value of N is obtained. The results are numerically illustrated to show the effect of the change of values of parameters. It is discovered that the cost function is separately convex in the variables s, S, and N. One can extend this model to a buffer of varying capacity and a retrial production inventory system.

References

1. Jayachitra, P., Albert, A.: Recent developments in queueing models under n-policy: A short survey. Int. J. Math. Arch. **5**, 227–233 (2014)
2. Jose, K.P., Reshmi, P.S.: A production inventory model with deteriorating items and retrial demands. Opsearch **58**(1), 71–82 (2020). https://doi.org/10.1007/s12597-020-00471-8
3. Jose, K.P., Nair, S.S.: Analysis of two production inventory systems with buffer, retrials and different production rates. J. Indust. Eng. Int. **13**(3), 369–380 (2017). https://doi.org/10.1007/s40092-017-0191-0
4. Jose, K.P., Salini, S.N.: A map/ph/1 production inventory model with varying service rates. Int. J. Pure Appl. Math. **117**, 373–381 (2017)
5. Krishnamoorthy, A., Narayanan, V.C., Deepak, T., Vineetha, P.: Control policies for inventory with service time. Stochastic Anal. Appl.**24**(4), 889–899 (2006). https://doi.org/10.1080/07362990600753635, https://doi.org/10.1080/07362990600753635
6. Rejitha, K.R., Jose, K.P.: A stochastic inventory system with two modes of service and retrial of customers. Opsearch **55**(1), 134–149 (2017). https://doi.org/10.1007/s12597-017-0322-9
7. Suganya, R., et al.: Perishable inventory system with n-policy, map arrivals, and impatient customers. Mathematics 9(13) (2021). https://doi.org/10.3390/math9131514, https://www.mdpi.com/2227-7390/9/13/1514
8. Tao, L., Zhang, L., Gao, S.: M/m/1 retrial queue with working vacation interruption and feedback under n-policy. J. Appl. Math. **2014**, 1–9 (04 2014). https://doi.org/10.1155/2014/414739
9. Thresiamma, N.J., Jose, K.P.: N-policy for a production inventory system with positive service time. Information Technologies and Mathematical Modelling. Queueing Theory and Application, pp. 52–66 (2022)
10. Tweedie, R.L.: Sufficient conditions for regularity, recurrence and ergodicity of markov processes. Math. Proc. Cambridge Philos. Soc. **78**(1), 125–136 (1975). https://doi.org/10.1017/S0305004100051562
11. Yadin, M., Naor, P.: Queueing systems with a removable service station. OR 14, 393–405 (1963). https://doi.org/10.2307/3006802
12. Yen, T.C., Wang, K.H., Chen, J.Y.: Optimization analysis of the n policy m/g/1 queue with working breakdowns. Symmetry **12**, 583 (04 2020). https://doi.org/10.3390/sym12040583

Research of Dynamic RQ System M/M/1 with Unreliable Server

N. M. Voronina[1](✉), S. V. Rozhkova[1,2], and E. A. Fedorova[2]

[1] National Research Tomsk Polytechnic University,
Lenin Avenue 30, Tomsk, Russia
{vnm,rozhkova}@tpu.ru

[2] National Research Tomsk State University,
Lenin Avenue 36, Tomsk, Russia

Abstract. The paper considers a single-line retrial queueing (RQ) system with an unreliable server controlled by a dynamic random multiple access protocol. A study of the prelimit probability distribution of the number of applications in orbit has been carried out. To study this system, the method of generating functions is used.

Keywords: Retrial queue · Dynamic random multiple access protocol · Unreliable server

1 Introduction

When designing or upgrading a data transmission network, it often becomes necessary to quantify network characteristics, such as the intensity of data flows over network communication lines, delays that occur at various stages of processing and transmitting packets. At the moment, such a network research tool is a protocol analyzer. But it does not give the opportunity to obtain probabilistic-temporal characteristics. Therefore, to study such systems, the apparatus of the theory of queuing is used, which makes it possible to build mathematical models of the data transmission network and find the main characteristics of the system.

A large number of works are devoted to the study of models of data transmission networks with various access protocols [1–8]. Various modifications of access protocols are proposed to solve the problems of repetitive applications. In [9–14], the authors investigate models with adaptive access protocols. The papers [15–23] consider the study of queuing systems with a dynamic access protocol. In this paper, we study a single-channel RQ system with an unreliable server controlled by a dynamic access protocol. The server is considered unreliable if it fails from time to time and requires restoration (repair). Only after that, the server resumes servicing new requests.

2 Description of the Mathematical Model

A prerequisite in data transmission networks over a communication line is the availability of a common resource. Any subscriber station, having generated requests, sends them to a common resource (server).

A. Dudin et al. (Eds.): ITMM 2022, CCIS 1803, pp. 212–224, 2023.
https://doi.org/10.1007/978-3-031-32990-6_18

If the server is free, then the customer is serviced. If the server fails during the service of the customer, then it is sent for repair, and the customer goes into orbit. To study such systems, consider a single-line RQ system with an unreliable server controlled by a dynamic access protocol.

Let's consider a single-server retrial queueing system with an unreliable server and the stationary Poisson flow of customers with parameter λ. A customer is serviced during random time distributed exponentially with parameter μ_1. We assume that the server is unreliable. An unreliable server may be in the following states: idle, busy or under repair. If the server is idle, and customer arrives, then the servicing immediately begins. If the server is busy at an arrival moment, then the customer goes into the orbit and waits for the opportunity to occupy the server at the next attempt. After a random time interval, a customer with intensity σ/i again tries to occupy the server for service, where i is the number of customers in orbit at time t (see Fig. 1). The working time is distributed exponentially with parameter γ_1, if server is idle and with parameter γ_2, if the server is busy. As soon as a breakdown occurs, the server is sent to repair and the servicing customer goes into the orbit. During repairing, all incoming customers go into the orbit. The recovery time is distributed exponentially with parameter μ_2. The goal of the research is to study such a system, as well as to determine its main characteristics and to find a stationary probability distribution of the number of customers in the orbit.

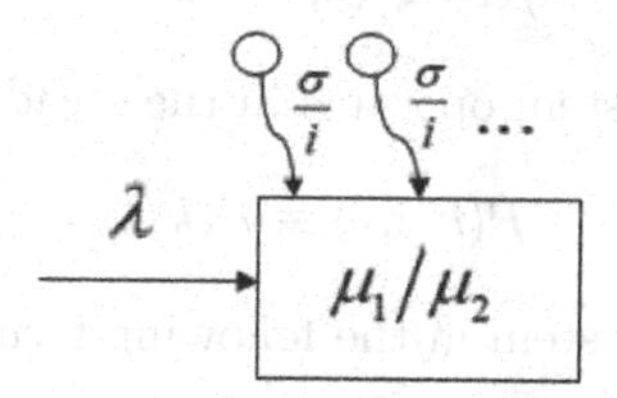

Fig. 1. Model of dynamic retrial queueing system M/M/1 with unreliable server

Let $i(t)$ be the number of customers in the orbit at time t and $k(t)$ determine the state of the server as follows:

$$k(t) = \begin{cases} 0, \text{if the server is idle,} \\ 1, \text{if the server is busy,} \\ 2, \text{if the server is under repair.} \end{cases}$$

3 Method of Generating Functions

Denote $P\{i(t) = i, k(t) = k\} = P(k, i, t)$ the probability that at time t the server is in state k and there are i customers in the orbit.

The probability distribution $P(k,i,t)$ satisfies the following system of equations:

$$\begin{cases} P(0,i,t+\Delta t) = (1-\lambda\Delta t)(1-\sigma\Delta t)(1-\gamma_1\Delta t)P(0,i,t)+ \\ +\mu_1\Delta t P(1,i,t)+\mu_2\Delta t P(2,i,t)+o(\Delta t), \\ P(1,i,t+\Delta t) = (1-\lambda\Delta t)(1-\mu_1\Delta t)(1-\gamma_2\Delta t)P(1,i,t)+ \\ +\lambda\Delta t P(0,i,t)+\sigma\Delta t P(0,i+1,t)+\lambda\Delta t P(1,i-1,t)+o(\Delta t), \\ P(2,i,t+\Delta t) = (1-\lambda\Delta t)(1-\mu_2\Delta t)P(2,i,t)+\gamma_1\Delta t P(0,i,t)+ \\ +\gamma_2\Delta t P(1,i-1,t)+\lambda\Delta t P(2,i-1,t)+o(\Delta t). \end{cases}$$

Let us compose a system of Kolmogorov differential equations for $i \geq 1$:

$$\begin{cases} \dfrac{\partial P(0,i,\ t)}{\partial t} = -(\lambda+\sigma+\gamma_1)P(0,i,\ t)+\mu_1 P(1,i,\ t)+ \\ +\mu_2 P(2,i,t), \\ \dfrac{\partial P(1,i,\ t)}{\partial t} = -(\lambda+\mu_1+\gamma_2)P(1,i,\ t)+\lambda P(0,i,\ t)+ \\ +\sigma P(0,i+1,t)+\lambda P(1,i-1,t), \\ \dfrac{\partial P(2,i,\ t)}{\partial t} = -(\lambda+\mu_2)P(2,i,\ t)+\gamma_1 P(0,i,\ t)+ \\ +\gamma_2 P(1,i-1,t)+\lambda P(2,i-1,t). \end{cases} \tag{1}$$

We assume that the system operates in the steady-state regime, i.e.

$$P(k,i,t) \equiv P(k,t).$$

Then we can rewrite System in the following form

$$\begin{cases} -(\lambda+\gamma_1)P(0,0)+\mu_1 P(1,0)+\mu_2 P(2,0)=0, i=0, \\ -(\lambda+\mu_1+\gamma_2)P(1,0)+\lambda P(0,0)+\sigma P(0,1)=0, i=0, \\ -(\lambda+\mu_2)P(2,0)+\gamma_1 P(0,0)=0, i=0, \\ -(\lambda+\sigma+\gamma_1)P(0,i)+\mu_1 P(1,i)+\mu_2 P(2,i)=0, i\ \geq\ 1, \\ -(\lambda+\mu_1+\gamma_2)P(1,i)+\lambda P(0,i)+\sigma P(0,i+1)+ \\ +\lambda P(0,i-1)=0, i\ \geq\ 1, \\ -(\lambda+\mu_2)P(2,i)+\gamma_1 P(0,i)+\gamma_2 P(1,i-1)+ \\ +\lambda P(2,i-1)=0, i\ \geq\ 1. \end{cases} \tag{2}$$

To find a solution of System (2), it is necessary to define the generating functions:

$$G(k,x)=\sum_{i=0}^{\infty} x^i P(k,i).$$

Partial generating function

$$G(k,x) = \sum_{i=0}^{\infty} x^i P(k,i) \neq Mx^i,$$

but

$$G(k,x) = Mx^i,$$

therefore

$$G(x) = \{G(0,x),\ G(1,x),\ G(2,x)\}$$

is a generating function, and we introduce its components $G(k,x)$ as partial generating functions.

Then we get the following system of equations

$$\begin{cases} -(\lambda+\sigma+\gamma_1)\, G(0,x) + \mu_1 G(1,x) + \mu_2 G(2,x) = -\sigma P(0,0), \\ \left(\lambda + \dfrac{\sigma}{x}\right) G(0,x) + (\lambda x - \lambda - \mu_1 - \gamma_2)\, G(1,x) = \dfrac{\sigma}{x} P(0,0), \\ \gamma_1 G(0,x) + \gamma_2 x G(1,x) + (\lambda x - \lambda - \mu_2)\, G(2,x) = 0. \end{cases} \tag{3}$$

Let us multiply the second equation of System (3) by x.

Then we have the following system of equation

$$\begin{cases} -(\lambda+\sigma+\gamma_1)\, G(0,x) + \mu_1 G(1,x) + \mu_2 G(2,x) = -\sigma P(0,0), \\ (\lambda x + \sigma)\, G(0,x) + \left(\lambda x^2 - \lambda x - \mu_1 x - \gamma_2 x\right) G(1,x) = \sigma P(0,0), \\ \gamma_1 G(0,x) + \gamma_2 x G(1,x) + (\lambda x - \lambda - \mu_2)\, G(2,x) = 0. \end{cases} \tag{4}$$

We will find a solution of System (4) by denoting

$$G(x) = G(0,x) + G(1,x) + G(2,x).$$

Then we have the following system of equation

$$\begin{aligned} &G(x) = P(0,0)\sigma((x-1)\lambda - \gamma_1 - \mu_2)((x-1)\lambda - \gamma_2 - \mu_1)/((x^2 - x)* \\ &* \lambda^3 + x((\sigma+\gamma_1)x - \sigma - \gamma_1 - \gamma_2 - \mu_2)\lambda^2 + (((-\gamma_2 - \mu_1 - \mu_2)\sigma - \\ &- \gamma_1(\gamma_2 + \mu_1))x + \sigma\mu_1)\lambda + \sigma\mu_2(\gamma_2 + \mu_1)). \end{aligned} \tag{5}$$

Taking into account the normalization condition, where

$$G(1) = 1,$$

we obtain an expression for $P(0,0)$:

$$P(0,0) = \frac{(-\gamma_2 - \mu_2)\lambda^2 - (\sigma\mu_2 + (\sigma+\gamma_1)\gamma_2 + \gamma_1\mu_1)\lambda + \sigma\mu_2(\gamma_2 + \mu_1)}{\sigma(\gamma_1 + \mu_2)(\gamma_2 + \mu_1)}.$$

By substituting $P(0,0)$ into the Eq. (5), we obtain the expression for the generating function:

$$G(x) = ((-\gamma_2 - \mu_2)\lambda^2 - ((\gamma_2 + \mu_2)\sigma + \gamma_1(\gamma_2 + \mu_1))\lambda + \sigma\mu_2(\gamma_2 + \mu_1))*$$
$$* ((x-1)\lambda - \gamma_1 - \mu_2)((x-1)\lambda - \gamma_2 - \mu_1)/((\gamma_1 + \mu_2)((x^2 - x)\lambda^3 +$$
$$+ x((\sigma + \gamma_1)x - \sigma - \gamma_1 - \gamma_2 - \mu_2)\lambda^2 + (((-\gamma_2 - \mu_1 - \mu_2)\sigma -$$
$$- \gamma_1(\gamma_2 + \mu_1))x + \sigma\mu_1)\lambda + \sigma\mu_2(\gamma_2 + \mu_1))(\gamma_2 + \mu_1)).$$

The values of the stationary distribution of server states R_k will look like this:

$$R_0 = G(0,1) = \frac{(-\lambda + \gamma_2 + \mu_1)\mu_2 - \lambda\gamma_2}{(\gamma_2 + \mu_1)(\gamma_1 + \mu_2)},$$

$$R_1 = G(1,1) = \frac{\lambda}{\gamma_2 + \mu_1},$$

$$R_2 = G(2,1) = \frac{(-\lambda + \gamma_2 + \mu_1)\gamma_1 + \lambda\gamma_2}{(\gamma_2 + \mu_1)(\gamma_1 + \mu_2)}.$$

In case of the probabilities $P(0,0)$ must be positive, the following inequalities must be true:

$$\frac{\lambda}{\mu_1} \leq \frac{\sigma\mu_2(\gamma_2 + \mu_1)}{(\lambda + \sigma)(\gamma_2\mu_1 + \mu_2\mu_1) + \gamma_1\mu_1(\gamma_2 + \mu_1)} = S, \tag{6}$$

where S is the throughput of the system under consideration.

Definition. Throughput is the upper limit of those load values $\rho = \frac{\lambda}{\mu_1}$, for which there is the steady-state regime.

The inequality (6) determines the condition for the existence of a steady-state regime for the considered dynamical system.

4 Method of Characteristic Functions

Introducing the partial characteristic function

$$H(k,u) = \sum_{i=0}^{\infty} e^{jui} P(k,i),$$

where $j = \sqrt{-1}$, System (2) can be rewritten as

$$\begin{cases} -(\lambda + \sigma + \gamma_1) H(0,u) + \mu_1 H(1,u) + \mu_2 H(2,u) = -\sigma P(0,0), \\ \left(\lambda + \frac{\sigma}{e^{ju}}\right) H(0,u) + \left(\lambda e^{ju} - \lambda - \mu_1 - \gamma_2\right) H(1,u) = \frac{\sigma}{e^{ju}} P(0,0), \\ \gamma_1 H(0,u) + \gamma_2 e^{ju} H(1,u) + \left(\lambda e^{ju} - \lambda - \mu_2\right) H(2,u) = 0. \end{cases} \tag{7}$$

We will find a solution of System (7) by denoting

$$H(u) = H(0,u) + H(1,u) + H(2,u).$$

Then we have the following system of equation

$$H(u) = P(0,0)\sigma((e^{ju}-1)\lambda-\gamma_1-\mu_2)((e^{ju}-1)\lambda-\gamma_2-\mu_1)/((e^{ju^2}- e^{ju})*\lambda^3+e^{ju}((\sigma+\gamma_1)e^{ju}-\sigma-\gamma_1-\gamma_2-\mu_2)\lambda^2+(((-\gamma_2-\mu_1-\mu_2)\sigma- -\gamma_1(\gamma_2+\mu_1))e^{ju}+\sigma\mu_1)\lambda+\sigma\mu_2(\gamma_2+\mu_1)). \quad (8)$$

By substituting $P(0,0)$ into the Eq. (8), we obtain

$$H(u) = ((-\gamma_2-\mu_2)\lambda^2-((\gamma_2+\mu_2)\sigma+\gamma_1(\gamma_2+\mu_1))\lambda+\sigma\mu_2(\gamma_2+\mu_1))* *((e^{ju}-1)\lambda-\gamma_1-\mu_2)((e^{ju}-1)\lambda-\gamma_2-\mu_1)/((\gamma_1+\mu_2)((e^{ju^2}-e^{ju})\lambda^3+ +e^{ju}((\sigma+\gamma_1)e^{ju}-\sigma-\gamma_1-\gamma_2-\mu_2)\lambda^2+(((-\gamma_2-\mu_1-\mu_2)\sigma- -\gamma_1(\gamma_2+\mu_1))e^{ju}+\sigma\mu_1)\lambda+\sigma\mu_2(\gamma_2+\mu_1))(\gamma_2+\mu_1)).$$

5 Asymptotic Analysis

Let us denoting

$$H(k,u) = \sum_{i=0}^{\infty} e^{jui} P(k,i),$$

where $j=\sqrt{-1}$ is the imaginary unit.

Thus System (2) can be rewritten as

$$\begin{cases} -(\lambda+\sigma+\gamma_1)H(0,u)+\mu_1 H(1,u)+\mu_2 H(2,u) = -\sigma P(0,0), \\ \left(\lambda+\frac{\sigma}{e^{ju}}\right)H(0,u)+\left(\lambda e^{ju}-\lambda-\mu_1-\gamma_2\right)H(1,u) = \frac{\sigma}{e^{ju}}P(0,0), \\ \gamma_1 H(0,u)+\gamma_2 e^{ju}H(1,u)+\left(\lambda e^{ju}-\lambda-\mu_2\right)H(2,u)=0. \end{cases} \quad (9)$$

We introduce a parameter

$$\rho = \frac{\lambda}{\mu_1},$$

that characterizes the system load.

Dividing all equations of System (9) by μ_1, we obtain the following system of equations

$$\begin{cases} -\left(\rho+\frac{\sigma}{\mu_1}+\frac{\gamma_1}{\mu_1}\right)H(0,u)+H(1,u)+\frac{\mu_2}{\mu_1}H(2,u) = \frac{\sigma}{\mu_1}P(0,0), \\ \left(\rho+\frac{\sigma}{\mu_1}e^{-ju}\right)H(0,u)+\left(\rho e^{ju}-\rho-1-\frac{\gamma_2}{\mu_1}\right)H(1,u) = \\ \qquad = \frac{\sigma}{\mu_1}e^{-ju}P(0,0), \\ \frac{\gamma_1}{\mu_1}H(0,u)+\frac{\gamma_2}{\mu_1}e^{ju}H(1,u)+\left(\rho e^{ju}-\rho-\frac{\mu_2}{\mu_1}\right)H(2,u)=0, \end{cases} \quad (10)$$

Denoting three-dimensional vectors

$$H(u) = \{H(0,u),\ H(1,u),\ H(2,u)\},$$

$$P(0) = \{P(0,0),\ P(1,0),\ P(2,0)\}$$

and matrices

$$A(ju,\rho) = \begin{bmatrix} -\left(\rho + \frac{\sigma}{\mu_1} + \frac{\gamma_1}{\mu_1}\right) & \left(\rho + \frac{\sigma}{\mu_1} e^{-ju}\right) & \frac{\gamma_1}{\mu_1} \\ 1 & \left(\rho e^{ju} - \rho - 1 - \frac{\gamma_2}{\mu_1}\right) & \frac{\gamma_2}{\mu_1} e^{ju} \\ \frac{\mu_2}{\mu_1} & 0 & \left(\rho e^{ju} - \rho - \frac{\mu_2}{\mu_1}\right) \end{bmatrix},$$

$$B(ju) = \begin{bmatrix} -\frac{\sigma}{\mu_1} & \frac{\sigma}{\mu_1} e^{-ju} & 0 \\ 0 & 0 & 0 \\ 0 & 0 & 0 \end{bmatrix},$$

we rewrite the System (10) in the form

$$H(u)\ A(ju,\rho) + P(0)\ B(ju) = 0. \tag{11}$$

To find the value of the throughput S, we will solve the Eq. (11) by the method of asymptotic analysis in the condition of a large load, denoting $\varepsilon = S - \rho$ and setting that $\varepsilon \to 0$.

Let us introduce the following substitutions

$$\rho = S - \varepsilon,\ u = \varepsilon w,\ H(u) = F(w,\varepsilon),\ P(0) = \varepsilon \Pi, \tag{12}$$

we get

$$F(w,\varepsilon) A(j\varepsilon w,\ S - \varepsilon) + \varepsilon \Pi B(j\varepsilon w) = 0, \tag{13}$$

where

$$A(j\varepsilon w,\ S-\varepsilon) = \sum_{n=0}^{\infty} \frac{(j\varepsilon w)^n}{n!} A_n(S) - \varepsilon \sum_{n=0}^{\infty} \frac{(j\varepsilon w)^n}{n!} A'_n(S). \tag{14}$$

Theorem 1. *The value S of throughput is equal to the value of the root of the equation*

$$R(S)A_1(S)E = 0, \tag{15}$$

where the vector R is determined by the equation $R(S)A_0(S) = 0$ and the normalization condition $RE = 1$, and equality $A_0(S) = A(0,S)$.

The characteristic function of the normalized number of customers in the server has the form

$$\Phi(w) = \lim_{\varepsilon \to 0} \{F(w,\varepsilon)E\} = \frac{\kappa}{\kappa - jw}, \tag{16}$$

where

$$\kappa = \frac{R(S)A'_1(S)E + f_2 A_1(S)E}{f_1 A_1(S)E + \frac{1}{2} R(S) A_2(S) E}. \tag{17}$$

Proof. Let us denote

$$\lim_{\varepsilon\to 0} A(j\varepsilon w,\ S-\varepsilon) = A(0,S) = A_0(S),$$

where $A_0(S)$ is determined from the Eq. (14). Then by performing the limit transition in Eq. (13), we obtain

$$F(w)A_0(S) = 0. \tag{18}$$

It follows from the form of the matrix $A_0(S)$ that its properties are similar to those of the matrix of infinitesimal characteristics, so the solution of the homogeneous System (13) can be written as

$$F(w) = \Phi(w)R(S),$$

where $R(S)$ is the probability distribution of chain values determined by the equation

$$R(S)A_0(S) = 0$$

and the normalization condition $RE = 1$.

Using matrix decomposition

$$A(j\varepsilon w,\ S-\varepsilon) = A_0(S) + j\varepsilon w A_1(S) - \varepsilon A'_0(S) + O(\varepsilon^2),$$
$$B(j\varepsilon w) = B_0 + O(\varepsilon^2),$$

we write the Eq. (13) in the following form

$$F(w,\varepsilon)(A_0(S) + j\varepsilon w A_1(S) - \varepsilon A'_0(S)) + \varepsilon \Pi B_0. \tag{19}$$

The solution of this equation is written in the form of decomposition

$$F(w,\varepsilon) = \Phi(w)R(S) + \varepsilon f(w) + O(\varepsilon^2), \tag{20}$$

Substitute the decomposition in the Eq. (19), we get

$$(\Phi(w)R(S) + \varepsilon f(w))(A_0(S) + j\varepsilon w A_1(S) - \varepsilon A'_0(S)) + \varepsilon \Pi B_0 =$$
$$= \Phi(w)R(S)(A_0(S) + j\varepsilon w A_1(S) - \varepsilon A'_0(S)) + \varepsilon f(w)A_0(S) + \varepsilon \Pi B_0 = O(\varepsilon^2).$$

Considering that $R(S)A_0(S) = 0$, then for the function $f(w)$ at $\varepsilon \to 0$ we can write the equation

$$f(w)A_0(S) + jw\Phi(w)R(S)A_1(S) - \Phi(w)R(S)A'_0(S) + \Pi B_0,$$

This equation is an inhomogeneous system of linear algebraic equations, so the solution of $f(w)$ can be written as

$$f(w) = \Phi(w)(jwf_1 - f_2) + f_3, \tag{21}$$

where vectors f_1, f_2, f_3 are solutions of systems:

$$f_1 A_0(S) + R(S)A_1(S) = 0, \tag{22}$$

$$f_2 A_0(S) + R(S){A'}_0(S) = 0, \tag{23}$$

$$f_3 A_0(S) + \Pi B_0 = 0.$$

The solution of the System (23) has the form

$$f_2 = R'(S).$$

To find the solution of System (22), the Eq. (15) must be satisfied. Thus, the Eq. (21) can be written as

$$f(w) = \Phi(w)(jwf_1 - R'(S)) + f_3, \tag{24}$$

So the Eq. (20) can be written as follows:

$$F(w,\varepsilon) = \Phi(w)R(S) + \varepsilon\Phi(w)(jwf_1 - R'(S)) + \varepsilon f_3 + O(\varepsilon^2). \tag{25}$$

To find the function $\Phi(w)$, we sum over k all the equations of System (13), then we obtain

$$F(w,\varepsilon)A(j\varepsilon w,\ S-\varepsilon)E + \varepsilon\Pi B(j\varepsilon w)E = 0. \tag{26}$$

For matrices A and B from this equation we write the decompositions

$$A(j\varepsilon w,\ S-\varepsilon) = A_0(S) + j\varepsilon w A_1(S) + \frac{(j\varepsilon w)^2}{2}A_2(S) - \varepsilon {A'}_0(S) +$$

$$+ \frac{\varepsilon^2}{2}{A''}_0(S) - j\varepsilon^2 w {A'}_1(S) + O(\varepsilon^2),$$

$$B(j\varepsilon w) = B_0 + j\varepsilon w B_1 + O(\varepsilon^2).$$

It follows from the form of matrices $A(ju)$ and $B(ju)$ that

$${A''}_0(S) = 0,\ A_0(S)E = 0,\ {A'}_0(S)E = 0,\ B_0 E = 0,$$

therefore

$$A(j\varepsilon w,\ S-\varepsilon)E = j\varepsilon w A_1(S)E + \frac{(j\varepsilon w)^2}{2}A_2(S)E - j\varepsilon^2 w {A'}_1(S) + O(\varepsilon^2),$$

$$B(j\varepsilon w)E = j\varepsilon w B_1 + O(\varepsilon^2),$$

Then the Eq. (26) will take the form

$$F(w,\varepsilon)(A_1(S) + \frac{jw\varepsilon}{2}A_2(S) - \varepsilon {A'}_1(S))E + \varepsilon\Pi B_1 E = O(\varepsilon^2).$$

Substituting the Eq. (25), we get the following form

$$(\Phi(w)R(S) + \varepsilon\Phi(w)(jwf_1 - R'(S)) + \varepsilon f_3)\times$$
$$\times\,(A_1(S)E + \frac{jw\varepsilon}{2}A_2(S)E - \varepsilon A'_1(S)E) + \varepsilon\Pi B_1 E + O(\varepsilon^2) =$$
$$= \Phi(w)R(S)\left[jw\varepsilon\, A_1(S)E + \frac{(jw\varepsilon)^2}{2}A_2(S)E - j\varepsilon^2 wA'_1(S)E\right] +$$
$$+\,\varepsilon(\Phi(w)(jwf_1 - R'(S)) + \varepsilon f_3)jw\varepsilon A_1(S)E + \varepsilon\Pi j\varepsilon w B_1 + O(\varepsilon^2).$$

Since the condition $R(S)A_1(S)E = 0$ is satisfied then at $\varepsilon \to 0$, we obtain

$$\Phi(w)\left[jw\,(f_1A_1(S)E + \frac{1}{2}R(S)A'_2(S)E) - (R(S)A'_1(S)E+\right.$$
$$\left.+\, f_2A_1(S)E)\right] + f_3A_1(S)E + \Pi B_1E = 0.$$

Then

$$\Phi(w) = \frac{f_3A_1(S)E + \Pi B_1E}{R(S)A'_1(S)E + f_2A_1(S)E - jw(f_1A_1(S)E + \frac{1}{2}R(S)A'_2(S)E)}.$$

Taking into account that $\Phi(0) = 1$, we get

$$\kappa = \frac{R(S)A'_1(S)E + f_2A_1(S)E}{f_1A_1(S)E + \frac{1}{2}R(S)A_2(S)E}.$$

So Theorem 1 is proved.

6 Numerical Example

In order to find the probability distribution $P(i)$, it suffices to apply the inverse Fourier transform to the characteristic function.

$$P(i) = \frac{1}{2\pi}\int_{-\pi}^{\pi} e^{-jui}H(u)du,.$$

where $H(u) = G(e^{ju}),\ j = \sqrt{-1}$.

In a numerical example, we take

$$\mu_1 = 5,\quad \mu_2 = 2,\quad \gamma_1 = 0.03,\quad \gamma_2 = 0.03,\quad \lambda = 1,\quad \sigma = 1.$$

The value of throughput for given parameters of a given RQ system $S = 0,34$.

Table 1 and Fig. 2 show the distribution of the number of customers in orbit for this system.

Table 1. The probability distribution of the number of customers in orbit, i=0, 1, 2,...

i	0	1	2	3	4	5
$P(i)$	0,48213	0,19421	0,12012	0,07523	0,04734	0,02985
i	6	7	8	9	10	11
$P(i)$	0,01883	0,01188	0,00750	0,00473	0,00299	0,00189
i	12	13	14	15	16	...
$P(i)$	0,00119	0,00075	0,00047	0,00030	0,00019	

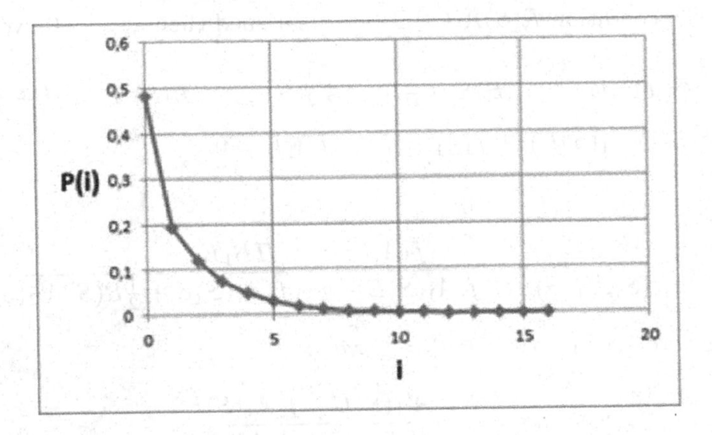

Fig. 2. The probability distribution of the number of customers

The values of the stationary distribution of server states R_k are:

$$R_0 = \frac{(-\lambda + \gamma_2 + \mu_1)\mu_2 - \lambda\gamma_2}{(\gamma_2 + \mu_1)(\gamma_1 + \mu_2)} = 0,68,$$

$$R_1 = \frac{\lambda}{\gamma_2 + \mu_1} = 0,19,$$

$$R_2 = \frac{(-\lambda + \gamma_2 + \mu_1)\gamma_1 + \lambda\gamma_2}{(\gamma_2 + \mu_1)(\gamma_1 + \mu_2)} = 0,13.$$

Table 2. Variation of S at different values σ

σ	1	5	10	50	100	500
S	0,342	0,689	0,788	0,891	0,906	0,919

Table 2 shows as the σ increases, the throughput S increases.

7 Conclusion

In this paper, we study the dynamic RQ-system M/M/1 with an unreliable server. As a result of the study, the generating and characteristic functions for the probability distribution of the number of applications in orbit are obtained. Further, the study was carried out by the method of asymptotic analysis under the condition of a large system load. The main characteristics of the system, the stationary distribution of server states, and the throughput of the system under consideration are found.

References

1. Lyubina T.V., Nazarov A.A.: Research of dynamic and adaptive RQ-systems with an incoming MMPP flow of applications. Bull. Tomsk State Univ. Manag. Comput. Technol. Inform. **3**(24), 104–112 (2013)
2. Lyubina T., Bublik Y.: Analysis of dynamic and adaptive retrial queue system with the incoming MAP-flow of requests. In: International Conference on Information Technologies and Mathematical Modelling, pp. 384–392 (2015)
3. Nazarov, A.A., Tsoy, S.A.: A general approach to the analysis of Markov models of data communication networks controlled by static random multiple access protocols. Autom. Control. Comput. Sci. **38**(4), 64–75 (2004)
4. Apostolopoulos, T.K., Protonotarios, E.N.: Queueing analysis of buffered slotted multiple access protocols. Comput. Commun. **8**(1), 9–21 (1985)
5. Ephremides A.: Analysis of protocols of multiple access. In: Baccelli, F., Fayolle, G. (eds.) Modelling and Performance Evaluation Methodology, vol. 60, pp. 563–575. Springer, Heidelberg (1984). https://doi.org/10.1007/BFb0005192
6. Nazarov, A.A., Nikitina, M.A.: Application of Markov chain ergodicity conditions to the study of the existence of stationary regimes in communication networks. Autom. Control. Comput. Sci. **37**(1), 50–55 (2003)
7. Gao, J., Rubin, I.: Analysis of a random-access protocol under long-range-dependent traffic. IEEE Trans. Veh. Technol. **52**(3), 693–700 (2003)
8. Nazarov, A.A., Sudyko, E.A.: Method of asymptotic semiinvariants for studying a mathematical model of a random access network. Probl. Inf. Transm. **46**(1), 86–102 (2010)
9. Nazarov A.A., Kuznetsov D.Yu.: Adaptive random access networks. Tomsk: TPU, p. 256 (2002)
10. Lyubina T.V., Nazarov A.A.: Study of an adaptive RQ-system with an incoming MMPP flow of applications using asymptotic analysis. In: Information technologies and mathematical modeling (ITMM-2012): materials of the XI All-Russian. Scientific-Practical. Conf. with International Participation (Anzhero-Sudzhensk, 23–24 November 2012). Kemerovo: Practice, Part 2, pp. 94–99 (2013)
11. Kuznetsov, D.Y., Nazarov, A.A.: Analysis of a communication network governed by an adaptive random multiple access protocol in critical load. Probl. Inf. Transm. **40**(3), 243–253 (2004)
12. Kuznetsov, D.Y., Nazarov, A.A.: Investigation of communication networks with adaptive random multiple access protocol under heavy load conditions for an infinite number of nodes. Autom. Control. Comput. Sci. **37**(3), 47–55 (2003)

13. Nazarov, A.A., Odyshev, Y.D.: Analysis of a communication network with the adaptive ALOHA protocol for a finite number of stations under overload. Problemy Peredachi Informatsii **36**(3), 83–93 (2000)
14. Anyugu Francis Lin, B., Ye, X., Hao, S.: Adaptive protocol for full-duplex two-way systems with the buffer-aided relaying. IET Commun. **13**(1), 54–58 (2019)
15. Shokhor, S.L.: Distribution of the number of messages in a communication network with channel reservation and a dynamic access protocol. Bull. Tomsk State Univ. **271**, 67–69 (2000)
16. Lyubina T.V., Nazarov A.A.: Investigation of a Markovian dynamical RQ-system with conflicts of requests. Bull. Tomsk State Univ. Manag. Comput. Technol. Inform. **3**(12), 73–84 (2010)
17. Nazarov, A.A., Shokhor, S.L.: Comparison of an asymptotic and prelimit model of a communication network with a dynamic protocol of random multiple access. In: Alexandrova, I.A., et al. (eds.) Mathematical Modeling and Probability Theory, Tomsk: Peleng, pp. 233–241 (1988)
18. Klimenok, V.I.: Optimization of dynamic management of the operating mode of data systems with repeat calls. Autom. Control. Comput. Sci. **24**(1), 23–28 (1990)
19. Gupur, G., Li, X.Z., Zhu, G.T.: The application of C 0 -semigroup theory to dynamic queueing systems. Semigroup Forum **62**, 205–216 (2001). https://doi.org/10.1007/s002330010030
20. Ouyang, H., Nelson, B.L.: Simulation-based predictive analytics for dynamic queueing systems. In: 2017 Winter Simulation Conference (WSC), pp. 1716–1727. IEEE (2017)
21. Mounce, R.: Existence of equilibrium in a continuous dynamic queueing model for traffic networks with responsive signal control. In: Lam, W., Wong, S., Lo, H. (eds.) Transportation and Traffic Theory 2009: Golden Jubilee, pp. 327–344. Springer, Boston (2009). https://doi.org/10.1007/978-1-4419-0820-9_16
22. Filipiak, J.: Dynamic routing in a queueing system with a multiple service facility. Oper. Res. **32**(5), 1163–1180 (1984)
23. Nazarov, A.A., Odyshev, Y.D.: Investigation of communications networks with dynamic synchronous ALOHA protocol under heavy load conditions. Avtomat. Vychisl. Tekhn. **35**(1), 77–84 (2001)

Estimating the Distribution Function Using Parametric Methods in Informative Model of Random Censorship from Both Sides

Abdushukurov Abdurakhim Akhmedovich[1] and Mansurov Dilshod Ravilovich[2](✉)

[1] Moscow State University, Tashkent Branch, Tashkent, Uzbekistan
a_abdushukurov@rambler.ru

[2] Navoi State Pedagogical Institute, Navoi, Uzbekistan
mathematicianmd@gmail.com

Abstract. This paper focuses on the investigation of two semi-parametric estimators for distribution functions in an informative model of random censorship from both sides. During the investigation of these estimators, we utilized the characterization properties of the informative model to gather insights. Additionally, we discussed the properties of these estimators using numerical modeling methods.

Keywords: Random censorship · Informative model · Proportional hazards · Exponential distribution · Gaussian process

Introduction

In the field of biomedical research and engineering tests for device reliability, there are situations where objects under observation are not observed immediately, but after some random amount of time. This phenomenon is called delayed entry or left random censoring. The random variable X represents the object's lifetime, and is only observed if $X \geq L$, where L is the moment when the object is placed under observation. Furthermore, X may be censored from the right by another random variable Y. Our interest lies in studying X when it is subject to random censorship from both sides by the random vector (L, Y). Assuming L, X, and Y are mutually independent with continuous distribution functions K, F, and G, respectively.

1 Informative Model of Random Censorship from Both Sides

The informative model of random censorship proposes that the survival distribution of the censoring variables is a power of the survival distribution of the

A. Dudin et al. (Eds.): ITMM 2022, CCIS 1803, pp. 225–237, 2023.
https://doi.org/10.1007/978-3-031-32990-6_19

lifetimes. It is also known as the proportional hazards model (PHM) because it assumes proportionality of corresponding hazard functions. The PHM is a special semiparametric model of random censorship that is both appealing and potentially useful. One of its great advantages is its ability to provide an easy and clear interpretation of results and conditions formulated in the general censorship model. The PHM was initially considered in the framework of right random censorship by the authors [16] and is often referred to as the Koziol-Green model, starting from [9]. Csörgő [10] provided a comprehensive overview of the estimation theory for many reliability functions based on the power-type estimator, often referred to as the ACL (Abdushukurov-Cheng-Lin) estimator. For more information, see [2–6,10–13,15,20,21].

This paper discusses the asymptotic characteristics of a power-type semiparametric statistics that estimates the survival function of a lifetime in an informative model of random censorship from both sides. The study establishes a weak convergence result for the empirical process and includes numerical modeling comparative results.

Let $\{(X_k, L_k, Y_k), k \geq 1\}$ be a sequence of independent realizations of the triple (X, L, Y) and

$$S^{(n)} = \{(\mathrm{Z}_i,\, \Delta_i),\, i = 1, \dots, n\}$$

-the observed sample, where $\mathrm{Z}_i = \max\{L_i, \min\{X_i, Y_i\}\}, \Delta_i = \left(\delta_i^{(0)}, \delta_i^{(1)}, \delta_i^{(2)}\right)$, with $\delta_i^{(0)} = I(\min(X_i,\, Y_i) < L_i), \delta_i^{(1)} = I(L_i \leq X_i < Y_i), \delta_i^{(2)} = I(L_i \leq Y_i < X_i)$ and $I(A)$ standing for indicator of the event A. Note that in sample $S^{(n)}$ the number of observed r.v.-s X_i is equal to $\delta_1^{(1)} + \dots + \delta_n^{(1)}$. The statistical task is consist in estimating of d.f. F from a sample $S^{(n)}$. However, such a general statement of the estimating problem, d.f.-s K and G are considered as a nuisance. In this paper, we will investigate the evaluation of d.f. F in the case of informative censoring from two sources, when d.f.-s K and G functionally depend on F. To describe such a model by H and N we denote d.f.-s of r.v.-s Z_i and $V_i = \min(X_i,\, Y_i)$. Then it is easy to see, that

$$H(x) = K(x)\, N(x),\;\; N(x) = 1 - (1 - F(x))\,(1 - G(x)),\;\; x \in \mathrm{R}^1. \quad (1)$$

Let us assume that there exist positive, unknown parameters θ and β such that the following representations hold for all $x \in R$:

$$\begin{cases} 1 - G(x) = (1 - F(x))^{\theta}, \\ K(x) = (N(x))^{\beta}, \end{cases} \quad (2)$$

Here, the parameter β controls the power of censoring from the left, while θ controls the power of censoring from the right. The smaller the values of these parameters, the weaker the corresponding censorship. This special model of random censorship on both sides was introduced in [3] and was later generalized for the case of competing risks in [4]. In this model, the parameters θ and β determine the level of censorship from both sides. By using formulas (1) and (2), we can obtain the following representation for the distribution function F:

$$1 - F(x) = \left[1 - (H(x))^{\lambda}\right]^{\gamma}, \; x \in \mathrm{R}^1, \tag{3}$$

where parameters $\lambda = \frac{1}{1+\beta}$ and $\gamma = \frac{1}{1+\theta}$ determine the degree of censoring. When these parameters are close to 1, the censoring is weak. By using the representation (3), we can estimate a triple $(H(x), \lambda, \gamma)$ and construct a semiparametric estimate for F over a sample $S^{(n)}$. We also use the characterization properties of the model to achieve this. In particular, we define subdistributions$\left\{T^{(m)}(x) = P\left(\mathrm{Z}_i \le x, \delta_i^{(m)} = 1\right), m = 0,1,2\right\}$ using the following formulas:

$$T^{(0)}(x) = \int_{-\infty}^{x} N(s)\, dK(s), \; T^{(1)}(x) = \int_{-\infty}^{x} K(s)(1 - G(s))\, dF(s),$$

$$T^{(2)}(x) = \int_{-\infty}^{x} K(s)(1 - F(s))\, dG(s).$$

It's easy to see that $T^{(0)}(x) + T^{(1)}(x) + T^{(2)}(x) = H(x), \; x \in \mathrm{R}^1$ and because of representation (3) in the model (2):

$$T_F = T_G = T_N = T_K = T_H = \inf\left\{x \in \mathrm{R}^1 : H(x) = 1\right\},$$
$$\tau_F = \tau_G = \tau_N = \tau_K = \tau_H = \sup\left\{x \in \mathrm{R}^1 : H(x) = 0\right\}.$$

It is worth noting that the model being considered is a generalization of the well-known proportional hazards model of Koziol-Green, which has been studied by authors [2–4, 8–13, 15–18, 20, 21]. This model is obtained from Eq. (2) when $\beta = 0$, meaning in the absence of random left-side censoring ($K(x) = 1$), and is a special case of the model from [5, 6] in the absence of a covariate. The following theorem characterizes the model given in Eq. (2). It was first introduced without proof in the paper [3].

Theorem 1. *Equalities* (2) *hold if and only if the r.v.-s* Z_i *and* Δ_i *are independent.*

Proof. If conditions (2) are satisfied, then the computation of subdistribution functions is straightforward and can be easily obtained

$$T^{(0)}(x) = (1 - \lambda) H(x), \; T^{(1)}(x) = \gamma\lambda H(x), \; T^{(2)}(x) = (1 - \gamma)\lambda H(x) \tag{4}$$

and under $x \to +\infty$ from (4) we have

$$P\left(\delta_i^{(0)} = 1\right) = 1 - \lambda, \; P\left(\delta_i^{(1)} = 1\right) = \gamma\lambda, \; P\left(\delta_i^{(2)} = 1\right) = (1 - \gamma)\lambda. \tag{5}$$

Now from relations (4) and (5) is easy follows independence of events $A = \{Z_i \le x\}$ and $\left\{B_m = \left\{\delta_i^{(m)} = 1\right\}, m = 0,1,2\right\}$ for all $x \in R^1$. This entails independence of $\overline{A}$ and $\{B_m, m = 0,1,2\}$, A and $\left\{\overline{B}_m, m = 0,1,2\right\}$ also $\overline{A}$ and $\left\{\overline{B}_m, m = 0,1,2\right\}$. Hence this denotes independence of r.v. Z_i and vector $(\delta_i^{(0)}, \delta_i^{(1)}, \delta_i^{(2)})$.

Conversely, let r.v. Z_i and indicators $\left\{\delta_i^{(0)},\ \delta_i^{(1)},\ \delta_i^{(2)}\right\}$ are independent. Then for $x > \tau_K$

$$K(x) = \exp\left\{-\int_x^{+\infty} \frac{dK(s)}{K(s)}\right\} = \exp\left\{-\int_x^{+\infty} \frac{dT^{(0)}(s)}{H(s)}\right\}$$
$$= \exp\left\{-P\left(\delta_i^{(0)} = 1\right)\int_x^{+\infty} \frac{dH(s)}{H(s)}\right\} = [H(x)]^{P\left(\delta_i^{(0)}=1\right)} = [K(x)\,N(x)]^{P\left(\delta_i^{(0)}=1\right)}.$$

From here we have $K(x) = [N(x)]^{\beta}$, where $\beta = P\left(\delta_i^{(0)} = 1\right)\left[P\left(\delta_i^{(0)} \neq 1\right)\right]^{-1}$, i.e. we obtain second relation in (2). Analogously, by independence of Z_i and indicators $\left\{\delta_i^{(1)}, \delta_i^{(2)}\right\}$, we get for $x < T_F$

$$1 - F(x) = \exp\left\{-\int_{-\infty}^{x} \frac{dF(s)}{1-F(s)}\right\} = \exp\left\{-\int_{-\infty}^{x} \frac{dT^{(1)}(s)}{K(s)(1-G(s))(1-F(s))}\right\}$$
$$= \exp\left\{-P\left(\delta_i^{(1)} = 1\right)\int_{-\infty}^{x} \frac{dH(s)}{K(s)-H(s)}\right\},$$

and also for $x < T_G$

$$1 - G(x) = \exp\left\{-\int_{-\infty}^{x} \frac{dG(s)}{1-G(s)}\right\} = \exp\left\{-\int_{-\infty}^{x} \frac{dT^{(2)}(s)}{K(s)(1-G(s))(1-F(s))}\right\}$$
$$= \exp\left\{-P\left(\delta_i^{(2)} = 1\right)\int_{-\infty}^{x} \frac{dH(s)}{K(s)-H(s)}\right\},$$

Now from last two expressions one can get equality

$$-\log(1 - G(x)) = -\theta \log(1 - F(x)), \quad x < \min(T_F, T_G),$$

where $\theta = P\left(\delta_i^{(2)} = 1\right)\left[P\left(\delta_i^{(1)} = 1\right)\right]$, i.e. we have first of representations (2) that complete the proof of Theorem 1.

Estimating the probabilities $p^{(m)} = P\left(\delta_i^{(m)} = 1\right)$, $m = 0, 1, 2$ by the corresponding frequencies $p_n^{(m)} = \frac{1}{n}\sum_{i=1}^{n} \delta_i^{(m)}$, $m = 0, 1, 2$ from formulas (5), we find the following estimates of the parameters λ and γ : $\lambda_n = 1 - p_n^{(0)}$, $\gamma_n = p_n^{(1)}\left(1 - p_n^{(0)}\right)^{-1}$. For d.f. $H(x)$ we use an empirical estimator

$$H_n(x) = \frac{1}{n}\sum_{i=1}^{n} I(Z_i \leq x), \quad x \in R^1.$$

Now, using (3), we can construct an estimator for the distribution function $F(x)$ using the plug-in method:

$$F_n(x) = 1 - \left[1 - (H_n(x))^{\lambda_n}\right]^{\gamma_n}, \quad x \in R^1. \tag{6}$$

We can define a normalized process constructed using the estimator (6):

$$\left\{Q_n(x) = \sqrt{n}\left(F_n(x) - F(x)\right), x \in R^1\right\}, \tag{7}$$

Select the interval $D = [\tau, T] \subset R^1$ so that $\tau_H < \tau \leq T < T_H$ and $q = \sup_{x \in D} \left[(H(x))^{p^{(0)}} - H(x) \right]^{-1} > 0$.

The following theorem provides a statement regarding the limiting Gaussian property of the sequence given in Eq. (7).

Theorem 2. *If $q > 0$, then the sequence of random processes $Q_n(x), x \in D$ converges weakly to a central Gaussian process $A(x), x \in D$ with a covariance structure for $x_1, x_2 \in D$:*

$$\begin{aligned} &cov\{A(x_1), A(x_2)\} = (1 - F(x_1))(1 - F(x_2)) \\ &\times \{a(x_1) a(x_2) [H(\min(x_1, x_2)) - H(x_1) H(x_2)] + b(x_1) b(x_2) \\ &\times p^{(0)} (1 - p^{(0)}) + c(x_1) c(x_2) p^{(1)} (1 - p^{(1)}) - 2b(\min(x_1, x_2)) c(\min(x_1, x_2)) \\ &\times p^{(0)} p^{(1)}\}, \end{aligned}$$

where

$$\begin{aligned} a(x) &= p^{(1)} \left[(H(x))^{p^{(0)}} - H(x) \right]^{-1}, \\ b(x) &= -\left[\frac{p^{(1)}}{(1-p^{(0)})} c(x) + \frac{a(x)}{(1-p^{(0)})} H(x) \log H(x) \right], \\ c(x) &= -\frac{1}{(1-p^{(0)})} \log \left[1 - (H(x))^{1-p^{(0)}} \right]. \end{aligned}$$

Proof. At first we establish the following asymptotic representation, which decomposes $F_n(x) - F(x)$ into normalized sum of independent and identically distributed random functions and a remainder term:

$$F_n(x) - F(x) = \frac{1}{n} \sum_{i=1}^{n} \Psi_x(Z_i, \Delta_i) + r_n(x), \tag{8}$$

where

$$\begin{aligned} &\Psi_x(Z_i, \Delta_i) = (1 - F(x)) \left\{ p^{(1)} \left[(H(x))^{p^{(0)}} - H(x) \right]^{-1} \right\} (I(Z_i \leq x) - H(x)) \\ &- \left[\frac{p^{(1)}}{(1-p^{(0)})^2} \log \left[1 - (H(x))^{1-p^{(0)}} \right] + \frac{p^{(1)}}{1-p^{(0)}} H(x) \log H(x) \left[(H(x))^{p^{(0)}} - H(x) \right]^{-1} \right] \\ &\times \left(\delta_i^{(0)} - p^{(0)} \right) - \frac{1}{1-p^{(0)}} \log \left[1 - (H(x))^{1-p^{(0)}} \right] \left(\delta_i^{(1)} - p^{(1)} \right), \end{aligned}$$

and

$$\sup_{x \in D} |r_n(x)| \overset{a.s.}{=} O\left(\frac{\log n}{n} \right). \tag{9}$$

Consider a function

$$f(u, y, z) = \left(1 - u^{1-y}\right)^{\frac{z}{1-y}}, \quad u, y, z \in (0, 1).$$

Then by (3) and (6),

$$1 - F(x) = f\left(H(x), p^{(0)}, p^{(1)}\right) = f_0, \ 1 - F_n(x) = f\left(H_n(x), p_n^{(0)}, p_n^{(1)}\right) = f_n.$$

also $F_n(x) - F(x) = -(f_n - f_0)$. By second order Taylor expansion

$$-(f_n - f_0) = \frac{1}{n}\sum_{i=1}^{n} \Psi_x(Z_i, \Delta_i) + q_n(x). \tag{10}$$

Here under $(u, y, z) = \left(H_n(x), p_n^{(0)}, p_n^{(1)}\right)$ and $(u_0, y_0, z_0) = \left(H(x), p^{(0)}, p^{(1)}\right)$ we have

$$\begin{aligned} q_n(x) &= \tfrac{1}{2}\left[f''_{uu}\cdot(u-u_0)^2 + f''_{yy}\cdot(y-y_0)^2 + f''_{zz}\cdot(z-z_0)^2\right] \\ &+ f''_{uy}\cdot(u-u_0)(y-y_0) + f''_{uz}\cdot(u-u_0)(z-z_0) + f''_{yz}\cdot(y-y_0)(z-z_0), \end{aligned}$$

where $f_{uy} = f_{yu}$, $f_{uz} = f_{zu}$, $f_{yz} = f_{zy}$ and second order derivatives calculated at intermediate point $(u_*, y_*, z_*) = \left(H_*(t), p_*^{(0)}, p_*^{(1)}\right)$. By using elementary inequality for $\alpha, \gamma > 0$

$$0 \le \vartheta^{\gamma}|\log\vartheta|^{\alpha} \le \left(\frac{\alpha}{\gamma}\right)e^{-\alpha}, 0 < \vartheta \le 1,$$

from (10) we obtain

$$\sup_{x\in D}|r_n(x)| \overset{a.s.}{=} \sum_{j=1}^{5} r_{jn}, \tag{11}$$

where

$$\begin{aligned} r_{1n} &= \mathrm{O}\left(\left(\sup_{x\in D}|H_n(x) - H(x)|\right)^2\right), \\ r_{2n} &= \mathrm{O}\left(\left(p_n^{(0)} - p^{(0)}\right)^2\right) + \mathrm{O}\left(\left(p_n^{(1)} - p^{(1)}\right)^2\right), \\ r_{3n} &= \mathrm{O}\left(\left|p_n^{(0)} - p^{(0)}\right| \cdot \sup_{x\in D}|H_n(x) - H(x)|\right), \\ r_{4n} &= \mathrm{O}\left(\left|p_n^{(1)} - p^{(1)}\right| \cdot \sup_{x\in D}|H_n(x) - H(x)|\right), \\ r_{5n} &= \mathrm{O}\left(\left|p_n^{(0)} - p^{(0)}\right| \cdot \left|p_n^{(1)} - p^{(1)}\right|\right). \end{aligned}$$

By Dvoretzky-Kiefer-Wolfowitz exponential inequality for empirical estimators [14,19], we have

$$\begin{aligned} &\sup_{x\in D}|H_n(x) - H(x)| \overset{a.s.}{=} \mathrm{O}\left(\left(\tfrac{\log n}{n}\right)^{1/2}\right), \\ &\left|p_n^{(m)} - p^{(m)}\right| \le \sup_{x\in D}\left|T_n^{(m)}(x) - T^{(m)}(x)\right| \overset{a.s.}{=} \mathrm{O}\left(\left(\tfrac{\log n}{n}\right)^{1/2}\right), \end{aligned} \tag{12}$$

where for $m = 0, 1$, $T_n^{(m)}(x) = \frac{1}{n}\sum_{i=1}^{n} I\left(Z_i \le x, \delta_i^{(m)} = 1\right)$. Now (8) and (9) follows from formulas (1.10)-(1.12). From (8) follows that the empirical process $\sqrt{n}(F_n(x) - F(x))$ is strong approximated by normalized sum $A_n(x) = n^{-\frac{1}{2}}\sum_{i=1}^{n}\Psi_x(Z_i, \Delta_i)$ with the rate $\mathrm{O}\left(n^{-\frac{1}{2}}\log n\right)$ a.s.

Hence, it sufficient to prove the weak convergence of the process $A_n(x)$ to $A(x)$, $x \in D$ (see, Theorem 4.1 in [7]). We first show convergence of the finite dimensional distributions, i.e. for any $N = 1, 2, \ldots$ and $\tau \le x_1 \le \ldots \le x_N \le T$: $(A_n(x_1), \ldots, A_n(x_N)) \xrightarrow{d} N(0; (\gamma_{x_i x_j}))$. Since $A_n(x_i) = \sum_{k=1}^{N} A_{nki}$ where $A_{nki} = n^{-\frac{1}{2}} \Psi_{x_i}(Z_k, \Delta_k)$, it suffices to check that (see, [1]):

$$\lim_{n \to \infty} \sum_{k=1}^{n} E A_{nki} A_{nkj} = \gamma_{x_i x_j}, \ (1 \le i, \ j \le N) \tag{13}$$

$$\lim_{n \to \infty} \sum_{k=1}^{n} \int_{\{|A_{nk}| > \varepsilon\}} |A_{nk}|^2 \, dP = 0, \tag{14}$$

for every $\varepsilon > 0$, where $|A_{nk}|^2 = \sum_{i=1}^{N} A_{nki}^2$. Now it is easy to see that

$$\sum_{k=1}^{n} E A_{nki} A_{nkj} = \frac{1}{n} \sum_{k=1}^{n} Cov\left(\Psi_{x_i}(Z_k, \delta_k), \Psi_{xj}(Z_k, \delta_k)\right) = \gamma_{x_i x_j},$$

i.e. (13) is hold.

Also, since the functions $\Psi_{x_i}(Z_k, \delta_k)$ are uniformly bounded, we have that $\max_{1 \le k \le n} |A_{nk}| \overset{a.s.}{=} O\left(n^{-\frac{1}{2}}\right)$ and hence,

$$\sum_{k=1}^{n} \int_{\{|A_{nk}| > \varepsilon\}} |A_{nk}|^2 \, dP \le \int_{\left\{\max\limits_{1 \le k \le n} |A_{nk}| > \varepsilon\right\}} \sum_{k=1}^{n} |A_{nk}|^2 \, dP$$
$$\le O(1) P\left(\max_{1 \le k \le n} |A_{nk}| > \varepsilon\right) = o(1),$$

i.e. (14) also hold. Now from (13) and (14) follows convergence of finite dimensional distributions. Lastly, we confirm the moment conditions for tightness, for all $\tau \le r \le s \le t \le T$:

$$E\left[(A_n(r) - A_n(s))^2 (A_n(s) - A_n(t))^2\right] \le (\gamma(r) - \gamma(s))(\gamma(s) - \gamma(t)) + c_n, \tag{15}$$

For some continuous and non-decreasing function $\gamma(s)$ and some sequence $c_n = o(1)$. Let's denote $\Psi_{x_i} = \Psi_x(Z_i, \Delta_i)$. Then, the left hand side of (15) equals

$$n^{-2} \sum_{i=1}^{n} \sum_{j=1}^{n} \sum_{k=1}^{n} \sum_{l=1}^{n} E\left[(\Psi_{ri} - \Psi_{si})(\Psi_{rj} - \Psi_{sj})(\Psi_{sk} - \Psi_{tk})(\Psi_{sl} - \Psi_{tl})\right]$$
$$= n^{-2} \{\sum_{i=1}^{n} E\left[(\Psi_{ri} - \Psi_{si})^2\right] \cdot \sum_{j=1, j \ne i}^{n} E\left[(\Psi_{sj} - \Psi_{tj})^2\right]$$
$$+2 \sum_{i=1}^{n} \sum_{j=1, j \ne i}^{n} E\left[(\Psi_{ri} - \Psi_{si})(\Psi_{si} - \Psi_{ti})\right] E\left[(\Psi_{rj} - \Psi_{sj})(\Psi_{sj} - \Psi_{tj})\right]$$

$$+\sum_{i=1}^{n} E\left[(\Psi_{ri}-\Psi_{si})^2(\Psi_{si}-\Psi_{ti})^2\right]\}$$

$$\leq 3n^{-2}\sum_{i=1}^{n} E\left[(\Psi_{ri}-\Psi_{si})^2\right]\sum_{i=1}^{n} E\left[(\Psi_{si}-\Psi_{ti})^2\right]+O\left(n^{-3}\right),$$

using the inequalities of Cauchy-Bunyakovsky and Hölder and the fact that the Ψ functions are uniformly bounded. At this point, we obtain.

$$n^{-2}\sum_{i=1}^{n} E\left[(\Psi_{ri}-\Psi_{si})^2\right]=n^{-2}\sum_{i=1}^{n}\left[D\Psi_{ri}+D\Psi_{si}-2Cov\left(\Psi_{ri},\Psi_{si}\right)\right].$$

Subsequently, with the appropriate value of $\gamma(s)$, Eq. (15) is derived, but further elaboration is excluded. It should be noted that the covariance structure of the limiting Gaussian process $\{A(x), x \in D\}$ is computed by utilizing the independence of the random variables Z_i and $\Delta_i = \left(\delta_i^{(0)}, \delta_i^{(1)}, \delta_i^{(2)}\right)$ in accordance with Theorem 1. Therefore, Theorem 2 is proved.

It should be noted that Theorem 2 was first announced in [18].

2 Investigations on How the Estimator Is Influenced by Unknown Parameters of Censorship

Numerical simulations were executed in Python to investigate the impact of parameters on the accuracy of the estimator $F_n(x)$, as well as its closeness to the estimated function $F(x)$. The standard normal and exponential distribution functions were utilized as $F(x)$, and simulated volume data with $n = 500$ was employed for this purpose. Figures 1, 2, 3, 4, 5, 6, 7, 8, 9, 10, 11 and 12 illustrate the findings, demonstrating that when $0 < \beta, , \theta \leq 1$, $F_n(x)$ approximates $F(x)$ well (Figs. 1, 2, 3 and 4). However, in other cases where censoring increases from both sides, the deviation between the estimator and the distribution function becomes noticeable.

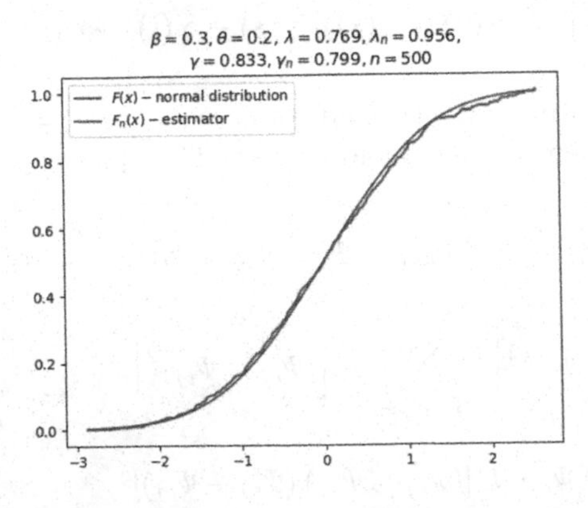

Fig. 1. $\beta = 0.3,\ \ \theta = 0.2,\ \ n = 500$

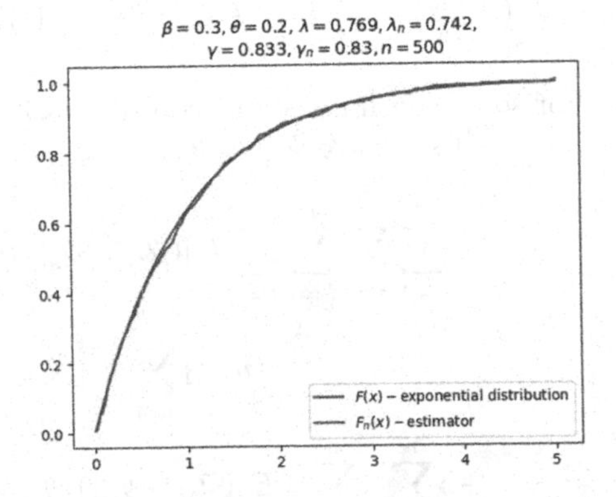

Fig. 2. $\beta = 0.3,\ \ \theta = 0.2,\ \ n = 500$

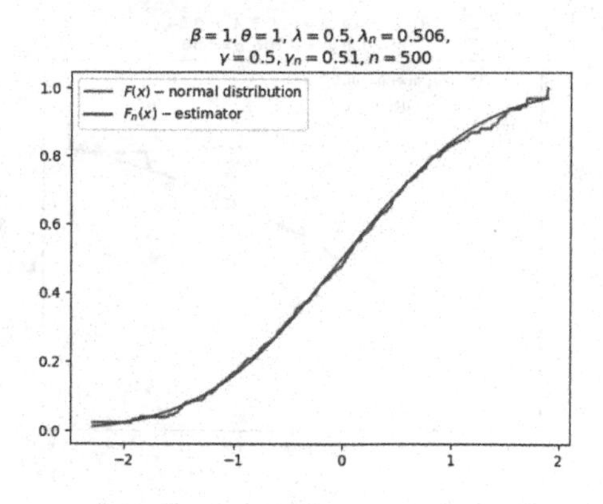

Fig. 3. $\beta = 1, \quad \theta = 1, \quad n = 500$

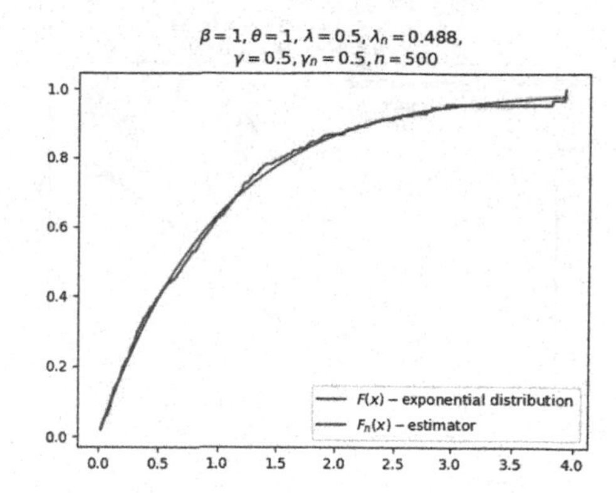

Fig. 4. $\beta = 1, \quad \theta = 1, \quad n = 500$

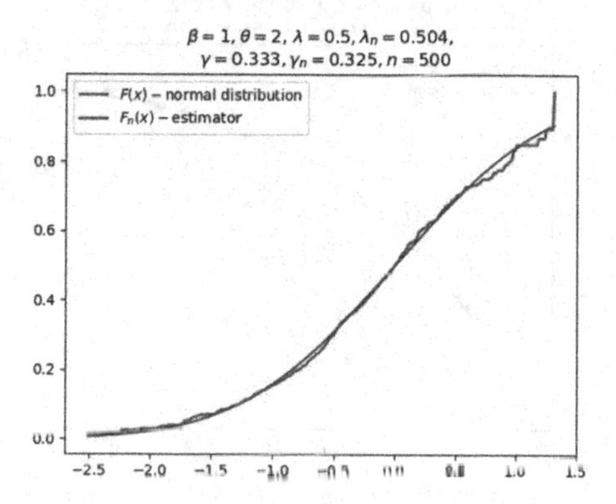

Fig. 5. $\beta = 1, \quad \theta = 2, \quad n = 500$

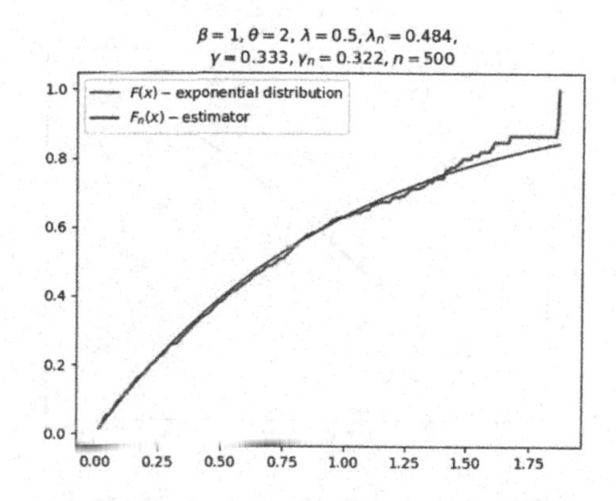

Fig. 6. $\beta = 1, \quad \theta = 2, \quad n = 500$

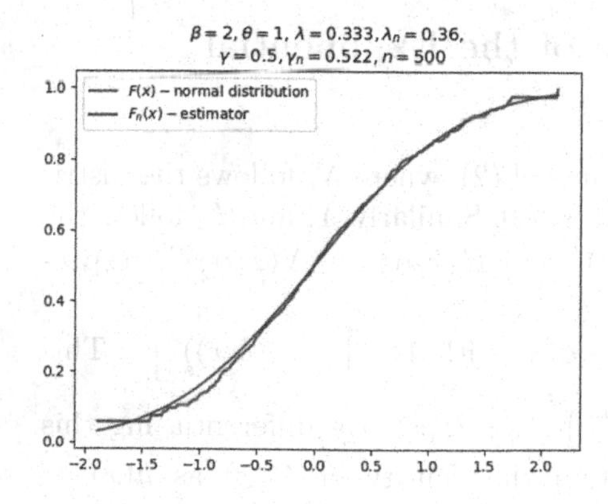

Fig. 7. $\beta = 2, \quad \theta = 1, \quad n = 500$

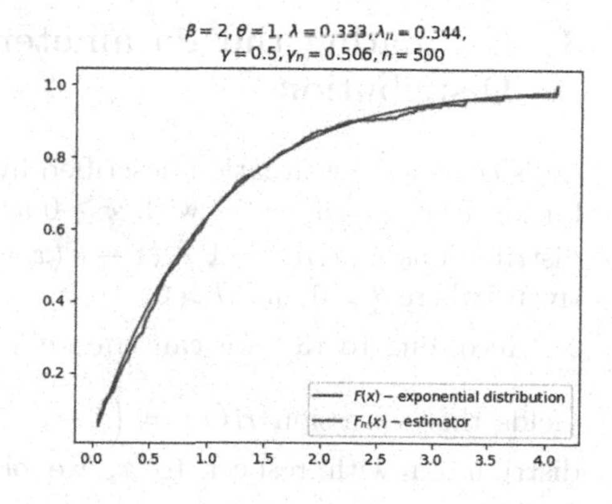

Fig. 8. $\beta = 2, \quad \theta = 1, \quad n = 500$

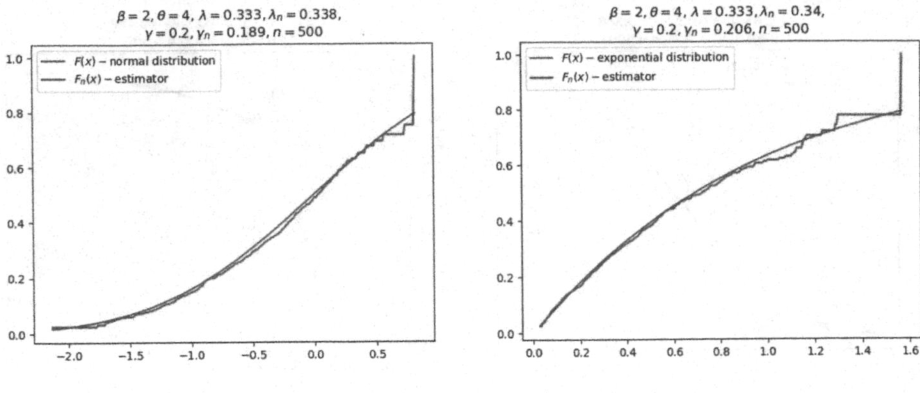

Fig. 9. $\beta = 2, \quad \theta = 4, \quad n = 500$

Fig. 10. $\beta = 2, \quad \theta = 4, \quad n = 500$

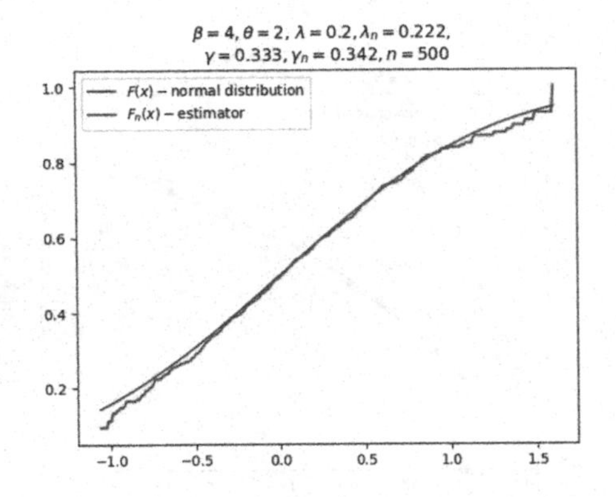

Fig. 11. $\beta = 4, \quad \theta = 2, \quad n = 500$

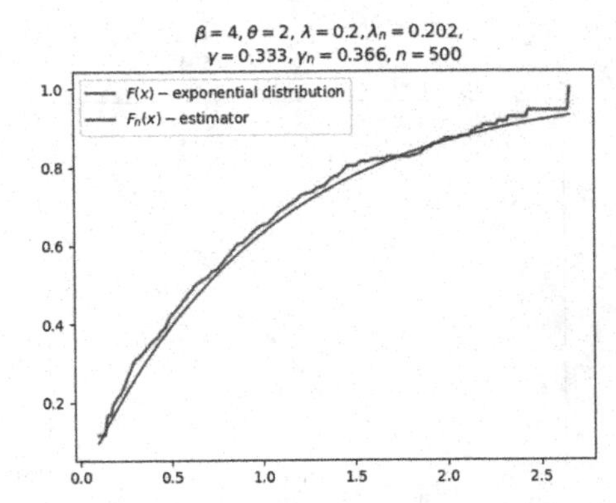

Fig. 12. $\beta = 4, \quad \theta = 2, \quad n = 500$

3 Estimating the Parameters of the Exponential Distribution

Let's consider the situation described by model (2), where X_i follows the distribution $F(x, \alpha) = 1 - e^{-\frac{x}{\alpha}}$ with $x \geq 0$ and $\alpha > 0$. Similarly, Y_i and L_i follow the distributions $G(x, \alpha) = 1 - (1 - F(x, \alpha))^{\theta}$ and $K(x, \alpha) = (N(x, \alpha))^{\beta}$, respectively, where $\theta > 0$ and $\beta > 0$.

According to (3), we can equate $1 - e^{-\frac{x}{\alpha}}$ with $1 - \left[1 - (H(x))^{\lambda}\right]^{\gamma}$. This yields the expression $H(x) = \left(1 - e^{-\frac{x}{\alpha \cdot \gamma}}\right)^{\frac{1}{\lambda}}$ for $H(x)$. By differentiating this distribution with respect to x, we obtain the density of $H(x)$ as $h(x) = \frac{1}{\alpha \lambda \gamma} e^{-\frac{x}{\alpha \gamma}} \left(1 - e^{-\frac{x}{\alpha \gamma}}\right)^{\frac{1-\lambda}{\lambda}}$.

Next, we can create a maximum likelihood function for $h(x)$ using a sample from Z_i:

$$L(Z,\alpha) =$$
$$\ln\left((\alpha\lambda\gamma)^{-n}\cdot e^{-\frac{\sum_{i=1}^{n} Z_i}{\alpha\gamma}}\left(\left(1-e^{-\frac{Z_1}{\alpha\gamma}}\right)\left(1-e^{-\frac{Z_1}{\alpha\gamma}}\right)\cdot\ \ldots\ \cdot\left(1-e^{-\frac{Z_n}{\alpha\gamma}}\right)\right)^{\frac{1-\lambda}{\lambda}}\right)$$
$$= -n\ln(\alpha\lambda\gamma) - \frac{\sum_{i=1}^{n} Z_i}{\alpha\gamma} + \frac{1-\lambda}{\lambda}\sum_{i=1}^{n}\ln\left(1-e^{-\frac{Z_i}{\alpha\gamma}}\right).$$

Now we calculate: $\frac{\partial L(Z,\alpha)}{\partial\alpha}$:

$$\frac{\partial L(Z,\alpha)}{\partial\alpha} = -\frac{n}{\alpha} + \frac{\sum_{i=1}^{n} Z_i}{\alpha^2\gamma} + \frac{1-\lambda}{\lambda}\sum_{i=1}^{n}\frac{-\frac{Z_i}{\alpha^2\gamma}e^{-\frac{Z_i}{\alpha\gamma}}}{1-e^{-\frac{Z_i}{\alpha\gamma}}}.$$

We solve the maximum likelihood equation $\frac{\partial L(Z,\alpha)}{\partial\alpha} = 0$ for α and obtain:

$$\alpha = \frac{1}{\lambda\gamma n}\cdot\sum_{i=1}^{n} Z_i\cdot\left(1-\frac{1-\lambda}{(H(Z_i))^{\lambda}}\right). \tag{16}$$

Using Eq. (16), we can obtain a semiparametric estimate for the parameter α, instead of supplying appropriate estimates for (H,λ,γ) as (H_n,λ_n,γ_n):

$$\alpha_n = \frac{1}{\lambda_n\gamma_n n}\cdot\sum_{i=1}^{n} Z_i\cdot\left(1-\frac{1-\lambda_n}{(H_n(Z_i))^{\lambda_n}}\right). \tag{17}$$

We obtain an estimate $F(x,,\alpha_n)$ for $F(x,,\alpha)$ using Eq. (17). It can be observed that in the absence of left censoring (i.e., $\lambda_n = 1$), estimator in Eq. (17) coincides with the well-known maximum likelihood estimator of the parameter α of the exponential distribution. Using numerical methods, we have created graphs to compare the estimate in Eq. (6) and the estimate constructed using Eq. (17). Based on these comparisons (Figs. 13, 14, 15 and 16), we conclude that the estimate $F(x,\alpha_n)$ is highly accurate.

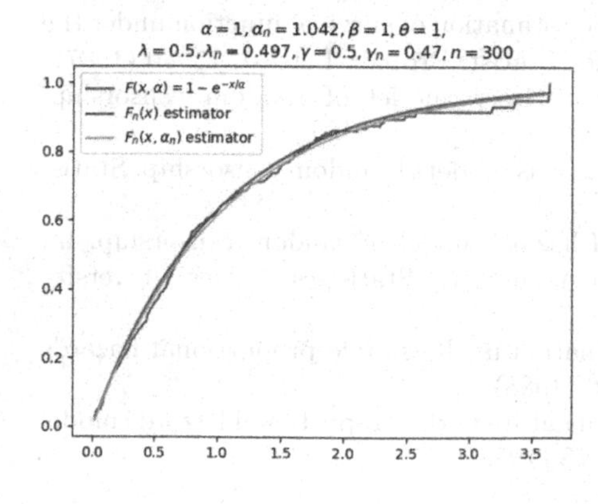

Fig. 13. $\beta = 1,\ \ \theta = 1,\ \ n = 300$

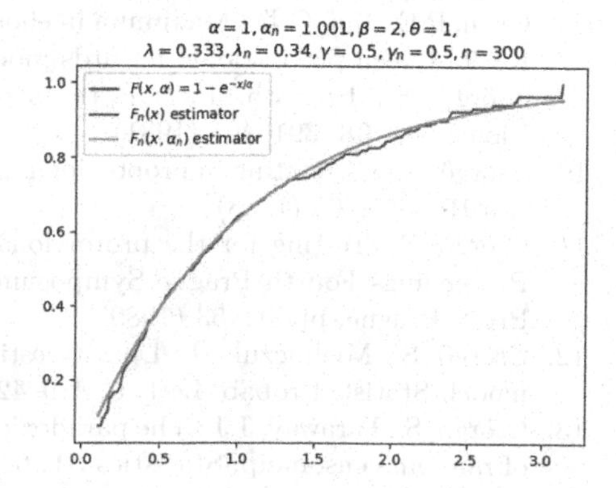

Fig. 14. $\beta = 2,\ \ \theta = 1,\ \ n = 300$

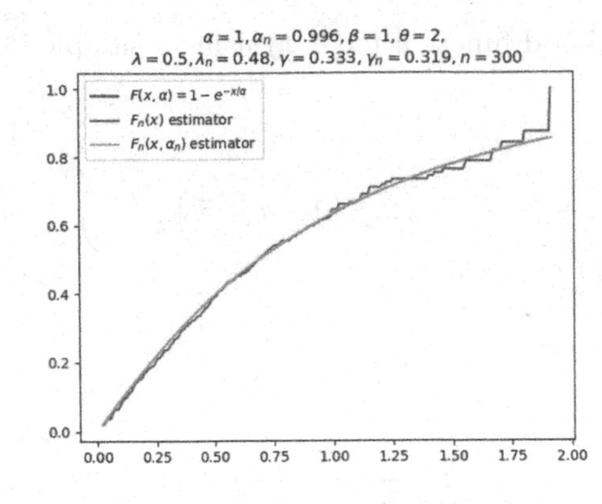

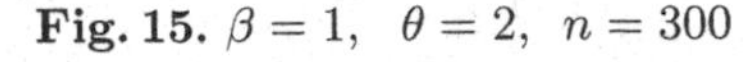
Fig. 15. $\beta = 1, \quad \theta = 2, \quad n = 300$

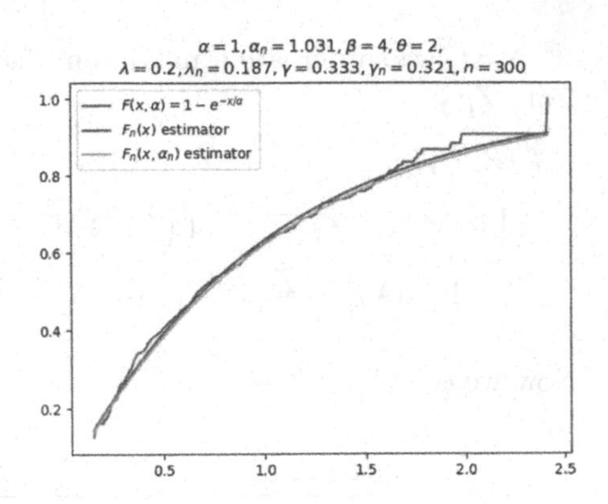

Fig. 16. $\beta = 4, \quad \theta = 2, \quad n = 300$

References

1. Araujo, A., Gine, E.: The Central Limit Theorem for Real and Banach Valued Random Variables. Willey, New York (1980)
2. Abdushukurov, A.A.: Estimation of probability density and intensity function of the Koziol-Green model of random cencoring. Sankhya: Indian J. Stat. Ser. A. **48**, 150–168 (1987)
3. Abdushukurov, A.A.: Random cencorship model from both sides and independence test for it. Report of Acad. Sci. Rep. Uz. Issue **11**, 8–9 (1994). (in Russian)
4. Abdushukurov, A.A.: Nonparametric estimation of the distribution function based on relative risk function. Commun. Statist. Th. Meth. **27**(8), 1991–2012 (1998)
5. Abdushukurov, A.A., Abdikalikov, F.A.: Semiparametric estimator of mean conditional residual life function under informative random censoring from both sides. Appl. Math. **6**, 319–325 (2015)
6. Abdikalikov, F.A., Abdushukurov, A.A.: Semiparametric estimation of conditional survival function in informative regression model of random censorship from both sides. Statisticheskie Metody Otsenivaniya i Proverki Gipotes. Perm. Russia. Perm State Univ. Press. Issue **23**, 145–162 (2012). (In Russian)
7. Billingsley, P.: Convergence of Probability Measures. Willey, New York (1968)
8. Chen, P.E., Lin, G.D.: Maximum likehood estimation of survival function under the Koziol-Green proportional hazards model. Statist. Probab. Lett. **5**, 75–80 (1987)
9. Csörgő, S., Horväth, L.: On the Koziol-Green model of random censorship. Biometrika **68**, 391–401 (1981)
10. Csörgő, S.: Estimating in proportional hazards model of random censorship. Statistics **19**, 437–463 (1988)
11. Csörgő, S.: Testing for the prorortional hazard model of random censorship. In: Proceedings Fourth Prague Symposium Asymptotic Statistics. Carles University Press. Prague, pp. 41–53 (1989)
12. Csörgő, S., Mielniczuk, J.: Density estimation in the simple proportional hazards model. Statist. Probab. Lett. **6**, 419–426 (1988)
13. Csörgő, S., Faraway, J.J.: The paradoxical nature of the proportional hazards model of random censorship. Statistics **31**, 67–78 (1998)

14. Dvoretzky, A., Kiefer, J., Wolfowitz, J.: Asymptotic minimax character of the sample distribution function and of the multinomil estimator. Ann. Math. Statist. **27**, 642–669 (1956)
15. Ghorai, J.: The asymptotic distribution of the suprema of the standardized empirical processes under the Koziol-Green model. Statist. Probab. Lett. **41**, 303–313 (1999)
16. Koziol, J.A., Green, S.B.: A Cramer-von Mises statistic for randomly censored data. Biometrika **63**(3), 465–476 (1976)
17. Hollander, M., Pena, E.: Families of confidence bands for the survival function under the general right censorship model and the Koziol-Green model. Canadian J. Statist. **17**(1), 59–74 (1989)
18. Mansurov, D.R.: Sequential empirical processes in informative models of incomplete observations. In: Materials of International Conference "Teoriya funcsiy odnogo i mnogich compleksnych peremennich", November 26–28. Nukus, pp. 165–168 (2020)
19. Massart, P.: The tight constant in the Dvoretzky-Kiefer-Wolfowitz inequality. Ann. Probab. **18**(3), 1269–1283 (1990)
20. Pawlitschko, J.: A comparison of survival function estimators in the Koziol-Green model. Statistics **32**, 277–291 (1999)
21. de Una-Álvares, J.: Kernel distribution function estimation under the Koziol-Green model. J. Stat. Plan. Infer. **87**, 199–219 (2000)

Author Index

A. Dudin et al. (Eds.): ITMM 2022, CCIS 1803, p. 239, 2023.
https://doi.org/10.1007/978-3-031-32990-6

Printed in the United States
by Baker & Taylor Publisher Services

Printed in the United States
by Baker & Taylor Publisher Services